Recent Advances in
Electromagnetic Devices

Recent Advances in Electromagnetic Devices

Guest Editor
Haejun Chung

Basel • Beijing • Wuhan • Barcelona • Belgrade • Novi Sad • Cluj • Manchester

Guest Editor
Haejun Chung
Department of Electronic Engineering
Hanyang University
Seoul
Korea, South

Editorial Office
MDPI AG
Grosspeteranlage 5
4052 Basel, Switzerland

This is a reprint of the Special Issue, published open access by the journal *Micromachines* (ISSN 2072-666X), freely accessible at: www.mdpi.com/journal/micromachines/special_issues/6DI5X3A7EM.

For citation purposes, cite each article independently as indicated on the article page online and using the guide below:

Lastname, A.A.; Lastname, B.B. Article Title. *Journal Name* **Year**, *Volume Number*, Page Range.

ISBN 978-3-7258-3200-2 (Hbk)
ISBN 978-3-7258-3199-9 (PDF)
https://doi.org/10.3390/books978-3-7258-3199-9

© 2025 by the authors. Articles in this book are Open Access and distributed under the Creative Commons Attribution (CC BY) license. The book as a whole is distributed by MDPI under the terms and conditions of the Creative Commons Attribution-NonCommercial-NoDerivs (CC BY-NC-ND) license (https://creativecommons.org/licenses/by-nc-nd/4.0/).

Contents

About the Editor vii

Chanik Kang and Haejun Chung
Recent Advances in Electromagnetic Devices: Design and Optimization
Reprinted from: *Micromachines* 2025, 16, 98, https://doi.org/10.3390/mi16010098 1

Zibin Weng, Chen Liang, Kaibin Xue, Ziming Lv and Xing Zhang
A Miniaturized Loaded Open-Boundary Quad-Ridge Horn with a Stable Phase Center for Interferometric Direction-Finding Systems
Reprinted from: *Micromachines* 2024, 16, 44, https://doi.org/10.3390/mi16010044 6

Shunlan Zhang, Weiping Cao, Jiao Wang, Tiesheng Wu, Yiying Wang and Yanxia Wang et al.
Reconfigurable Multifunctional Metasurfaces for Full-Space Electromagnetic Wave Front Control
Reprinted from: *Micromachines* 2024, 15, 1282, https://doi.org/10.3390/mi15111282 19

Binyi Ma, Jing Li, Yu Chen, Yuheng Si, Hongyan Gao and Qiannan Wu et al.
A 60 GHz Slotted Array Horn Antenna for Radar Sensing Applications in Future Global Industrial Scenarios
Reprinted from: *Micromachines* 2024, 15, 728, https://doi.org/10.3390/mi15060728 32

Chuan Zhao, Qinwei Zhang, Wenzhe Pei, Junjie Jin, Feng Sun and Hongkui Zhang et al.
Design and Analysis of 5-DOF Compact Electromagnetic Levitation Actuator for Lens Control of Laser Cutting Machine
Reprinted from: *Micromachines* 2024, 15, 641, https://doi.org/10.3390/mi15050641 45

Alfredo Gomes Neto, Jefferson Costa e Silva, Joabson Nogueira de Carvalho and Custódio Peixeiro
Planar Printed Structures Based on Matryoshka Geometries: A Review
Reprinted from: *Micromachines* 2024, 15, 469, https://doi.org/10.3390/mi15040469 61

Yanhu Huang, Jiajun Liang, Zhao Wu and Qian Chen
Design of 2.45 GHz High-Efficiency Rectifying Circuit for Wireless RF Energy Collection System
Reprinted from: *Micromachines* 2024, 15, 340, https://doi.org/10.3390/mi15030340 111

Fayyadh H. Ahmed, Rola Saad and Salam K. Khamas
A Novel Compact Broadband Quasi-Twisted Branch Line Coupler Based on a Double-Layered Microstrip Line
Reprinted from: *Micromachines* 2024, 15, 142, https://doi.org/10.3390/mi15010142 124

Bikash Ranjan Behera, Mohammed H. Alsharif and Abu Jahid
Investigation of a Circularly Polarized Metasurface Antenna for Hybrid Wireless Applications
Reprinted from: *Micromachines* 2023, 14, 2172, https://doi.org/10.3390/mi14122172 141

Fuwang Li, Yi-Feng Cheng, Gaofeng Wang and Jiang Luo
A Novel High-Isolation Dual-Polarized Patch Antenna with Two In-Band Transmission Zeros
Reprinted from: *Micromachines* 2023, 14, 1784, https://doi.org/10.3390/mi14091784 158

Tarek S. Abdou and Salam K. Khamas
A Multiband Millimeter-Wave Rectangular Dielectric Resonator Antenna with Omnidirectional Radiation Using a Planar Feed
Reprinted from: *Micromachines* 2023, 14, 1774, https://doi.org/10.3390/mi14091774 167

Xukun Hu, Guozhi Zhang, Guangyu Deng and Xuyu Li
Experimental Study on the Compatibility of PD Flexible UHF Antenna Sensor Substrate with SF6/N2
Reprinted from: *Micromachines* **2023**, *14*, 1516, https://doi.org/10.3390/mi14081516 . . . . **188**

Yongxin Zhan, Yu Chen, Honglei Guo, Qiannan Wu and Mengwei Li
Design of a Ka-Band Five-Bit MEMS Delay with a Coplanar Waveguide Loaded U-Shaped Slit
Reprinted from: *Micromachines* **2023**, *14*, 1508, https://doi.org/10.3390/mi14081508 . . . . **201**

Xiaoyan Wang, Yanfei Liu, Yilin Jia, Ningning Su and Qiannan Wu
Ultra-Wideband and Narrowband Switchable, Bi-Functional Metamaterial Absorber Based on Vanadium Dioxide
Reprinted from: *Micromachines* **2023**, *14*, 1381, https://doi.org/10.3390/mi14071381 . . . . **214**

About the Editor

Haejun Chung

Dr. Haejun Chung has been an Assistant Professor at Hanyang University since 2022, working in the area of inverse design of electromagnetic devices. He graduated with a B. S. degree in Electrical Engineering from Illinois Institute of Technology in 2010 and then obtained M. S. (2013) and Ph. D. (2017) degrees, both from Purdue University. He joined Applied Physics at Yale University as a Postdoctoral Research Associate, developing a fast inverse design algorithm and novel metasurfaces. In 2020, he joined MIT Mechanical Engineering as a Postdoctoral Research Associate to develop large-area metalenses and tunable metasurfaces.

Editorial

Recent Advances in Electromagnetic Devices: Design and Optimization

Chanik Kang [1] and Haejun Chung [1,2,*]

1. Department of Artificial Intelligence, Hanyang University, Seoul 04763, Republic of Korea; chanik@hanyang.ac.kr
2. Department of Electronic Engineering, Hanyang University, Seoul 04763, Republic of Korea
* Correspondence: haejun@hanyang.ac.kr

Electromagnetic devices are a continuous driving force in cutting-edge research and technology, finding applications in diverse fields such as optics [1–3], photonics [4], RF waves [5], and many others [6–8]. The design and optimization of electromagnetic devices have become essential to meet the demanding performance requirements for high efficiency, high power density, and reduced form factors. Over the years, various methods, such as analytic designs [9], evolutionary-based designs [10], gradient-based optimization [11,12], and neural network-based design techniques [13,14], have been employed to improve the design of electromagnetic devices.

Underpinning these design and optimization strategies are the fundamental principles of optics, photonics, and electromagnetics, which collectively drive modern technological advancements across a broad spectrum of applications. The field of optics focuses on the behavior and manipulation of light, including reflection, refraction, and diffraction, while photonics focuses on generating, controlling, and detecting photons for use in devices such as lasers, optical fibers, and imaging systems. Electromagnetics, the overarching discipline that unifies electric and magnetic phenomena, provides a theoretical framework for understanding how electromagnetic waves propagate and interact with various materials. Together, these fields enable innovations in high-speed communication systems [15], optical computing [16], and photonic integrated circuits [17], among others [18,19]. Continued research into advanced materials [20], fabrication processes [21], and simulation techniques [22–24] drives further progress, leading to the development of more compact, energy-efficient, and high-performance devices that continue to push the boundaries of what is possible in both fundamental science and practical applications.

This Special Issue features a diverse collection of recent advances in electromagnetic devices, offering insights and practical approaches that will be of significant value to researchers.

Zhang et al. [25] introduced a multifunctional metasurface designed for full-space electromagnetic wavefront control, which holds promise for applications in 6G communications. This work exemplifies how reconfigurability and compact designs can meet the growing demands for versatile electromagnetic systems. The authors achieved polarization conversion and reflection-beam pattern tuning, demonstrating a robust combination of theoretical modeling, simulation, and experimental validation.

Ma et al. [26] presented a 60 GHz slotted array horn antenna optimized for radar sensing in industrial scenarios. This work demonstrates the potential of millimeter-wave technologies in next-generation radar systems by achieving a high gain and wide bandwidth. Their method involved meticulous radiation-band structure design and array optimization to achieve a high gain and a wide impedance bandwidth, validated through fabrication and testing.

Received: 2 January 2025
Accepted: 14 January 2025
Published: 16 January 2025

Citation: Kang, C.; Chung, H. Recent Advances in Electromagnetic Devices: Design and Optimization. *Micromachines* **2025**, *16*, 98. https://doi.org/10.3390/mi16010098

Copyright: © 2025 by the authors. Licensee MDPI, Basel, Switzerland. This article is an open access article distributed under the terms and conditions of the Creative Commons Attribution (CC BY) license (https://creativecommons.org/licenses/by/4.0/).

Zhao et al. [27] explored a five-degree-of-freedom electromagnetic levitation actuator for laser cutting machines. Their work offers a compelling solution for achieving high-speed, high-precision control, which is crucial for advanced manufacturing. They employed nonlinear analytical modeling, finite element simulations, and PID-based centralized control to achieve high-speed and high-precision lens manipulation in laser cutting applications.

Huang et al. [28] proposed a high-efficiency 2.45 GHz rectifying circuit for RF energy collection systems. This innovative approach could pave the way for sustainable energy solutions in IoT and low-power devices. Their approach focused on suppressing harmonic components and optimizing the rectifier's DC-RF conversion efficiency through detailed simulation and experimental comparisons.

Ahmed et al. [29] and Abdou et al. [30] introduced compact and high-performance devices for 5G and beyond. Ahmed et al. focus on a quasi-twisted branch-line coupler, while Abdou et al. present a multiband millimeter-wave dielectric resonator antenna with omnidirectional radiation capabilities. By utilizing a double-layered microstrip line structure with a slow-wave design, Ahmed et al. achieved a significant size reduction and enhanced bandwidth, supported by their simulation and fabrication results. Abdou et al. utilized the excitation of specific electromagnetic modes and validated their design through simulation and measurements.

Behera et al. [31] investigated circularly polarized metasurface antennas tailored for hybrid wireless applications. This study emphasizes energy-efficient designs that cater to IoT and smart sensor networks. Behera et al. adopted AI-driven surrogate model-assisted optimization to design a polarization-reconfigurable metasurface antenna. Their study combined the use of smart metasurfaces with reconfigurable monopole antennas to achieve high gain and broad bandwidth.

Li et al. [32] contributed a novel dual-polarized patch antenna with enhanced isolation, showcasing its utility in modern wireless communication systems, where signal clarity and separation are paramount. Their design methodology incorporated equivalent circuit modeling and rigorous design formulas to enhance isolation and bandwidth.

Hu et al. [33] examined the compatibility of flexible UHF antenna sensors with SF6/N2 gas mixtures. Their findings support the development of reliable sensors for high-voltage applications. Their experimental approach combined Fourier-transform infrared spectroscopy, scanning electron microscopy, and X-ray photoelectron spectroscopy to analyze material interactions.

Zhan et al. [34] proposed a Ka-band MEMS delay with low insertion loss and high accuracy. This design has significant implications for phased-array radar and communication systems. Their method involved optimizing the structure to minimize insertion loss and enhance delay accuracy through system simulations.

Neto et al. [35] explored the development, applications, and potential of planar printed structures inspired by Matryoshka geometries. These structures leverage the nesting principles of Matryoshka dolls to achieve compact, multi-resonance, and wideband configurations. This study demonstrates various applications of planar printed circuit technology, focusing on frequency-selective surfaces (FSSs), filters, antennas, and sensors.

Wang et al. [36] introduced a terahertz metamaterial absorber based on vanadium dioxide (VO_2) that achieves switchable ultra-wideband and ultra-narrowband absorption by leveraging the material's phase-transition properties. The proposed absorber comprises a multilayer structure with VO_2 as the topmost layer, supported by an insulating Topas layer, a PMI dielectric layer, and a gold reflector.

Weng et al. [37] proposed a miniaturized loaded open-boundary quad-ridge horn (LOQRH) antenna engineered for interferometric direction-finding systems. By optimizing the ridge structure and incorporating resistive loading and a self-balanced feed, they effectively suppressed common-mode currents, ensuring radiation pattern symmetry and minimizing phase center fluctuations. This work underscores the potential of compact LOQRH antennas to improve accuracy and efficiency in multiprobe interferometric applications.

In conclusion, this Special Issue focused on the potential of recent progress in electromagnetic devices to address critical challenges across various applications. The papers included in this Special Issue demonstrate the diversity and impact of cutting-edge research in this field, from advanced wireless communication systems and energy-harvesting technologies to precision manufacturing and high-performance antennas. A key theme emerging from these works is the emphasis on innovative design methods and multidisciplinary approaches. For example, the development of reconfigurable metasurfaces, compact and high-efficiency components for 5G and beyond, and advanced actuators for laser machining illustrates the continuous push for solutions that combine functionality, precision, and scalability. Similarly, efforts to enhance energy-harvesting efficiency, improve antenna isolation, and enable omnidirectional radiation patterns reflect the field's commitment to addressing practical needs in real-world applications. Another notable aspect is the range of applications explored, from 6G communications, radar sensing, and hybrid wireless systems to high-voltage power monitoring and phased-array systems.

Funding: This work was supported by the National Research Foundation of Korea (NRF) grant funded by the Korean government (MSIT) under the following grant numbers: (RS-2024-00338048), and (RS-2024-00414119). It was also supported by the Global Research Support Program in the Digital Field (RS-2024-00412644) under the supervision of the Institute of Information and Communications Technology Planning & Evaluation (IITP), and by the Artificial Intelligence Graduate School Program (RS-2020-II201373, Hanyang University), also supervised by the IITP. Additionally, this research was supported by the Artificial Intelligence Semiconductor Support Program (RS-2023-00253914), funded by the IITP, and by the Korea government (MSIT) grant (RS-2023-00261368). This work received support from the Culture, Sports, and Tourism R&D Program through a grant from the Korea Creative Content Agency, funded by the Ministry of Culture, Sports and Tourism (RS-2024-00332210).

Conflicts of Interest: The authors declare no conflicts of interest.

References

1. So, S.; Mun, J.; Park, J.; Rho, J. Revisiting the design strategies for metasurfaces: Fundamental physics, optimization, and beyond. *Adv. Mater.* **2023**, *35*, 2206399. [CrossRef] [PubMed]
2. Elsawy, M.M.; Lanteri, S.; Duvigneau, R.; Fan, J.A.; Genevet, P. Numerical optimization methods for metasurfaces. *Laser Photonics Rev.* **2020**, *14*, 1900445. [CrossRef]
3. Seo, J.; Jo, J.; Kim, J.; Kang, J.; Kang, C.; Moon, S.W.; Lee, E.; Hong, J.; Rho, J.; Chung, H. Deep-learning-driven end-to-end metalens imaging. *Adv. Photonics* **2024**, *6*, 066002. [CrossRef]
4. Cui, T.J.; Zhang, S.; Alù, A.; Wegener, M.; Pendry, J.; Luo, J.; Lai, Y.; Wang, Z.; Lin, X.; Chen, H.; et al. Roadmap on electromagnetic metamaterials and metasurfaces. *J. Phys. Photonics* **2024**, *6*, 032502. [CrossRef]
5. Pérez-López, D.; Gutierrez, A.; Sánchez, D.; López-Hernández, A.; Gutierrez, M.; Sánchez-Gomáriz, E.; Fernández, J.; Cruz, A.; Quirós, A.; Xie, Z.; et al. General-purpose programmable photonic processor for advanced radiofrequency applications. *Nat. Commun.* **2024**, *15*, 1563. [CrossRef]
6. Park, J.; Kim, S.; Nam, D.W.; Chung, H.; Park, C.Y.; Jang, M.S. Free-form optimization of nanophotonic devices: From classical methods to deep learning. *Nanophotonics* **2022**, *11*, 1809–1845. [CrossRef]
7. Lee, S.; Hong, J.; Kang, J.; Park, J.; Lim, J.; Lee, T.; Jang, M.S.; Chung, H. Inverse design of color routers in CMOS image sensors: Toward minimizing interpixel crosstalk. *Nanophotonics* **2024**, *13*, 3895–3914. [CrossRef]
8. Cho, M.; Jung, J.; Kim, M.; Lee, J.Y.; Min, S.; Hong, J.; Lee, S.; Heo, M.; Kim, J.U.; Joe, I.S.; et al. Color arrestor pixels for high-fidelity, high-sensitivity imaging sensors. *Nanophotonics* **2024**, *13*, 2971–2982. [CrossRef]

9. Tang, R.J.; Lim, S.W.D.; Ossiander, M.; Yin, X.; Capasso, F. Time reversal differentiation of fdtd for photonic inverse design. *ACS Photonics* **2023**, *10*, 4140–4150. [CrossRef]
10. Jafar-Zanjani, S.; Inampudi, S.; Mosallaei, H. Adaptive genetic algorithm for optical metasurfaces design. *Sci. Rep.* **2018**, *8*, 11040. [CrossRef]
11. Chung, H.; Miller, O.D. High-NA achromatic metalenses by inverse design. *Opt. Express* **2020**, *28*, 6945–6965. [CrossRef] [PubMed]
12. Chung, H.; Miller, O.D. Tunable metasurface inverse design for 80% switching efficiencies and 144 angular deflection. *ACS Photonics* **2020**, *7*, 2236–2243. [CrossRef]
13. Kang, C.; Seo, D.; Boriskina, S.V.; Chung, H. Adjoint method in machine learning: A pathway to efficient inverse design of photonic devices. *Mater. Des.* **2024**, *239*, 112737. [CrossRef]
14. Kang, C.; Seo, J.; Jang, I.; Chung, H. Adjoint Method-based Fourier Neural Operator Surrogate Solver for Wavefront Shaping in Tunable Metasurfaces. *iScience* **2024**, *28*, 111545. [CrossRef]
15. Ji, R.; Wang, S.; Liu, Q.; Lu, W. High-speed visible light communications: Enabling technologies and state of the art. *Appl. Sci.* **2018**, *8*, 589. [CrossRef]
16. Kazanskiy, N.L.; Butt, M.A.; Khonina, S.N. Optical computing: Status and perspectives. *Nanomaterials* **2022**, *12*, 2171. [CrossRef]
17. Bogaerts, W.; Pérez, D.; Capmany, J.; Miller, D.A.; Poon, J.; Englund, D.; Morichetti, F.; Melloni, A. Programmable photonic circuits. *Nature* **2020**, *586*, 207–216. [CrossRef]
18. Park, J.S.; Lim, S.W.D.; Amirzhan, A.; Kang, H.; Karrfalt, K.; Kim, D.; Leger, J.; Urbas, A.; Ossiander, M.; Li, Z.; et al. All-glass 100 mm diameter visible metalens for imaging the cosmos. *ACS Nano* **2024**, *18*, 3187–3198. [CrossRef]
19. Haim, O.; Boger-Lombard, J.; Katz, O. Image-guided computational holographic wavefront shaping. *Nat. Photonics* **2024**, *19*, 44–53. [CrossRef]
20. Chen, G.; Li, N.; Ng, J.D.; Lin, H.L.; Zhou, Y.; Fu, Y.H.; Lee, L.Y.T.; Yu, Y.; Liu, A.Q.; Danner, A.J. Advances in lithium niobate photonics: Development status and perspectives. *Adv. Photonics* **2022**, *4*, 034003. [CrossRef]
21. Shekhar, S.; Bogaerts, W.; Chrostowski, L.; Bowers, J.E.; Hochberg, M.; Soref, R.; Shastri, B.J. Roadmapping the next generation of silicon photonics. *Nat. Commun.* **2024**, *15*, 751. [CrossRef] [PubMed]
22. Mao, C.; Lupoiu, R.; Dai, T.; Chen, M.; Fan, J.A. Towards General Neural Surrogate Solvers with Specialized Neural Accelerators. *arXiv* **2024**, arXiv:2405.02351.
23. Seo, J.; Kang, C.; Seo, D.; Chung, H. Wave Interpolation Neural Operator: Interpolated Prediction of Electric Fields Across Untrained Wavelengths. *arXiv* **2024**, arXiv:2408.02971.
24. Xue, W.; Zhang, H.; Gopal, A.; Rokhlin, V.; Miller, O.D. Fullwave design of cm-scale cylindrical metasurfaces via fast direct solvers. *arXiv* **2023**, arXiv:2308.08569.
25. Zhang, S.; Cao, W.; Wang, J.; Wu, T.; Wang, Y.; Wang, Y.; Zhou, D. Reconfigurable Multifunctional Metasurfaces for Full-Space Electromagnetic Wave Front Control. *Micromachines* **2024**, *15*, 1282. [CrossRef]
26. Ma, B.; Li, J.; Chen, Y.; Si, Y.; Gao, H.; Wu, Q.; Li, M. A 60 GHz Slotted Array Horn Antenna for Radar Sensing Applications in Future Global Industrial Scenarios. *Micromachines* **2024**, *15*, 728. [CrossRef]
27. Zhao, C.; Zhang, Q.; Pei, W.; Jin, J.; Sun, F.; Zhang, H.; Zhou, R.; Liu, D.; Xu, F.; Zhang, X.; et al. Design and Analysis of 5-DOF Compact Electromagnetic Levitation Actuator for Lens Control of Laser Cutting Machine. *Micromachines* **2024**, *15*, 641. [CrossRef]
28. Huang, Y.; Liang, J.; Wu, Z.; Chen, Q. Design of 2.45 GHz High-Efficiency Rectifying Circuit for Wireless RF Energy Collection System. *Micromachines* **2024**, *15*, 340. [CrossRef]
29. Ahmed, F.H.; Saad, R.; Khamas, S.K. A Novel Compact Broadband Quasi-Twisted Branch Line Coupler Based on a Double-Layered Microstrip Line. *Micromachines* **2024**, *15*, 142. [CrossRef]
30. Abdou, T.S.; Khamas, S.K. A Multiband Millimeter-Wave Rectangular Dielectric Resonator Antenna with Omnidirectional Radiation Using a Planar Feed. *Micromachines* **2023**, *14*, 1774. [CrossRef]
31. Behera, B.R.; Alsharif, M.H.; Jahid, A. Investigation of a circularly polarized metasurface antenna for hybrid wireless applications. *Micromachines* **2023**, *14*, 2172. [CrossRef] [PubMed]
32. Li, F.; Cheng, Y.F.; Wang, G.; Luo, J. A Novel High-Isolation Dual-Polarized Patch Antenna with Two In-Band Transmission Zeros. *Micromachines* **2023**, *14*, 1784. [CrossRef]
33. Hu, X.; Zhang, G.; Deng, G.; Li, X. Experimental Study on the Compatibility of PD Flexible UHF Antenna Sensor Substrate with SF_6/N_2. *Micromachines* **2023**, *14*, 1516. [CrossRef]
34. Zhan, Y.; Chen, Y.; Guo, H.; Wu, Q.; Li, M. Design of a Ka-band five-bit MEMS delay with a coplanar waveguide loaded U-shaped slit. *Micromachines* **2023**, *14*, 1508. [CrossRef]
35. Neto, A.G.; Silva, J.C.e.; Carvalho, J.N.d.; Peixeiro, C. Planar Printed Structures Based on Matryoshka Geometries: A Review. *Micromachines* **2024**, *15*, 469. [CrossRef]

36. Wang, X.; Liu, Y.; Jia, Y.; Su, N.; Wu, Q. Ultra-wideband and narrowband switchable, bi-functional metamaterial absorber based on vanadium dioxide. *Micromachines* **2023**, *14*, 1381. [CrossRef]
37. Weng, Z.; Liang, C.; Xue, K.; Lv, Z.; Zhang, X. A Miniaturized Loaded Open-Boundary Quad-Ridge Horn with a Stable Phase Center for Interferometric Direction-Finding Systems. *Micromachines* **2024**, *16*, 44. [CrossRef]

Disclaimer/Publisher's Note: The statements, opinions and data contained in all publications are solely those of the individual author(s) and contributor(s) and not of MDPI and/or the editor(s). MDPI and/or the editor(s) disclaim responsibility for any injury to people or property resulting from any ideas, methods, instructions or products referred to in the content.

Article

A Miniaturized Loaded Open-Boundary Quad-Ridge Horn with a Stable Phase Center for Interferometric Direction-Finding Systems

Zibin Weng *, Chen Liang, Kaibin Xue, Ziming Lv and Xing Zhang

The National Key Laboratory of Radar Detection and Sensing, Xidian University, Xi'an 710071, China; liangchen@stu.xidian.edu.cn (C.L.); kbxue@stu.xidian.edu.cn (K.X.); 23021211267@stu.xidian.edu.cn (Z.L.); monicastarstar@163.com (X.Z.)
* Correspondence: zibinweng@mail.xidian.edu.cn

Abstract: In order to achieve high accuracy in interferometric direction-finding systems, antennas with a stable phase center in the working bandwidth are required. This article proposes a miniaturized loaded open-boundary quad-ridge horn (LOQRH) antenna with dimensions of 40 mm × 40 mm × 49 mm. First, to stabilize the phase center of the antenna, the design builds on the foundation of a quad-ridge horn antenna, where measures such as optimizing the ridge structure and introducing resistive loading were implemented to achieve size reduction. Second, electrically small-sized antennas are more susceptible to the effects of common-mode currents (CMCs), which can reduce the symmetry of the radiation pattern and the stability of the phase center. To avoid the generation of common-mode currents during operation, a self-balanced feed structure was introduced into the proposed antenna design. This structure establishes a balanced circuit and routes the feedline at the voltage null point, effectively suppressing the common-mode current. As a result, the miniaturization of the LOQRH antenna was achieved while ensuring the suppression of the common-mode current, thereby maintaining the stability of the antenna's electromagnetic performance. The measured results show that the miniaturized antenna has a small phase center change of less than 20.3 mm within 2–18 GHz, while the simulated phase center fluctuation is only 14.6 mm. In addition, when taking 18.5 mm in front of the antenna's feed point as the phase center, the phase fluctuation is less than 22.5° within the required beam width. Along with the desired stable phase center, the miniaturized design makes the proposed antenna suitable for interferometric direction-finding systems.

Keywords: common-mode current (CMC); stable phase center; radio interferometry; ultrawideband antennas

Academic Editor: Haejun Chung

Received: 17 December 2024
Revised: 27 December 2024
Accepted: 28 December 2024
Published: 30 December 2024

Citation: Weng, Z.; Liang, C.; Xue, K.; Lv, Z.; Zhang, X. A Miniaturized Loaded Open-Boundary Quad-Ridge Horn with a Stable Phase Center for Interferometric Direction-Finding Systems. *Micromachines* **2025**, *16*, 44. https://doi.org/10.3390/mi16010044

Copyright: © 2024 by the authors. Licensee MDPI, Basel, Switzerland. This article is an open access article distributed under the terms and conditions of the Creative Commons Attribution (CC BY) license (https://creativecommons.org/licenses/by/4.0/).

1. Introduction

Accurate positioning and localization are essential in modern technologies such as radar detection, communication systems [1,2], and electronic warfare. Traditional methods like time of arrival (TOA) [3–5], frequency-based estimation, and velocity-based techniques have been widely studied for localization. While these methods have their merits, they often face challenges in complex environments, such as susceptibility to errors from signal timing, mobility, or the need for extensive hardware and signal processing.

In contrast, Direction of Arrival (DOA) estimation [6–8] has gained significant attention due to its ability to provide precise localization through the measurement of signal angles. Using multiple antennas, DOA estimation is highly effective for applications like satellite navigation, automotive systems, and communication networks.

Figure 1 shows the simplified interferometric model for DOA estimation, which ignores factors such as differences between receiving channels and the coupling between antenna elements.

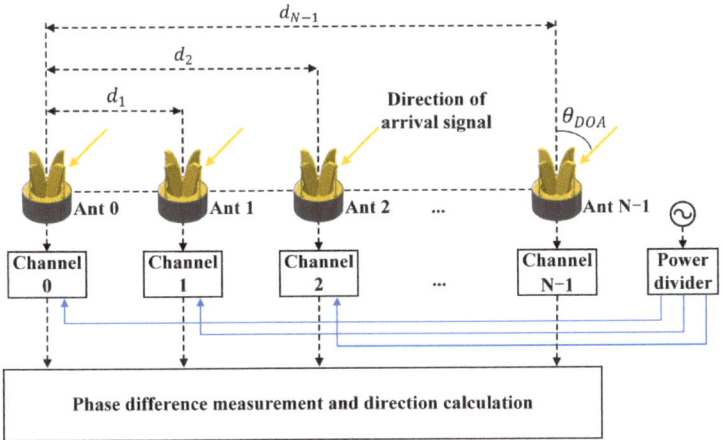

Figure 1. Simplified schematic of an interferometric system.

The calculation process of the simplified interferometric model is shown in (1)–(3). Table 1 shows the detailed explanation of each parameter in Equations (1)–(3).

$$\theta_{DOA} = \arcsin\left(\frac{\lambda}{2\pi}\sum_{k=1}^{N-1} d_k \phi_{W,k} \Big/ \sum_{k=1}^{N-1} d_k^2\right) \quad (1)$$

$$\begin{cases} \phi_{W,k+1} = \phi_{Mea,k+1} + \gamma_{Mea,k+1} + \sigma \\ \gamma_{Mea,k} = 2\pi \cdot floor\left(\frac{\alpha_k \phi_{W,k}}{2\pi}\right), \alpha_k = \frac{d_{k+1}}{d_k}, \phi_{W,1} = \phi_{Mea,1} \end{cases} \quad (2)$$

$$\sigma = \begin{cases} 0 & \phi_{Mea,k+1} + \gamma_{Mea,k+1} - \alpha_k \phi_{W,k} \in [-\pi, \pi) \\ 2\pi & \phi_{Mea,k+1} + \gamma_{Mea,k+1} - \alpha_k \phi_{W,k} < -\pi \\ -2\pi & \phi_{Mea,k+1} + \gamma_{Mea,k+1} - \alpha_k \phi_{W,k} \geq \pi \end{cases} \quad (3)$$

Table 1. Symbol table.

Symbol	Explanation
θ_{DOA}	Minimum mean square error solution for the incoming wave angle
d_k	Length of the interferometer baseline between each unit antenna N and the reference antenna 0
α_k	Ratio of the lengths of the adjacent baseline
$\phi_{W,k}$	Phase deblurring result of $\phi_{W,k}$
$\gamma_{Mea,k}/2\pi$	Number of fuzzy baseline k-measurement directions
$\phi_{Mea,k}$	Measured phase difference between channels k and 0
$\Delta\phi$	The phase measurement error
$floor(\cdot)$	The downward rounding function

As electromagnetic field detectors in the interferometer direction-finding system, the performance of antennas is crucial for receiving signals, particularly phase center stability. Since the DOA is calculated based on the phase difference between different channels and the baseline length (distance between unit antennas), any phase measurement error can significantly impact detection accuracy.

According to the reciprocity characteristic of the antenna, the phase center is also the location of the antenna reference point (ARP), where the antenna receives signals [9,10]. There is a unique phase center for an ideal antenna, but this is impossible in practice: different signal frequencies will cause phase center offset (PCO), and different signal incident angles also cause phase center variations (PCVs) [11,12]. Fluctuating phase centers will cause changes in the baseline of the interferometer, which will inevitably introduce deviations in the test results.

In the case of only considering the phase detection error $\Delta\phi$, which is caused by the fluctuation of the phase center, the baseline interferometry error $\Delta\theta_i$ can be given by Equation (4):

$$\Delta\theta_i = \frac{\lambda_{\min} \cdot \Delta\phi}{2\pi d_i \cos\theta_{DOA}} \quad (4)$$

Since the phase center of most antennas varies with azimuth and is challenging to solve analytically, there are two main methods to correct the errors caused by phase center fluctuations. The first method is to ensure consistency across the antenna units. When all units operating in the same frequency band are consistent, PCO and PCV will only cause a small shift in the baseline, which will not significantly affect the test results. The second method involves making sufficient corrections. If PCO and PCV fluctuations lead to large changes in the baseline length, repeated measurements are needed to identify the pattern, allowing for corrections to eliminate this error [13–20].

However, when antenna units are without choking [21,22] structures, a part of the feed current will flow on the outer surface of the coaxial cable, generating the common-mode current (CMC). For electrically large-sized antennas, the influence of the CMC is often ignored. However, the influence must be addressed for many electrically small-sized antennas. Impacted by the CMC, the feeding structures easily become a part of the radiation structure, largely affecting the radiation pattern symmetry and phase center stability. The CMC will distort the radiation characteristics of the antenna units, and its influence on the phase center is complex to evaluate [23,24]. In addition, there will inevitably be mutual coupling between the units, which will also make the PCO and PCV change irregularly and affect the measurement accuracy.

To address these challenges, this article proposes a miniaturized loaded open-boundary quad-ridge horn (LOQRH) antenna, designed to maintain stable phase center characteristics while minimizing the impact of CMC. The proposed antenna incorporates several key design techniques, including optimizing the ridge curve and incorporating resistive loading, which reduces surface current flow and achieves compact dimensions of $0.27\lambda_L \times 0.27\lambda_L \times 0.33\lambda_L$ (where λ_L is the free-space wavelength at the lowest operating frequency). Furthermore, by conducting an odd–even mode analysis [25,26], this article gives point N as the lead-out position for the coaxial feedline. When the feedline is led from point N, there is no energy flowing on the outer surface of the feedline, which also means no CMC. These key design techniques enhance phase center stability and improve radiation pattern symmetry, making the LOQRH antenna ideal for DOA estimation systems.

2. Antenna Geometry and Design

2.1. Structure of the Proposed Antenna

Figure 2a shows the whole 3D structure of the antenna unit, which includes three main components: the ridged horn, the loaded resistor, and the self-balance feeding structure.

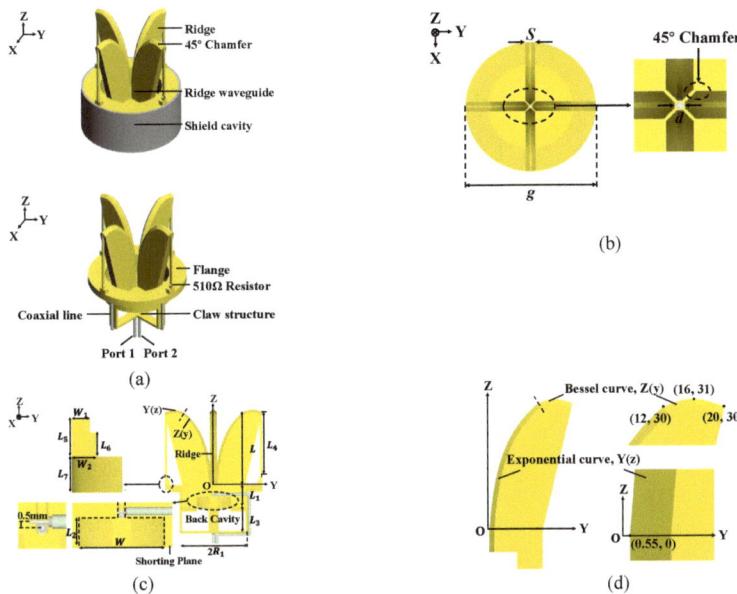

Figure 2. Geometry of the proposed antenna. (**a**) 3D view. (**b**) Top view. (**c**) Section view. (**d**) Detail of the ridge. $S = 3$ mm, $L = 30$ mm, $L_1 = 9$ mm, $L_2 = 4$ mm, $L_3 = 10$ mm, $L_4 = 26$ mm, $L_5 = 3.2$ mm, $L_6 = 2$ mm, $L_7 = 3$ mm, $R_1 = 13$ mm, $h_r = 8$ mm, $W = 14$ mm, $W_1 = 1.5$ mm, $W_2 = 2$ mm, $g = 40$ mm, $d = 1.1$ mm.

The ridge is critical in the impedance transformation from the feed point to the aperture. The proposed antenna combines an additional Bessel curve with the exponential curve to optimize the ridge more flexibly.

L indicates the length of the exponential curve in the z direction, and h_r indicates the length of the Bessel curve in the y direction. $Y'(L)$, g, and d are the slope at the end of $Y(z)$, the diameter of the aperture, and the ridge spacing, respectively. What is more, the smallest ridge spacing can lower the ridge waveguide's cut-off frequency and make the equivalent impedance closer to 50 Ω.

As shown in Figure 2b, we perform 45° chamfering treatments on ridges, which prevents the radiation performance from being affected by too large ridge spacing and avoids the collision of ridges caused by too small ridge spacing.

To improve the low-frequency characteristics of the proposed LOQRH, a resistor is loaded between each ridge and the fixed flange to reduce the ridge length and absorb the energy not radiated. The VSWR performance of the antenna loaded with resistors of different resistance values and an unloaded resistor is given in Figure 3. It can be seen that the VSWR of the antenna is significantly decreased at 2–5 GHz after loading the resistor compared to when no resistor is loaded, with little change at high frequencies. Integrating the proposed antenna's S_{11} and gain performance, the resistors are chosen as 510 Ω.

As shown in Figure 2c, two 047 50 Ω semi-rigid coaxial probes are used for feeding to facilitate processing and miniaturizing. The two probes are placed orthogonally for dual polarization and better isolation between the ports. For the VSWR of the ports to be as similar as possible, it is necessary to make the two probes as close as possible without contact. Finally, the distance between them is 0.5 mm.

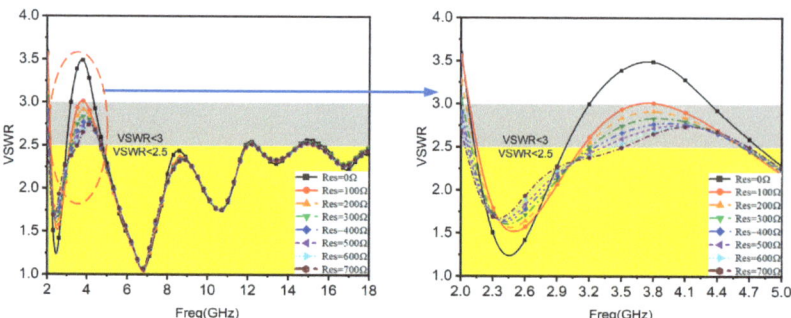

Figure 3. Comparison of VSWR performance of antennas before and after loading resistors.

There is also a stepped cavity with a shorting plane at the bottom of the ridge waveguide, as shown in Figure 2c. To make the shorting plane realize high impedance characteristics in the entire frequency band, the height L_2 and length W of the back cavity steps can be adjusted, thereby inhibiting the backward transmission of electromagnetic waves.

In addition, the coaxial feedline will be distributed along the claw structure and led out at the bottom center of the claw structure, as shown in Figure 2c. The detailed mechanism will be discussed later.

In summary, the proposed miniaturization design combines an exponential curve and a Bessel curve for the ridge profile, ensuring a smooth transition of the antenna's characteristic impedance from the feed point to the bell mouth surface, with chamfered ridge edges for enhanced performance. Additionally, resistors are loaded at the ends of each ridge to absorb the energy that is not radiated due to the reduced vertical size of the ridge at low frequencies, effectively improving low-frequency performance. Compared to the traditional four-ridge horn antenna, the antenna proposed in this paper achieves a reduction in both the vertical dimension L and aperture size g. To attenuate higher-order modes in the horn, the opening section length is typically greater than 0.5λ (where λ corresponds to the lowest frequency). For 2 GHz, this value is 75 mm, whereas the proposed antenna achieves a length of 30 mm. Furthermore, as the vertical size L of the antenna decreases, the aperture size g also reduces. Therefore, the proposed four-ridge horn antenna successfully reduces both the vertical dimension L and aperture size g, leading to a compact design.

2.2. Common-Mode Current Suppression

The fluctuation of the antenna's phase center is often related to the antenna's dimensions and also weakens as the antenna's dimensions decrease. Therefore, interferometric direction-finding antennas can stabilize their phase center once their dimensions are small enough. Also, benefitting from the limited dimensions, interferometric direction-finding antennas can reduce coupling between themselves. In this way, each unit can be a "similar point source", which is essential for ensuring the antennas' performance and improving interferometric direction-finding systems' accuracy.

Due to the potential difference between the inner and outer surfaces of the outer conductor of the coaxial line, a part of the current on the inner surface flows back to the source along the outer surface. This part of the current is the CMC. The miniaturized antenna is more susceptible to the influence of the CMC. The CMC affects the feed balance, causing the pattern to deviate, intensifying phase center fluctuations, and even introducing unnecessary cross-polarization [27–29].

For the feeding structure, the inner conductor of the coaxial line is connected to a ridge, and the outer conductor is connected to the opposite ridge. To suppress the CMC, the proposed antenna added a claw structure after the shorting plane, as shown in Figure 2c. The coaxial feedline will be distributed along the claw structure and led out at the bottom center of the claw structure.

The self-balanced feeding structure's equivalent model of the odd and even modes is shown in Figure 4. A_1 and A_2 are the input electric field at points M_1 and M_2, respectively. The electrical length from point M_1 to point N is θ_1, and the electrical length from point M_1 to point M_2 is θ_T. Then, the odd-mode input $A_o = (A_1 - A_2)/2$, and even-mode input $A_e = (A_1 + A_2)/2$. The function of electric field distribution and electric length can be expressed by $e^{-\alpha\theta}e^{j\beta\theta}$, where α represents the attenuation factor of the electric field as a function of electrical length, and β is the propagation factor. At point N, where the coaxial feedline is led out, the electric field B can be expressed by Equation (9):

$$B = (A_e + A_o) \cdot e^{-\alpha\theta_1} \cdot e^{j\beta\theta_1} + (A_e - A_o) \cdot e^{-\alpha(\theta_T - \theta_1)} \cdot e^{j\beta(\theta_T - \theta_1)} \quad (5)$$

Let B = 0, then

$$\theta_1 = \frac{1}{2}\theta_T + \frac{1}{2(-\alpha + j\beta)}\ln(-\frac{A_1}{A_2}) \quad (6)$$

Near the midpoint, there must be a point where the electric field intensity is 0. When leading from this point, no energy flows on the outer layer of the coaxial line, thereby suppressing the occurrence of the CMC from its source. This point is taken as the lead-out point, point N.

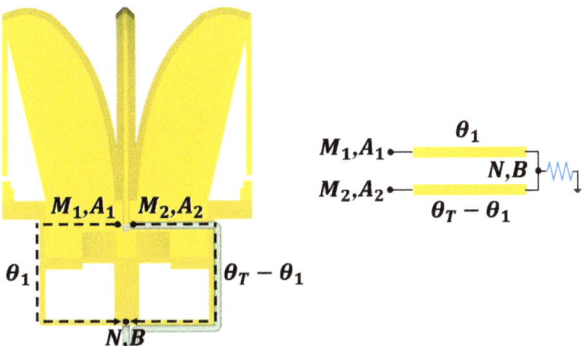

Figure 4. The equivalent odd and even mode analysis model of the self-balanced feeding structure.

Figure 5 shows the surface current distribution and the normalized radiation patterns at some frequency points. When feeding without CMC suppression, there will be a CMC on the coaxial line, and the pattern of the antenna will have obvious distortion accordingly; when feeding with CMC suppression, there will be no CMC on the coaxial line, and the pattern of the antenna also maintains good symmetry.

The specific normalized radiation pattern distortion is shown in the gray ovals in Figure 5, while the specific details of common-mode current suppression are shown in the red ovals in Figure 5.

At the same time, since the common-mode current suppression effect is frequency related, we show the effect of the CMC suppression structure at high frequencies in Figure 6. Combining Figures 5 and 6, we can clearly see that for antennas with electrically small-sized antennas, the common-mode current has a large effect. Due to the effect of common-mode current, the radiation pattern is severely distorted. Thus, for our proposed LOQRH antenna, the effect of the common-mode current is much larger in the low-frequency case than in the high-frequency case.

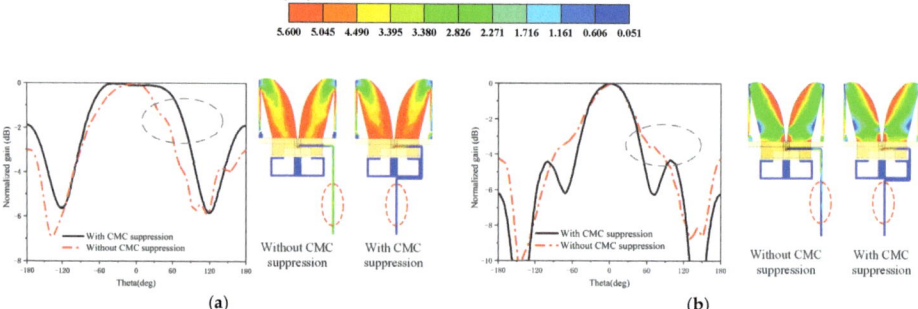

Figure 5. Normalized radiation patterns and surface current distribution of the proposed antenna with or without CMC suppression in the XOZ plane. (**a**) 2 GHz. (**b**) 4 GHz.

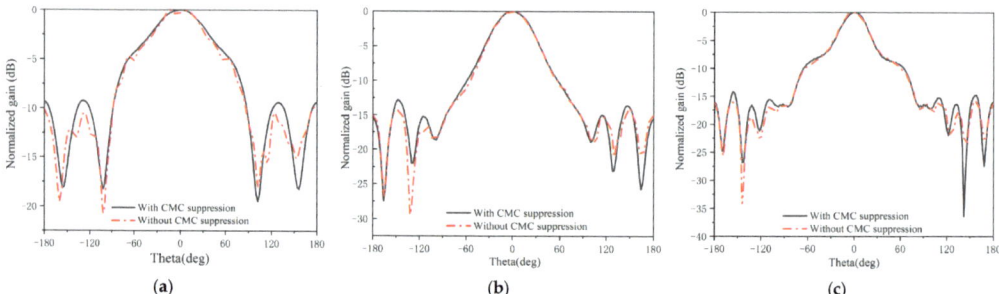

Figure 6. The effect of the CMC suppression at high frequencies. (**a**) 6 GHz, (**b**) 10 GHz, (**c**) 14 GHz.

Therefore, the CMC suppression structure can effectively suppress the CMC on the coaxial line and reduce the antenna pattern distortion caused by the CMC.

2.3. Phase Center Stability

This article solves the phase centers of the XOZ and YOZ planes, respectively, and then performs an arithmetic average on them to obtain the equivalent phase center in the entire space. As to the proposed antenna, it can be assumed that (x_0, y_0, z_0) is its phase center in a particular plane. We have simulated in CST to verify that x_0 and y_0 can be considered almost zero due to the fact that the proposed antenna is symmetric to both the XOZ plane and the YOZ plane. Taking the XOZ plane as an example, z_0 can be obtained by Equations (7)–(9).

$$\Psi(\theta) = \vec{k} \cdot \vec{r}_0 + \Psi_0 \tag{7}$$

$$\vec{r}_0 = x_0\hat{x} + y_0\hat{y} + z_0\hat{z} \tag{8}$$

$$z_0 = \frac{c_0}{\pi^2 f} \int_0^\pi \Psi(\theta) \cos\theta \, d\theta \tag{9}$$

where $\Psi(\theta)$ is the measured far-field phase pattern, c_0 is the speed of light, and f is the working frequency.

The optimal position of the proposed antenna, which is selected based on the equivalent phase centers at all frequency points, is determined to be 18.5 mm in front of the feed point of LOQRH. At 2 GHz, the phase center of the proposed antenna and the antenna in [30] deviates farthest from the optimal position. The deviation is 10.7 mm for the proposed antenna and 24.6 mm in [30]. Furthermore, compared to the antenna referenced in [30] at 2 GHz, the stabilization of the phase center in our proposed antenna contributes to an enhanced success rate in direction finding for incoming signals.

3. Measurement and Analysis

The radiation characteristics of the proposed antenna are measured by the Anritsu MS46322A VNA (which is manufactured by Anritsu Corporation, headquartered in Atsugi, Kanagawa, Japan) in the anechoic chamber. A broadband horn antenna (A-INFO LB-10180) produced by A-INFO Inc., based in Chengdu, China, was used as the probe antenna to perform far-field measurements of the main polarization and cross-polarization of the LOQRH antenna.

Laser alignment was employed to ensure proper main polarization alignment, and the LOQRH antenna mounted on a turntable was rotated 360 degrees to obtain the radiation pattern for the main polarization. Simultaneously, the broadband horn antenna was rotated by 90 degrees, and the turntable was similarly rotated 360 degrees to capture the cross-polarization radiation pattern. Figure 7 shows the simulated and measured radiation patterns between 2 and 18 GHz. As seen in Figure 7, measured patterns have good symmetry, showing that the CMC has been well suppressed.

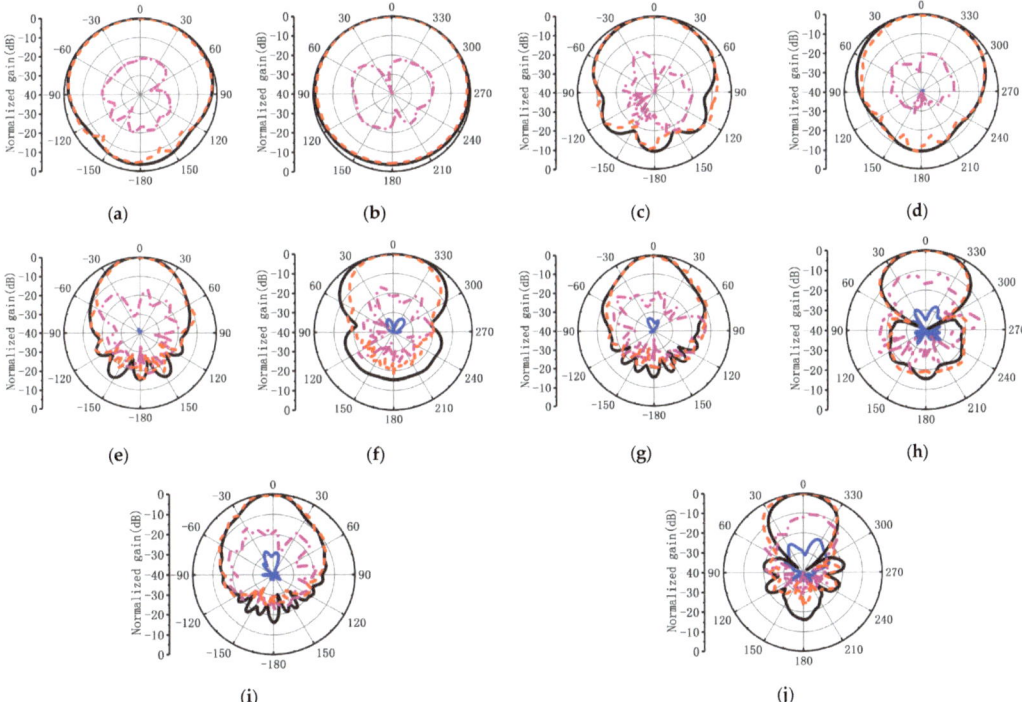

Figure 7. Simulated and measured patterns of the proposed antenna. (**a**) *XOZ* plane at 2 GHz. (**b**) *YOZ* plane at 2 GHz. (**c**) *XOZ* plane at 6 GHz. (**d**) *YOZ* plane at 6 GHz. (**e**) *XOZ* plane at 10 GHz. (**f**) *YOZ* plane at 10 GHz. (**g**) *XOZ* plane at 14 GHz. (**h**) *YOZ* plane at 14 GHz. (**i**) *XOZ* plane at 18 GHz. (**j**) *YOZ* plane at 18 GHz.

Table 2 shows the values of the simulated and measured phase fluctuations in the XOZ plane and the YOZ plane at some frequencies when taking the optimal position, 18.5 mm in front of the feed point. During the test, we normalized the measured phase values by the measured phase value at 0° to avoid unnecessary slight changes.

Table 2. Phase fluctuation values in the XOZ plane and YOZ plane.

Frequency (GHz)	XOZ Plane Phase Fluctuation (°)		YOZ Plane Phase Fluctuation (°)	
	Simulated	Measured	Simulated	Measured
2	10.2	21.2	7.6	12.0
6	17.4	22.2	11.3	15.8
10	4.3	8.5	18.6	22.5
14	8.4	12.5	12.9	14.5
18	11.4	16.1	14.1	17.8

If the HPBW of the proposed antenna is less than 90° at a specific frequency point, then we focus on the phase fluctuation within the HPBW; otherwise, we focus on the phase fluctuation within ±45°. It can be seen that within the required range, the phase fluctuation of the proposed antenna within 2–18 GHz does not exceed 22.5° [31] in Table 2, which means that the proposed antenna has a stable phase center within 2–18 GHz.

It can also be found that the measured phase fluctuation is larger than the simulated one in Table 2, which may be caused by the fact that the rotation axis and the phase center of the antenna do not entirely coincide due to the constraints of the measurement conditions.

Figure 8 presents the simulated and measured S parameters and gain. As Figure 8 shows, the measured $|S_{21}|$ between the ports is greater than 25 dB, possibly because the two ports were not strictly orthogonal during the test but still met the requirement. The measured $|S_{11}|$ is less than −6 dB from 2 to 18 GHz. Affected by the resistors, the antenna's gain is low when working at 2 GHz. As the frequency increases, the proposed antenna becomes less affected by the resistors, and its gain rises accordingly.

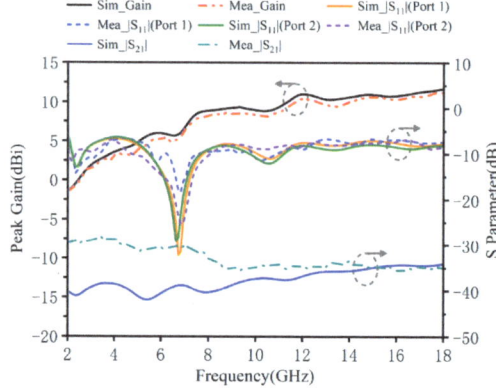

Figure 8. Simulated and measured S parameter and peak gain of the proposed antenna. The direction of the arrows indicates which y-axis is referenced for each data set.

Figure 9 shows the difference between the simulated and measured phase center change in the full space.

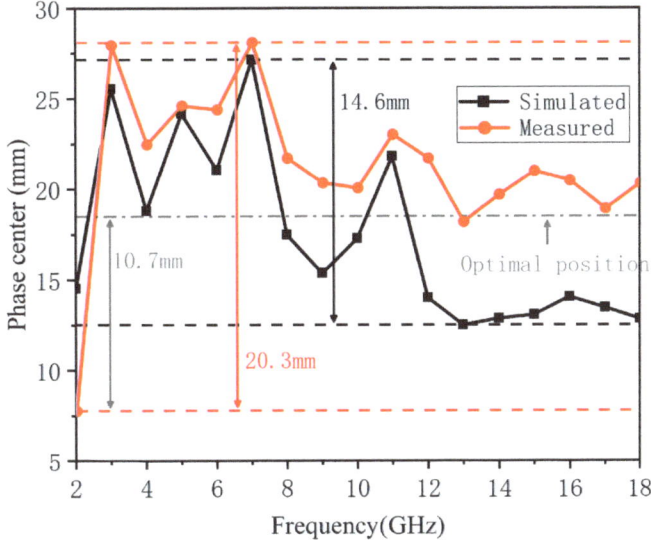

Figure 9. Simulated and measured phase center in the entire space.

The average of the *XOZ* plane and *YOZ* plane phase centers is usually used as the full-space phase center. As shown in Figure 9, the measured full-space phase center change fluctuates less than 20.3 mm in the working frequency band. If 18.5 mm is used as the optimal phase center, the maximum phase center change is 10.7 mm at 2 GHz.

Figure 10 shows a prototype of the proposed antenna.

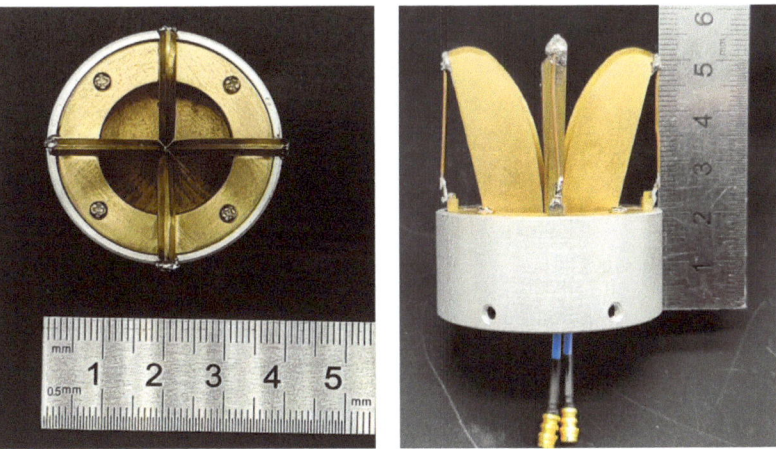

Figure 10. Photographs of the antenna prototype.

Table 3 compares the proposed antenna with previously published works. It can be observed that the proposed design achieves miniaturization while maintaining the stability of the phase center and a wide operating bandwidth.

Table 3. Comparison of main results with previously published works.

Ref. Work	BW (f_{max}/f_{min})	Phase Center Change (mm)	Profile ($\lambda_{f\,min}$)
[30]	6:1	<30 (2~12 GHz)	$1.40 \times 1.40 \times 1.07$
[32]	1.125:1	\	$1.61 \times 1.61 \times 0.30$
[33]	2.25:1	\	$2.00 \times 2.00 \times 1.43$
[34]	5.2:1	\	$1.61 \times 1.61 \times 1.46$
[35]	6:1	<50 (2~12 GHz)	$1.14 \times 1.14 \times 1.04$
[36]	3.3:1	<40 (8~18 GHz)	$1.29 \times 1.29 \times 3.16$
[37]	8:1	<50 (1.5~12 GHz)	$0.42 \times 1.42 \times 1.32$
This work	9:1	<20.3 (2~18 GHz)	$0.27 \times 0.27 \times 0.33$

4. Conclusions

Based on CMC suppression, a miniaturized, broadband LOQRH antenna for interferometric direction-finding systems is proposed in this article. By introducing the self-balanced feeding structure, the proposed LOQRH can not only achieve small dimensions of $0.27\,\lambda_L \times 0.27\,\lambda_L \times 0.33\,\lambda_L$ but also effectively suppress the CMC and stabilize the phase center. The measured results illustrate that the proposed antenna has a 2 to 18 GHz bandwidth. Within the working bandwidth, the measured phase center of the proposed LOQRH fluctuates less than 20.3 mm, while the simulated phase center fluctuation is only 14.6 mm. Therefore, the proposed antenna is very suitable for interferometric direction-finding systems, especially for those tiny, multiprobe ones.

Author Contributions: All authors have significantly contributed to the research presented in this manuscript. Conceptualization, Z.W., C.L. and K.X.; investigation, Z.W. and C.L.; writing, C.L. and K.X.; Z.W., Z.L. and X.Z. reviewed and revised the manuscript. All authors have read and agreed to the published version of the manuscript.

Funding: This work was supported by ZTE Industry–University–Institute Cooperation Funds under grant No. IA20240710018.

Data Availability Statement: All data generated or analyzed during this study are included in this manuscript. There are no additional data or datasets beyond what is presented in the manuscript.

Conflicts of Interest: The authors declare no conflicts of interest.

References

1. Cheng, L.; Qin, S.; Feng, G. Learning-based admission control for low-earth-orbit satellite communication networks. *ZTE Commun.* **2023**, *21*, 54–62.
2. Fan, G.T.; Wang, Z.B. Intelligent antenna attitude parameters measurement based on deep learning SSD model. *ZTE Commun.* **2022**, *20*, 36–43.
3. Chen, Z.; Wang, L.; Zhang, M. Virtual Antenna Array and Fractional Fourier Transform-Based TOA Estimation for Wireless Positioning. *Sensors* **2019**, *19*, 638. [CrossRef]
4. Zou, Y.; Fan, J.; Wu, L.; Liu, H. Fixed Point Iteration Based Algorithm for Asynchronous TOA-Based Source Localization. *Sensors* **2022**, *22*, 6871. [CrossRef]
5. Luo, R.; Yan, L.; Deng, P.; Kuang, Y. Hybrid TOA/AOA Virtual Station Localization Based on Scattering Signal Identification for GNSS-Denied Urban or Indoor NLOS Environments. *Appl. Sci.* **2022**, *12*, 12157. [CrossRef]
6. Cheong, P.; Tu, M.; Choi, W.-W.; Wu, K. An Ingenious Multiport Interferometric Front-End for Concurrent Dual-Band Transmission. *IEEE Trans. Microw. Theory Tech.* **2022**, *70*, 1725–1731. [CrossRef]
7. Abdulkawi, W.M.; Alqaisei, M.A.; Sheta, A.-F.A.; Elshafiey, I. New Compact Antenna Array for MIMO Internet of Things Applications. *Micromachines* **2022**, *13*, 1481. [CrossRef]
8. Liu, M.; Hu, J.; Zeng, Q.; Jian, Z.; Nie, L. Sound Source Localization Based on Multi-Channel Cross-Correlation Weighted Beamforming. *Micromachines* **2022**, *13*, 1010. [CrossRef]

9. IEEE Standard 145-2013; IEEE Standard for Definitions of Terms for Antennas. IEEE: New York, NY, USA, 2014.
10. IEEE 149-1979 (R2008); IEEE Standard Test Procedures for Antennas. IEEE: New York, NY, USA, 2008.
11. Hu, Z.Y.; Li, Z.; Gang, O.; Bin, Z. Research on antenna phase center anechoic chamber calibration method. In Proceedings of the International Conference on Microwave and Millimeter Wave Technology, Chengdu, China, 8–11 May 2010; pp. 1522–1524.
12. Yao, Y.; Zhang, L.; Chen, W.; Yu, C.; Dong, D. Phase Center Variation Modification on BDS High-Precision Baseline Solution. In Proceedings of the 4th International Conference on Information Science and Control Engineering, Changsha, China, 21–23 July 2017; pp. 1591–1595.
13. Padilla, P.; Pousi, P.; Tamminen, A.; Mallat, J.; Ala-Laurinaho, J.; Sierra-Castaner, M.; Raisanen, A.V. Experimental Determination of DRW Antenna Phase Center at mm-Wavelengths Using a Planar Scanner: Comparison of Different Methods. IEEE Trans. Antennas Propag. 2011, 59, 2806–2812. [CrossRef]
14. Chen, Y.; Vaughan, R.G. Determining the three-dimensional phase center of an antenna. In Proceedings of the 31th URSI General Assembly and Scientific Symposium (URSI GASS), Beijing, China, 16–23 August 2014; pp. 1–4.
15. Menudier, C.; Chantalat, R.; Thevenot, M.; Monediere, T.; Dumon, P.; Jecko, B. Phase Center Study of the Electromagnetic Band Gap Antenna: Application to Reflector Antennas. IEEE Antennas Wirel. Propag. Lett. 2007, 6, 227–231. [CrossRef]
16. Waidelich, D. The phase centers of aperture antennas. IEEE Trans. Antennas Propag. 1980, 28, 263–264. [CrossRef]
17. Esposito, C.; Gifuni, A.; Perna, S. Measurement of the Antenna Phase Center Position in Anechoic Chamber. IEEE Antennas Wirel. Propag. Lett. 2018, 17, 2183–2187. [CrossRef]
18. Choni, Y.I. Hodograph of Antenna's Local Phase Center: Computation and Analysis. IEEE Trans. Antennas Propag. 2015, 63, 2819–2823. [CrossRef]
19. Harke, D.; Garbe, H.; Chakravarty, P. A new method to calculate phase center locations for arbitrary antenna systems and scenarios. In Proceedings of the IEEE International Symposium on Electromagnetic Compatibility (EMC), Ottawa, ON, Canada, 25–29 July 2016; pp. 674–678.
20. Harke, D.; Garbe, H.; Chakravarty, P. A system-independent algorithm for phase center determination. In Proceedings of the International Symposium on Electromagnetic Compatibility EMC Europe, Angers, France, 4–7 September 2017; pp. 1–5.
21. Fukushima, T.; Michishita, N.; Morishita, H.; Fujimoto, N. Coaxially Fed Monopole Antenna with Choke Structure Using Left-Handed Transmission Line. IEEE Trans. Antennas Propag. 2017, 65, 6856–6863. [CrossRef]
22. Zhang, Z.-Y.; Zhao, Y.; Zuo, S.; Yang, L.; Ji, L.-Y.; Fu, G. A Broadband Horizontally Polarized Omnidirectional Antenna for VHF Application. Trans. Antennas Propag. 2018, 66, 2229–2235. [CrossRef]
23. Zheng, Y.; Weng, Z.; Qi, Y.; Fan, J.; Li, F.; Yang, Z.; Drewniak, J.L. Calibration Loop Antenna for Multiple Probe Antenna Measurement System. IEEE Trans. Instrum. Meas. 2020, 69, 5745–5754. [CrossRef]
24. Zheng, Y.; Lin, B.; de Paulis, F.; Violette, M.; Ye, X.; Qi, Y. Loop Antennas for Accurate Calibration of OTA Measurement Systems: Review, Challenges, and Solutions. IEEE Trans. Instrum. Meas 2022, 71, 1–13. [CrossRef]
25. Jones, E.M.T. Coupled-Strip-Transmission-Line Filters and Directional Couplers. IRE Trans. Microw. Theory Technol. 1956, 4, 75–81. [CrossRef]
26. Speciale, R.A. Even- and Odd-Mode Waves for Nonsymmetrical Coupled Lines in Nonhomogeneous Media. IEEE Trans. Microw. Theory Technol. 1975, 23, 897–908. [CrossRef]
27. Cai, Z.; Weng, Z.; Qi, Y.; Fan, J.; Zhuang, W. A High-Performance Standard Dipole Antenna Suitable for Antenna Calibration. IEEE Trans. Antennas Propag. 2021, 69, 8878–8883. [CrossRef]
28. Fukasawa, T.; Yoneda, N.; Miyashita, H. Investigation on current reduction effects of baluns for measurement of a small antenna. IEEE Trans. Antennas Propag. 2019, 67, 4323–4329. [CrossRef]
29. Nguyen, V.-A.; Park, B.-Y.; Park, S.-O.; Yoon, G. A planar dipole for multiband antenna systems with self-balanced impedance. IEEE Antennas Wirel. Propag. Lett. 2014, 13, 1632–1635. [CrossRef]
30. Beukman, T.S.; Meyer, P.; Ivashina, M.V.; Maaskant, R. Modal-Based Design of a Wideband Quadruple-Ridged Flared Horn Antenna. IEEE Trans. Antennas Propag. 2016, 64, 1615–1626. [CrossRef]
31. Balanis, C.A. Antenna Theory and Design; Wiley: Hoboken, NJ, USA, 2005; pp. 566–568.
32. Moy-Li, H.C.; Sanchez-Escuderos, D.; Antonino-Daviu, E.; Ferrando-Bataller, M. Low-profile radially corrugated horn antenna. IEEE Antennas Wirel. Propag. Lett. 2017, 16, 3180–3183. [CrossRef]
33. Abbas-Azimi, M.; Mazloumi, F.; Behnia, F. Design of broadband constant-beamwidth conical corrugated-horn antenna. IEEE Trans. Antennas Propag. 2009, 51, 109–114. [CrossRef]
34. Dong, B.; Yang, J.; Dahlström, J.; Flygare, J.; Pantaleev, M.; Billade, B. Optimization and Realization of Quadruple-Ridge Flared Horn with New Spline-Defined Profiles as a High-Efficiency Feed From 4.6 GHz to 24 GHz. IEEE Trans. Antennas Propag. 2019, 67, 585–590. [CrossRef]
35. Akgiray, A.; Weinreb, S.A.; Imbriale, W.; Beaudoin, C. Circular quadruple-ridged flared horn achieving near-constant beamwidth over multioctave bandwidth: Design and measurements. IEEE Trans. Antennas Propag. 2013, 61, 1099–1108. [CrossRef]

36. Manshari, S.; Koziel, S.; Leifsson, L. A Wideband Corrugated Ridged Horn Antenna with Enhanced Gain and Stable Phase Center for X- and Ku-Band Applications. *IEEE Antennas Wirel. Propag. Lett.* **2019**, *18*, 1031–1035. [CrossRef]
37. Manshari, S.; Koziel, S.; Leiffson, L.; Glazunov, A.A. High-Performance Wideband Horn Antenna for Direction Finding Arrays. In Proceedings of the 2020 14th EuCAP, Copenhagen, Denmark, 15–20 March 2020; pp. 1–5.

Disclaimer/Publisher's Note: The statements, opinions and data contained in all publications are solely those of the individual author(s) and contributor(s) and not of MDPI and/or the editor(s). MDPI and/or the editor(s) disclaim responsibility for any injury to people or property resulting from any ideas, methods, instructions or products referred to in the content.

Reconfigurable Multifunctional Metasurfaces for Full-Space Electromagnetic Wave Front Control

Shunlan Zhang [1,*], Weiping Cao [1,*], Jiao Wang [1], Tiesheng Wu [1], Yiying Wang [1], Yanxia Wang [2] and Dongsheng Zhou [2]

1. School of Information and Communication, Guilin University of Electronic Technology, Guilin 541004, China
2. Hebei Jinghe Electronic Technology Incorporated Company, Shijiazhuang 050200, China
* Correspondence: zhslan@guet.edu.cn (S.Z.); weipingc@guet.edu.cn (W.C.)

Abstract: In order to implement multiple electromagnetic (EM) wave front control, a reconfigurable multifunctional metasurface (RMM) has been investigated in this paper. It can meet the requirements for 6G communication systems. Considering the full-space working modes simultaneously, both reflection and transmission modes, the flexible transmission-reflection-integrated RMM with p-i-n diodes and anisotropic structures is proposed. By introducing a 45°-inclined H-shaped AS and grating-like micro-structure, the polarization conversion of linear to circular polarization (LP-to-CP) is achieved with good angular stability, in the transmission mode from top to bottom. Meanwhile, reflection beam patterns can be tuned by switching four p-i-n diodes to achieve a 1-bit reflection phase, which are embedded in the bottom of unit cells. To demonstrate the multiple reconfigurable abilities of RMMs to regulate EM waves, the RMMs working in polarization conversion mode, transmitted mode, reflected mode, and transmission-reflection-integrated mode are designed and simulated. Furthermore, by encoding two proper reflection sequences with 13 × 13 elements, reflection beam patterns with two beams and four beams can be achieved, respectively. The simulation results are consistent with the theoretical method. The suggested metasurface is helpful for radar and wireless communications because of its compact size, simple construction, angular stability, and multi-functionality.

Keywords: metasurface; multifunctional; reconfigurable; transmission-reflection integrated; full-space

1. Introduction

A metasurface (MS) is a surface with periodic or aperiodic structures, consisting of subwavelength elements, which possess the unique ability to control the amplitude [1,2], phase [2,3], and polarization states [4,5] of incident electromagnetic (EM) waves. They have some advantages such as a lower profile, low insertion loss, and easy integration with other circuits [6,7]. With the help of innovative techniques for modulating electromagnetic waves and a variety of useful applications, including diffusion, anomalous reflection and refraction [8,9], radar cross-section reduction [10,11], beam scanning [12–14], focusing [15,16], polarization conversion [17–19], and holography [20], researchers have flexibly designed MSs based on the generalized Snell's law [21]. Nonetheless, passive unit cells make up the majority of these designs. Their functions are set once they are produced, and they can only be used for a limited number of pre-planned uses. As a result, they are unable to meet the growing needs for communications and multifunctional devices.

Compared with passive MSs whose functions are fixed [22], reconfigurable MSs possess a stronger superiority [23,24]. Versatile features are expected because of their dynamic status changes. Therefore, the reconfigurable multifunctional metasurfaces (RMMs) can be controlled by electrical, optical, mechanical, and thermal means, which have been developed. These RMMs control EM waves from microwave to terahertz bands; furthermore, they are useful in a variety of applications, including antenna design [25], polarization conversion [26–29], and beam steering [30–32]. In the meantime, tunable MSs have been designed using functional materials such as liquid crystals [33], graphene [34,35], and vanadium dioxide (VO2) [36]. For example, using the insulator-to-metal transition feature of VO2, the researchers in [36] created a switchable MS that can accomplish broadband absorption and reflection. However, these RMMs mostly operate in either the transmission or reflection mode to regulate EM waves in half-space, and the other space is unutilized. Nonetheless, RMMs in full-space will possess a wider prospect, particularly with the growing demands for highly integrated and more powerful devices in 6G communication systems.

Driven by the imperative demand for both design integration and miniaturization for 6G applications, based on the earlier research in [37], we further present a flexible transmission-reflection-integrated RMM with p-i-n diodes and an anisotropic structure (AS), which can convert the x-polarized EM waves to circular polarization (CP) in the transmission mode from top to bottom, and control reflection beam patterns of the y-polarized EM waves in the reflection mode from bottom to top. The suggested unit cell for the MS is made up of five delicately designed metal patterns separated by four substrate layers. The upper three metal patterns complete the polarization conversion of linear-to-right hand circular polarization (LP-to-CP). By integrating four p-i-n diodes into metal layer 5, 1-bit tunable reflection phases are realized, which can control the y-polarized EM wave reflection patterns from bottom to top. To verify its adjustable property, a 13×13 array is designed, which can reflect beam patterns with two beams and four beams in reflection mode through the modulation of the switchable status on p-i-n diodes. Unlike previous research, our designs provide a feasible way of realizing adjustable multifunctional MSs operating in full space, which possess good angular stability and can result in many fascinating applications in wireless communications and radar.

2. Metasurface Unit Cell Design

Through rigorous structural design, numerous functionalities can be extracted from a single geometric structure, evoking unique reactions, as described in Figure 1. The RMM unit cell structure is depicted in Figure 2, where five metal layers, divided by four substrates, make up the RMM unit cell. Metal layers 1 to 3 are etched onto the Arlon AD255A (tm) substrate with $\varepsilon_r = 2.55$ and $\tan\delta = 0.0015$. Additionally, the F4B substrate is etched with the patterns of metal layers 4 and 5, which exhibit $\varepsilon_r = 2.65$ and $\tan\delta = 0.0015$. In order to produce orthogonally polarized waves when the incoming waves are linearly polarized, the element structure must be asymmetric along the incident wave's polarized direction. Meanwhile, the structure needs to be symmetrical in the direction along a 45° angle, which can enhance the polarization conversion ratio [38]. As described in Figure 2c, the pattern of metal layer 2 is a 45°-inclined H-shape, which completes the polarization conversion of LP-to-CP. As depicted in Figure 2b,d, the polarization grids on metal layers 1 and 3 are along the y- and x-directions, respectively, which improve the conversion attributes of the x-polarized EM waves and enhance polarization conversion purity. In conjunction with the metal grating on metal layer 3, four open trapezoid patches on metal layer 5 work as an artificial magnetic conductor (AMC). The two dc bias signal lines are designed on metal layers 4 and 5, as presented in Figure 2e,f. To isolate the high-frequency signals, two crescent-distributed capacitors and symmetrically distributed inductances are integrated into the bias layer of metal layer 4. In the meantime, the bias layer on metal layer 5 adopts the four inductors with $L = 270$ nH to choke RF currents. Through four metallized via-holes, the bottom trapezoid patches and the bias layer of metal layer 4 are connected. Four red square components on metal layer 5 are p-i-n diodes, which connect with a pair of symmetric trapezoids, respectively, as shown in Figure 2f. The proposed RMM is simulated and analyzed using the software Ansys HFSS 2018, the Floquet port, and periodic boundary conditions. The unit designated in Figure 2 has its optimal parameters shown in Table 1.

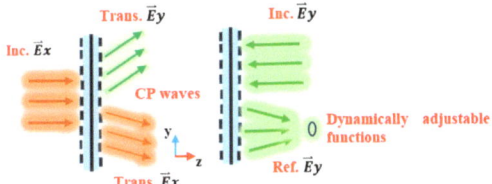

Figure 1. The schematic diagram of the suggested RMM working for full-space.

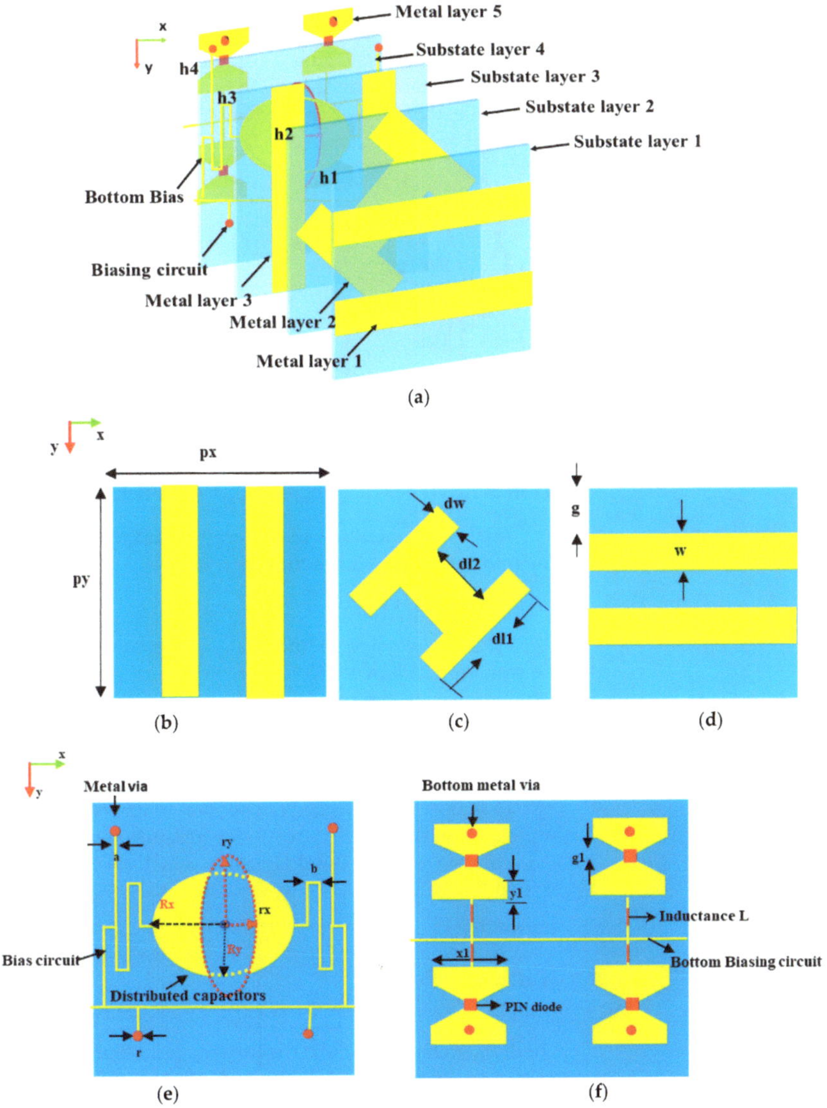

Figure 2. The RMM unit cell structure. (**a**) Schematic of the unit cell; (**b**–**f**) structures of five metal patterns from 1 to 5.

Table 1. Dimensions of the proposed unit cell (unit: mm).

Parameter	px	py	g	w	g1	x1	y1
Value	18	18	4.05	1.35	0.33	5.5	1
Parameter	h1	h2	h3	h4	dl1	dl2	dw
Value	3.175	0.813	0.254	1.524	9	9	2.7
Parameter	a	b	r	Rx	rx	Ry	ry
Value	0.1	1	0.1	4.2	1.66	2.5	3.46

The p-i-n diode used is MADP-000907-14020 from MACOM, Lowell, Massachusetts, USA. According to the data sheet, the forward-biased diode can be analogous to a series RL circuit with resistance $R = 7.8\ \Omega$ and inductance L = 30 pH, as described in Figure 3a. Meanwhile, the reverse-biased diode can be analogous to a series LC circuit with resistance L = 30 pH and capacitance C = 28 fF, as depicted in Figure 3b.

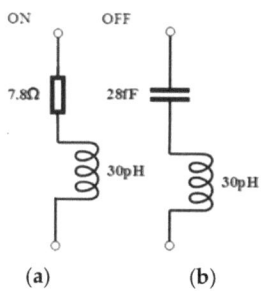

Figure 3. Analogous circuit of p-i-n diodes in different states: (**a**) ON state, (**b**) OFF state.

Analysis of the RMM in transmitted and reflected modes is carried out independently. The parts below provide a thorough analysis.

3. Transmission Mode

When the x-polarized incoming waves are impinged on the RMM from top to bottom, the transmitted wave will be converted in polarization due to matching electric fields being excited. In the meantime, the RMM operates in transmission mode, and works as a polarizer.

3.1. Working Principle

To realize the polarized conversion of LP-to-CP with an x-polarized incoming wave impinging on the RMM, the element structure should be asymmetric along the x-direction and embrace a 45°-inclined H-shape, which can translate the x-polarized wave into two perpendicularly polarized waves and produce different accumulations of phases. As depicted in Figure 4, the x-polarized incoming wave $\vec{E}_i$ can be divided into the tangential (u) and normal (v) directions at the center point of the inclined H-shape, and can be written as

$$\vec{E}_i = \hat{x} E_{ix} e^{-jkz} = \hat{u} E_{iu} e^{-jkz} + \hat{v} E_{iv} e^{-jkz} \tag{1}$$

where $E_{iu} = E_{iv} = E_{ix}/\sqrt{2}$; k is the wavenumber; and $\hat{x}$, $\hat{u}$, and $\hat{v}$ are the unit vectors with respect to the x-axis and u- and v-directions, respectively. The transmitted fields can be written using

$$\vec{E}_t = [\hat{u}(T_{uu} E_{iu} + T_{uv} E_{iv}) + \hat{v}(T_{vu} E_{iu} + T_{vv} E_{iv})] e^{-jkz} \tag{2}$$

where $T_{uu} = |T_{uu}| e^{j\varphi_{uu}}$, $T_{vu} = |T_{vu}| e^{j\varphi_{vu}}$, $T_{uv} = |T_{uv}| e^{j\varphi_{uv}}$, and $T_{vv} = |T_{vv}| e^{j\varphi_{vv}}$ are transmission coefficients.

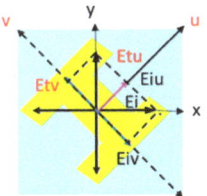

Figure 4. The electric field along the x-axis splitting into the u and v components, where the right-hand coordinate system is assumed, and the z axis is directed towards the reader.

When $|T_{vu}| = |T_{uv}| = 0, |T_{uu}| = |T_{vv}|$, and $\Delta\varphi = \varphi_{uu} - \varphi_{vv} = \pi$,

$$\vec{E}_t = (\hat{v} - \hat{u})T_{vv}\left(E_{ix}/\sqrt{2}\right)e^{-jkz} = \hat{y}E_{ty}e^{-jkz} \quad (3)$$

which implies that the EM waves are converted from the x-polarized waves to the y-polarized ones. When the phase differences $\Delta\varphi = \varphi_{uu} - \varphi_{vv} = 0$, the polarization of the transmission wave is still along the x-direction. If $\Delta\varphi = \pm 90°$, the incoming linearly polarized waves are transformed into circularly polarized waves, which can be described as follows:

$$\vec{E}_t = (\hat{x}E_{tx} + \hat{y}E_{ty})e^{-jkz} = \left(\hat{x}|T_{xx}|e^{j\varphi_{xx}} + \hat{y}|T_{yx}|e^{j\varphi_{xx}\pm\pi/2}\right)E_{ix}e^{-jkz} \quad (4)$$

where $T_{xx} = |E_{tx}/E_{ix}|e^{j\varphi_{xx}}$ and $T_{yx} = |E_{ty}/E_{ix}|e^{j(\varphi_{xx}\pm\pi/2)}$ are transmission coefficients of co- and cross-polarized components under the x-polarized incident wave. It is clear from Equation (4) that the phase differences $\Delta\varphi$ and the co-polarized and cross-polarized transmission coefficients (T_{xx} and T_{yx}) can be used to identify the state of polarization for the transmitted field. When $|T_{xx}| = |T_{yx}|$, if $\Delta\varphi \approx 90°$ and $\Delta\varphi \approx -90°$, a polarized conversion of LP-to-LHCP and LP-to-RHCP is achieved, respectively, whereas when $|T_{xx}| \neq |T_{yy}|$, an elliptic polarization wave is generated.

From Equation (4), it is concluded that controlling polarization can be accomplished through altering $\Delta\varphi$, which is mostly decided by the structure of resonant unit cells. $\Delta\varphi$ is a function of frequency when metasurfaces are unreconfigurable structures, meaning that distinct polarized conversion functions will be generated at distinct frequencies [37].

3.2. Simulation Results

We further discuss circular polarization conversion through the axial ratio (AR). The transmission coefficients of two waves with orthogonal polarization under an x-polarized incoming wave can be used to determine the AR parameter, which is crucial for assessing the electromagnetic waves' level of circular polarization.

$$AR = \frac{\sqrt{\left[T_{xx}^2 + T_{yx}^2 + \sqrt{T_{xx}^4 + T_{yx}^4 + 2(T_{xx}T_{yx})^2\cos(2\Delta\varphi)}\right]}}{\sqrt{\left[T_{xx}^2 + T_{yx}^2 - \sqrt{T_{xx}^4 + T_{yx}^4 + 2(T_{xx}T_{yx})^2\cos(2\Delta\varphi)}\right]}} \quad (5)$$

From the previous analysis, we are aware that in order to produce circularly polarized waves, the two orthogonal components of the transmitted electric field must have a phase difference of $\Delta\varphi = \pm 90°$ and equal amplitude, i.e., $|T_{xx}| = |T_{yx}|$. If the AR is less than 3 dB, it is roughly circular polarization in practice.

The full-wave simulations are used to examine the RMM's transmission magnitudes and phase discrepancies in order to verify our design. In Figure 5, the simulated co-polarized and cross-polarized transmission magnitudes and phase differences are displayed when the diode is forward-biased. It is evident that $|T_{xx}|$ and $|T_{yx}|$ are over -10 dB and nearly equal within the working frequency ranges of 7.65–7.7 GHz and 6.1–6.6 GHz, and that the phase differences are close to 270°, which indicates the transmission of RHCP waves. When the diode is reversebiased, Figure 6 displays the simulated co-polarized and cross-polarized transmission magnitudes and phase discrepancies. Thus, an excellent circularly polarized transmission wave is achieved in the two bands of 6.5–6.6 GHz and 7.65–7.7 GHz in both the ON and OFF states.

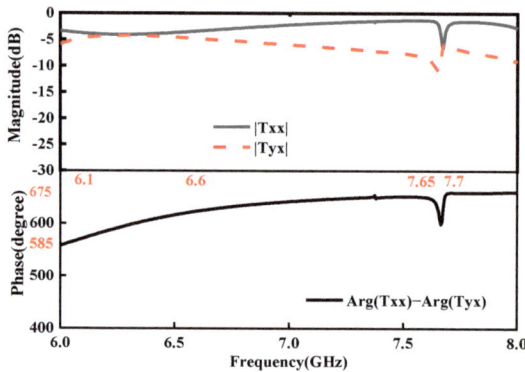

Figure 5. Simulated co-polarized and cross-polarized transmission magnitudes and phase differences under the x-polarized EM wave in the ON state.

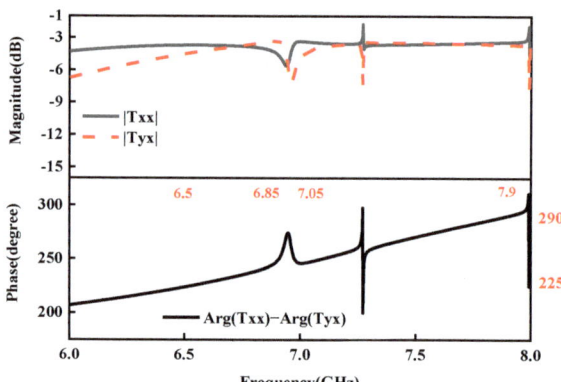

Figure 6. Simulated co-polarized and cross-polarized transmission magnitudes and phase differences under the x-polarized EM wave in the OFF state.

In order to look into the impact of oblique incidence on transmission performance, the variation in transmission performances is investigated for different obliques under the x-polarized EM waves from top to bottom. Figures 7 and 8 present the co- and cross-polarized transmitted magnitudes and phase differences versus operating frequency at oblique incidence with the incidence angles $\theta = 0°$, 15°, and 30° in the ON and OFF states, respectively. It is obvious from Figures 7 and 8 that the curved lines of the simulated co- and cross-polarization transmission magnitudes and phase differences are nearly coincident with those under normal incidence in the ON and OFF states, respectively, which shows the great angular stability in transmission mode in the ON and OFF states.

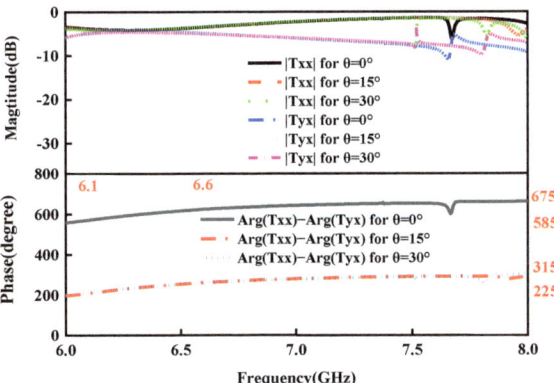

Figure 7. Simulated co- and cross-polarization transmission magnitudes and phase differences for different incident angles under the *x*-polarized EM waves in the ON state.

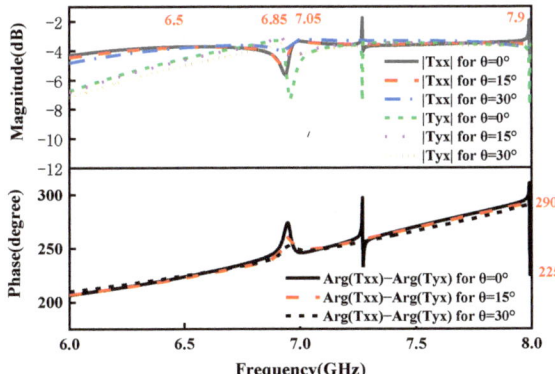

Figure 8. Simulated co- and cross-polarization transmission magnitudes and phase differences for different incident angles under the *x*-polarized EM waves in the OFF state.

4. Reflection Mode

When the y-polarization incident wave is impinged on the RMM from bottom to top, by switching the p-i-n states and designing the discontinuous reflection phases, the wavefront of EM waves can be freely tailored, opening a wide range of new phenomena and applications, including beam patterns with two beams and four beams. Meanwhile, the RMM works in reflection mode and acts as a reflector.

4.1. Performance Analysis

According to the analysis above, we can determine that four open trapezoidal patches on metal layer 5 serve as an AMC in conjunction with the metal grating on metal layer 3, which acts as the ground of the AMC. To verify this, the simulated reflection performances are analyzed. An AMC can be generated by impinging y-polarized EM waves on the MS from bottom to top almost in the frequency range of 14–16 GHz, which can be used for stealth materials in radars [39] or high-gain antennas [40]. Figure 9 presents reflection coefficients, phases, and phase differences of the proposed flexible regulated MS under the y-polarized normal incidence in the ON and OFF states within a bandwidth of 14–16 GHz. As shown in Figure 9a, folded reflected phases are almost in the range of $-90° \sim +90°$, and most of the incident waves are reflected with reflected coefficients ($|R_{yy}| = |E_{ry}/E_{iy}|$)

greater than −10 dB in each state in the operating band, which indicates that the proposed MS works as an AMC. Furthermore, it is obvious from Figure 9b that the reflection phases are discernibly distinct in every state, and the phase discrepancies between adjacent states fall in the range of (180° − 10°, 180° + 10°) around 15.15 GHz, 15.54 GHz, and 15.8 GHz, by switching simultaneously the four p-i-n diodes on metal layer 5 to operate in the ON/OFF states. Therefore, 1-bit coding elements can be obtained at 15.15 GHz, 15.54 GHz, and 15.8 GHz, meaning that the gradient reflection phase distribution of θ and $\theta + 180°$ can be produced by the suggested metasurface working from state 1 to state 2. As a result, the element configuration in each state can be thought of as a fundamental digital element. Two different element configurations yield 1 bit, simulating the state 1 and state 0 units, respectively.

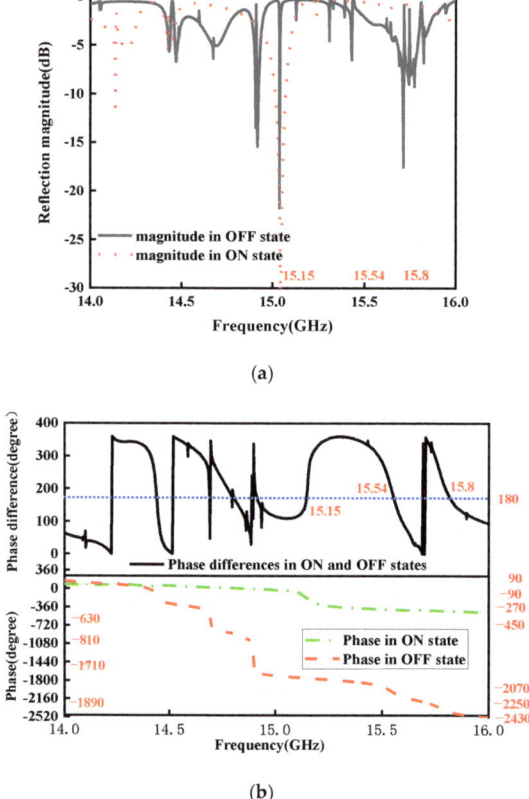

Figure 9. Simulated reflection coefficients of the presented MS in the ON and OFF states: (a) reflection coefficient; (b) reflection phase and reflection phase differences between p-i-n diode ON and OFF states.

Subsequently, the performance of reflection is analyzed for various oblique incidences in reflection modes. The reflection magnitudes and phases versus operating frequency with variable incident angles θ in the ON and OFF states are shown in Figures 10 and 11. From these figures, it is concluded that the reflection performance is almost stable up to 30° in the ON state and 20° in the OFF state in the band of 14–16 GHz, which implies that the proposed RMM possesses better stability at oblique incidence in the reflection mode. The above findings have conclusively demonstrated that the suggested RMM can still operate as a digital coding MS at 20 degrees of oblique illumination in each state. Greater incident angles cause the EM wave to take an extra path between the ground (metal layer 3) and the bottom, which increases the phase difference, resulting in destructive interference [38],

$$\Delta\phi = \phi_{oblique} - \phi_{normal} = 2\sqrt{\varepsilon_r}kd\left(\frac{1}{\sqrt{1-\frac{\sin^2\theta}{\varepsilon_r}}} - 1\right) \tag{6}$$

where θ represents the incident angle with respect to normal incidence, k signifies the vector of wave propagation in free space, d indicates the thickness of the substrate, and $\varepsilon_r = 2.65$. Reflection performance degrades severely for incidence angles more than 30° in the ON state and 20° in the OFF state. Folded reflection phases are out of −90°–90°. It is obvious from Equation (6) that higher-order modes or grating lobes are the cause of the performance deterioration. It can be concluded from the above analysis that the overall trend in the reflection performance under oblique incidence also meets the majority of large-angle incidence design criteria despite not being as stable as the transmission mode.

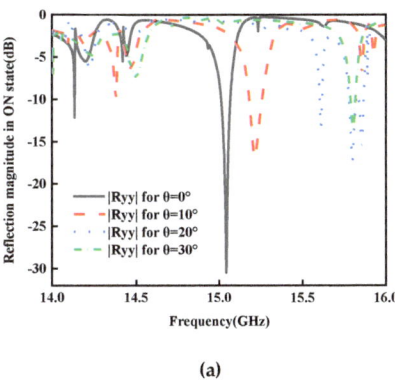
(a)

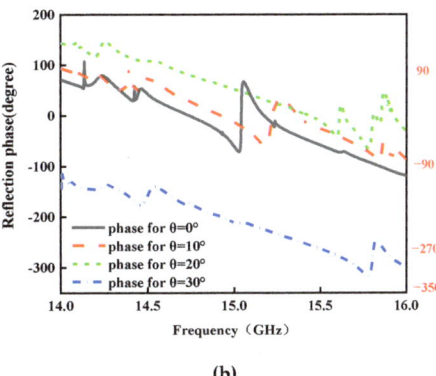
(b)

Figure 10. Simulated reflection magnitudes and phases for different incident angles θ in the ON states under the y-polarized wave from bottom to top: (**a**) magnitudes, (**b**) phases.

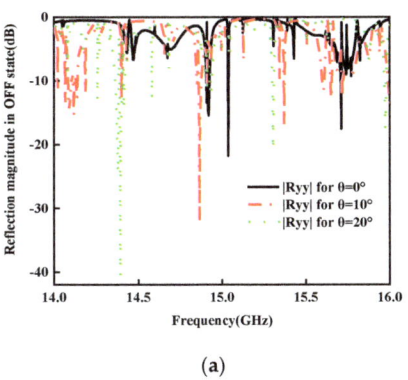
(a)

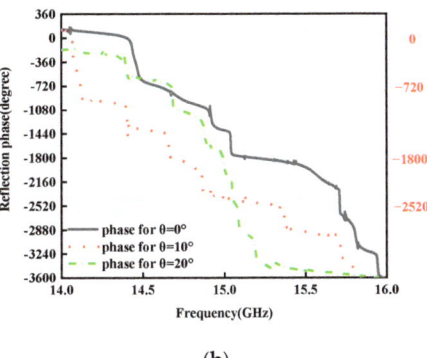
(b)

Figure 11. Simulated reflection magnitudes and phases for different incident angles θ in the OFF states under the y-polarized wave from bottom to top: (**a**) magnitudes, (**b**) phases.

4.2. Reconfigurable Reflection Coding MSs

These coding particles could be arranged to form a coding metasurface with various encoding sequences in two dimensions, which possesses different reflection patterns. According to array theory, the radiation pattern of a given encoding sequence, which is made up of $N \times N$ equal-sized unit cells with dimension D, can be analytically calculated. In the far-field area, the specific characteristic of each coding particle becomes hazy because of the subwavelength characteristic of the digital particle with a reflection phase of $\varphi(m,n)$ (either θ or $\theta + 180°$ in the 1-bit case) for the mnth element. The far-field-function-reflected metasurface at the normal incidence of plane waves is written as

$$f(\theta, \varphi) = f_e(\theta, \varphi) \sum_{n=1}^{N} \exp\{-i\{\varphi(m,n) + kD\sin\theta[(m-1/2)\cos\varphi + (n-1/2)\sin\varphi]\}\} \quad (7)$$

where the lattice pattern function is denoted by $f_e(\theta, \varphi)$ and the elevation and azimuth angles are represented by θ and φ, respectively. The relative phase of the "0" element has been assumed to be zero for simplicity's sake, and the term $f_e(\theta, \varphi)$ in Equation (7) has been neglected because it becomes ambiguous in the far-field. Using any given encoding sequence, we may determine the directivity function $Dir(\theta, \varphi)$ from Equation (7), which can be represented as

$$Dir(\theta, \varphi) = \frac{4\pi |f(\theta, \varphi)|^2}{\int_0^{2\pi} \int_0^{\pi/2} |f(\theta, \varphi)|^2 \sin\theta d\theta d\varphi} \quad (8)$$

In order to investigate the aforementioned beam modulation performance, the 1-bit digital MS is designed, which consists of 13×13 unit cells in the total size of 234 mm × 234 mm. The various reflection fields will be produced by various coding sequences in the reflection mode, and each particle can be independently controlled. Full-wave numerical simulations are used to finish all numerical simulations. Figure 12 presents the reflection pattern of the proposed RMM with the 0001111000111 coding sequence, under the y-normal illumination from bottom to top. The coding sequence is shown in Figure 12a. From Figure 12b, it is evident that normal-incidence plane waves are redirected in two symmetrical directions, which is consistent with the theoretical predictions in Equation (7). Figure 13 depicts the reflection pattern of the chess-board periodic coding MS under the y-normal illumination from bottom to top. The chess-board coding sequence with 01/10 is shown in Figure 13a. From Figure 13b, it can be obviously found that the normal-incidence plane waves are scattered in four symmetrical directions, which further verifies the theory derivation results in Equation (7). The comparison of the designed RMM with some existing multifunction MSs for full space are listed in Table 2. It can be observed that the proposed RMM possesses real-time multifunctional adjustability and a low profile. Further, transmission and reflection of the designed RMM have good angular stability compared to other existing multiple MSs.

Table 2. Comparison of the proposed RMM with some recently reported full-space MSs.

Refs.	Real-Time Tunable Phase	Active Element	Layer Number	Bias Circuit Design	Profile
[41]	Yes	P-i-n	9	No	High
[42]	No	-	5	-	Low
[43]	Yes	P-i-n	11	Yes	Low
[44,45]	No	-	7	-	Low
This work	Yes	P-i-n	9	Yes	Low

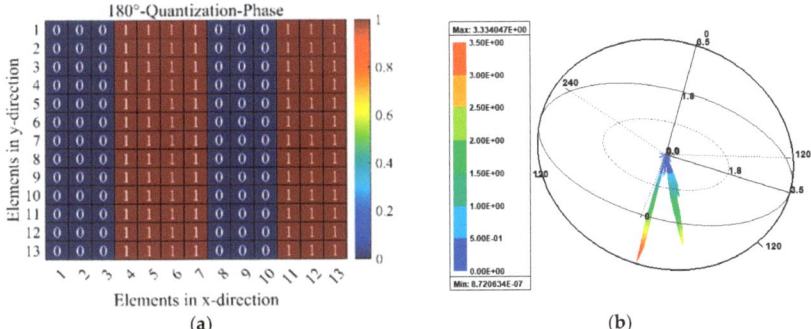

Figure 12. Schematic illustration of 0001111...... coding MSs under the y-normal illumination from bottom to top with dual-beam pattern: (**a**) coding phase profile, (**b**) simulated 3D far-field patterns.

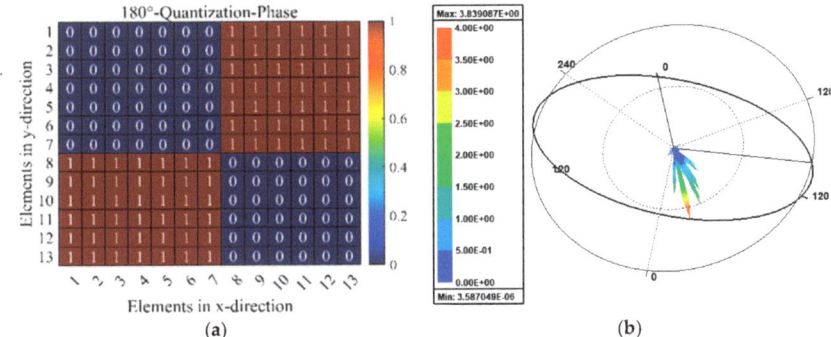

Figure 13. Schematic illustration of chess-board periodic coding metasurface under the y-normal illumination with quad-beam pattern: (**a**) coding phase profile, (**b**) simulated 3D far-field patterns.

5. Conclusions

In this paper, an important design has been implemented: a flexible transmission-reflection-integrated RMM with p-i-n diodes and an AS, which can realize the EM wave polarization conversion from the LP to CP in the transmission mode. As a good structure from top to bottom, it can modulate reflection beam patterns by switching the p-i-n diodes embedded into metal 5. Taking the transmitted mode as an example, the physical mechanism of multifunctionalities was developed. An RMM with 13×13 elements was designed and simulated, which can scatter reflection beam patterns with two beams and four beams by switching p-i-n diodes. The efficiency of the suggested design is confirmed by the simulation results, which agree with the theoretical predictions. Our design provides a new method for building an RMM in full space and possesses enormous potential applications in radar and wireless communications.

Author Contributions: All authors have significantly contributed to the research presented in this manuscript; S.Z. presented the main idea and wrote the manuscript; W.C. and J.W. edited the manuscript; T.W. and Y.W. (Yiyin Wang) reviewed and revised the manuscript. Y.W. (Yanxia Wang) and D.Z. finished some simulations. All authors have read and agreed to the published version of the manuscript.

Funding: This work is supported by the Major Science and Technology Programs for Universities in Hebei under Grant 241130447A and by the Natural Science Foundation of Guangxi Province (2024GXNSFAA010178).

Data Availability Statement: The original contributions presented in the study are included in the article, further inquiries can be directed to the corresponding author.

Conflicts of Interest: Authors Yanxia Wang and Dongsheng Zhou were employed by Hebei Jinghe Electronic Technology Incorporated Company. The remaining authors declare that the research was conducted in the absence of any commercial or financial relationships that could be construed as a potential conflict of interest.

References

1. Demetre, J.S.; Smy, T.J.; Gupta, S. Static Metasurface Reflectors With Independent Magnitude and Phase Control Using Coupled Resonator Configuration. *IEEE Trans. Antennas Propag.* **2023**, *71*, 3536–3545. [CrossRef]
2. Yang, J.; Yang, W.X.; Qu, K.; Zhao, J.M.; Jiang, T.; Chen, K.; Feng, Y.J. Active polarization-converting metasurface with electrically controlled magnitude amplification. *Opt. Express* **2023**, *31*, 28979–28986. [CrossRef] [PubMed]
3. Sleasman, T.; Duggan, R.; Awadallah, R.S.; Shrekenhamer, D. Dual-Resonance Dynamic Metasurface for Independent Magnitude and Phase Modulation. *Phys. Rev. Appl.* **2023**, *20*, 014004. [CrossRef]
4. Mueller, J.P.B.; Rubin, N.A.; Devlin, R.C.; Groever, B.; Capasso, F. Metasurface Polarization Optics: Independent Phase Control of Arbitrary Orthogonal States of Polarization. *Phys. Rev. Lett.* **2017**, *118*, 113901. [CrossRef] [PubMed]
5. Rudakova, N.V.; Bikbaev, R.G.; Tyryshkina, L.E.; Vetrov, S.Y.; Timofeev, I.V. Tuning Q-Factor and Perfect Absorption Using Coupled Tamm States on Polarization-Preserving Metasurface. *Photonics* **2023**, *10*, 1391. [CrossRef]
6. Smith, D.R.; Padilla, W.J.; Vier, D.C.; Nemat-Nasser, S.C.; Schultz, S. Composite medium with simultaneously negative permeability and permittivity. *Phys. Rev. Lett.* **2000**, *84*, 4184–4187. [CrossRef]
7. Pendry, J.B. Negative refraction makes a perfect lens. *Phys. Rev. Lett.* **2000**, *85*, 3966–3969. [CrossRef]
8. Li, Y.C.; Wan, S.C.; Zhao, R.Q.; Zhu, Z.; Li, W.J.; Guan, C.Y.; Yang, J.; Bogdanov, A.; Belov, P.; Shi, J.H. Huygens' metasurface: From anomalous refraction to reflection. *Opt. Commun.* **2024**, *565*, 130648. [CrossRef]
9. Aieta, F.; Genevet, P.; Yu, N.F.; Kats, M.A.; Gaburro, Z.; Capasso, F. Out-of plane reflection and refraction of light by anisotropic optical antenna metasurfaces with phase discontinuities. *Nano Lett.* **2012**, *12*, 1702–1706. [CrossRef]
10. Song, Y.C.; Ding, J.; Guo, C.J.; Ren, Y.H.; Zhang, J.K. Ultra-broadband backscatter radar cross section reduction based on polarization-insensitive metasurface. *IEEE Antennas Wirel. Propag.* **2016**, *15*, 329–331. [CrossRef]
11. Zhang, H.; Wang, X.; Liu, Y.; Liu, C.; Dong, H. A Combined Subdomain Method of Moments and Asymptotic Waveform Evaluation for Fast Wideband Bistatic Radar Cross Section Prediction. *IEEE T Antennas Propag.* **2024**, *72*, 5900–5909. [CrossRef]
12. Li, Y.B.; Wan, X.; Cai, B.G.; Cheng, Q.; Cui, T.J. Frequency-controls of electromagnetic multi-beam scanning by metasurfaces. *Sci. Rep.* **2014**, *4*, 6921. [CrossRef] [PubMed]
13. Chen, T.T.; Song, L.Z. A Simplified Dual-Polarized Beam-Scanning Reflectarray Using Novel Wheel-Rudder Elements. *IEEE Antennas Wirel. Propag.* **2024**, *23*, 2311–2315. [CrossRef]
14. Yu, H.; Zhang, Z.Y.; Su, J.X.; Qu, M.J.; Li, Z.R.; Xu, S.H.; Yang, F. Quad-Polarization Reconfigurable Reflectarray With Independent Beam-Scanning and Polarization Switching Capabilities. *IEEE Trans. Antennas Propag.* **2023**, *71*, 7285–7298. [CrossRef]
15. Hu, J.H.; Guo, Z.Y.; Shi, J.Y.; Jiang, X.; Chen, Q.M.; Chen, H.; He, Z.X.; Song, Q.H.; Xiao, S.M.; Yu, S.H.; et al. A metasurface-based full-color circular auto-focusing Airy beam transmitter for stable high-speed underwater wireless optical communications. *Nat. Commun.* **2024**, *15*, 2944. [CrossRef]
16. Shrestha, S.; Overvig, A.C.; Lu, M.; Stein, A.; Yu, N.F. Broadband achromatic dielectric metalenses. *Light. Sci. Appl.* **2018**, *7*, 85. [CrossRef]
17. Grady, N.K.; Heyes, J.E.; Chowdhury, D.R.; Zeng, Y.; Reiten, M.T.; Azad, A.K.; Taylor, A.J.; Dalvit, D.A.R.; Chen, H.T. Terahertz metamaterials for linear polarization conversion and anomalous refraction. *Science* **2013**, *340*, 1304–1307. [CrossRef]
18. Yuan, Y.Y.; Wu, Q.; Burokur, S.N.; Zhang, K. Chirality-Assisted Phase Metasurface for Circular Polarization Preservation and Independent Hologram Imaging in Microwave Region. *IEEE Trans. Microw. Theory* **2023**, *71*, 3259–3272. [CrossRef]
19. Gao, X.; He, L.Y.; Yin, S.J.; Xue, C.H.; Wang, G.F.; Xie, X.M.; Xiong, H.; Cheng, Q.; Cui, T.J. Ultra-Wideband Low-RCS Circularly Polarized Antennas Realized by Bilayer Polarization Conversion Metasurfaces and Novel Feeding Networks. *IEEE Trans. Antennas Propag.* **2024**, *72*, 1959–1964. [CrossRef]
20. Hu, Y.Q.; Luo, X.H.; Chen, Y.Q.; Liu, Q.; Li, X.; Wang, Y.S.; Liu, N.; Duan, H.G. 3D integrated metasurfaces for full-colour holography. *Light. Sci. Appl.* **2019**, *8*, 86. [CrossRef]
21. Yu, N.F.; Genevet, P.; Kats, M.A.; Aieta, F.; Tetienne, J.P.; Capasso, F.; Gaburro, Z. Light propagation with phase discontinuities: Generalized laws of reflection and refraction. *Science* **2011**, *334*, 333–337. [CrossRef] [PubMed]
22. Zhang, S.L.; Cao, W.P.; Wu, T.S.; Wang, J.; Wei, Y. The Design of a Multifunctional Coding Transmitarray with Independent Manipulation of the Polarization States. *Micromachines* **2024**, *15*, 1014. [CrossRef] [PubMed]
23. Yin, T.; Ren, J.; Chen, Y.J.; Xu, K.D.; Yin, Y.Z. Highly Integrated Reconfigurable Shared-Aperture EM Surface With Multifunctionality in Transmission, Reflection, and Absorption. *IEEE Trans. Antennas Propag.* **2024**, *72*, 6789–6794. [CrossRef]
24. Xing, H.R.; Tang, C.; Li, Z.W.; Wang, M.J.; Fan, C.; Zheng, H.X.; Li, E.R. Three-Polarization-Reconfigurable Antenna Array Implemented by Liquid Metal. *IEEE Antennas Wirel. Propag.* **2024**, *23*, 374–378. [CrossRef]

25. Afsari, A.; Van Veen, B.D.; Behdad, N. An Electronically Reconfigurable Matching and Decoupling Network for Two-Element HF Antenna Arrays. *IEEE Trans. Antennas Propag.* **2024**, *72*, 6332–6347. [CrossRef]
26. Guo, Q.X.; Hao, F.S.; Qu, M.J.; Su, J.X.; Li, Z.R. Multiband Multifunctional Polarization Converter Based on Reconfigurable Metasurface. *IEEE Antennas Wirel. Propag.* **2024**, *23*, 1241–1245. [CrossRef]
27. Ratni, B.; de Lustrac, A.; Piau, G.P.; Burokur, S.N. Electronic control of linear-to-circular polarization conversion using a reconfigurable metasurface. *Appl. Phys. Lett.* **2017**, *111*, 214101. [CrossRef]
28. Huang, C.; Zhang, C.L.; Yang, J.N.; Sun, B.; Zhao, B.; Luo, X.G. Reconfigurable metasurface for multifunctional control of electromagnetic waves. *Adv. Opt. Mater.* **2017**, *5*, 1700485. [CrossRef]
29. Lor, C.; Phon, R.; Lim, S. Reconfigurable transmissive metasurface with a combination of scissor and rotation actuators for independently controlling beam scanning and polarization conversion. *Microsyst. Nanoeng.* **2024**, *10*, 40. [CrossRef]
30. Ning, Y.F.; Zhu, S.C.; Chu, H.B.; Zou, Q.; Zhang, C.; Li, J.X.; Xiao, P.; Li, G.S. 1-bit Low-Cost Electronically Reconfigurable Reflectarray and Phased Array Based on p-i-n Diodes for Dynamic Beam Scanning. *IEEE T Antennas Propag.* **2024**, *72*, 2007–2012. [CrossRef]
31. Zhang, X.Y.; Kong, X.K.; Zhou, S.C.; Liu, P.Q.; Zou, Y.K.; Cheng, J.L.; Pang, X.Y.; Gao, S. High-accuracy beam generation and scanning using reconfigurable coding metasurface. *J. Phys. D Appl. Phys.* **2024**, *57*, 385108. [CrossRef]
32. Fu, X.J.; Yang, F.; Liu, C.X.; Wu, X.J.; Cui, T.J. Terahertz beam steering technologies: From phased arrays to field-programmable metasurfaces. *Adv. Opt. Mater.* **2020**, *8*, 190062. [CrossRef]
33. Sobczyk, A.; Mędoń, K.F.; Kowalczyk, M.A.; Suszek, J.; Zdrojek, M.; Sypek, M. Ultra-low reflection of sub-THz waves in graphene-based composite with metasurface for stealth technology. *Opt. Express* **2024**, *18*, 32118–32127. [CrossRef]
34. Torabi, E.S.; Fallahi, A.; Yahaqhi, A. Evolutionary optimization of graphene-metal metasurfaces for tunable broadband terahertz absorption. *IEEE T Antennas Propag.* **2017**, *65*, 1464–1467. [CrossRef]
35. Xing, B.B.; Liu, Z.G.; Lu, W.B.; Chen, H.; Zhang, Q.B. Wideband microwave absorber with dynamically tunable absorption based on graphene and random metasurface. *IEEE Antennas Wirel. Propag. Lett.* **2019**, *18*, 2602–2606. [CrossRef]
36. Yang, W.C.; Zhou, C.Y.; Xue, Q.; Wen, Q.Y.; Che, W.Q. Millimeter-Wave Frequency-Reconfigurable Metasurface Antenna Based on Vanadium Dioxide Films. *IEEE T Antennas Propag.* **2021**, *69*, 4359–4369. [CrossRef]
37. Zhang, S.L.; Cao, W.P.; Wu, T.S.; Wang, J.; Li, H.; Duan, Y.L.; Rong, H.Y.; Zhang, Y.L. Transmission-Reflection-Integrated Multifunctional Passive Metasurface for Entire-Space Electromagnetic Wave Manipulation. *Materials* **2023**, *16*, 4242. [CrossRef]
38. Yang, H.; Wang, S.C.; Li, P.; He, Y.; Zhang, Y.J. A Broadband Multifunctional Reconfigurable Polarization Conversion Metasurface. *IEEE Trans. Antennas Propag.* **2024**, *71*, 5759–5767. [CrossRef]
39. Galarregui, J.C.I.; Pereda, A.T.; de Falcon, J.L.M.; Ederra, I.; Gonzalo, R.; de Maagt, P. Broadband Radar Cross-Section Reduction Using AMC Technology. *IEEE Trans. Antennas Propag.* **2013**, *61*, 6136–6143. [CrossRef]
40. Su, R.; Gao, P.; Wang, R.D.; Wang, P. High-gain broadside dipole planar AMC antenna for 60 GHz applications. *Electron. Lett.* **2018**, *5*, 407–408. [CrossRef]
41. Qin, Z.; Li, Y.F.; Wang, H.; Li, C.C.; Liu, C.; Zhu, Z.B.; Yuan, Q.; Wang, J.F.; Qu, S.B. Transmission Reflection Integrated Programmable Metasurface for Real-Time Beam Control and High Efficiency Transmission Polarization Conversion. *Ann. Phys.* **2022**, *535*, 202200368. [CrossRef]
42. Sun, S.; Ma, H.F.; Chen, Y.F.; Cui, T.J. Transmission-Reflection-Integrated Metasurface with Simultaneous Amplitude and Phase Controls of Circularly Polarized Waves in Full Space. *Laser Photonics Rev.* **2024**, *18*, 2300945. [CrossRef]
43. Zhang, C.; Gao, J.; Cao, X.Y.; Li, S.J.; Yang, H.H.; Li, T. Multifunction Tunable Metasurface for Entire-Space Electromagnetic Wave Manipulation. *IEEE Trans. Antennas Propag.* **2020**, *68*, 3301–3306. [CrossRef]
44. Wang, J.Y.; Zhang, Z.J.; Huang, C.; Pu, M.B.; Lu, X.J.; Ma, X.L.; Guo, Y.H.; Luo, J.; Luo, X.G. Transmission–Reflection-Integrated Quadratic Phase Metasurface for Multifunctional Electromagnetic Manipulation in Full Space. *Adv. Opt. Mater.* **2022**, *10*, 2102111. [CrossRef]
45. Cai, T.; Wang, G.M.; Tang, S.W.; Xu, H.X.; Duan, J.W.; Guo, H.J.; Guan, F.X.; Sun, S.L.; He, Q.; Zhou, L. High-Efficiency and Full-Space Manipulation of Electromagnetic Wave Fronts with Metasurfaces. *Phys. Rev. Appl.* **2017**, *8*, 034033. [CrossRef]

Disclaimer/Publisher's Note: The statements, opinions and data contained in all publications are solely those of the individual author(s) and contributor(s) and not of MDPI and/or the editor(s). MDPI and/or the editor(s) disclaim responsibility for any injury to people or property resulting from any ideas, methods, instructions or products referred to in the content.

Article

A 60 GHz Slotted Array Horn Antenna for Radar Sensing Applications in Future Global Industrial Scenarios

Binyi Ma [1,2,3,4], Jing Li [1,2,3,4], Yu Chen [1,2,3,4], Yuheng Si [1,2,3,4], Hongyan Gao [2,5], Qiannan Wu [2,3,4,5,*] and Mengwei Li [1,2,3,4,6,*]

1. School of Instrument and Electronics, North University of China, Taiyuan 030051, China; mabinyi99@163.com (B.M.); lijing@tit.edu.cn (J.L.); cy13833876975@163.com (Y.C.); 18834372826@163.com (Y.S.)
2. School of Instrument and Intelligent Future Technology, North University of China, Taiyuan 030051, China; 13781790790@163.com
3. Academy for Advanced Interdisciplinary Research, North University of China, Taiyuan 030051, China
4. Center for Microsystem Intergration, North University of China, Taiyuan 030051, China
5. School of Semiconductors and Physics, North University of China, Taiyuan 030051, China
6. Key Laboratory of Dynamic Measurement Technology, North University of China, Taiyuan 030051, China
* Correspondence: qiannanwoo@nuc.edu.cn (Q.W.); lmwprew@163.com (M.L.)

Abstract: This paper presents the design of a 60 GHz millimeter-wave (MMW) slot array horn antenna based on the substrate-integrated waveguide (SIW) structure. The novelty of this device resides in the achievement of a broad impedance bandwidth and high gain performance by meticulously engineering the radiation band structure and slot array. The antenna demonstrates an impressive impedance bandwidth of 14.96 GHz (24.93%), accompanied by a remarkable maximum reflection coefficient of −39.47 dB. Furthermore, the antenna boasts a gain of 10.01 dBi, showcasing its outstanding performance as a high-frequency antenna with a wide bandwidth and high gain. To validate its capabilities, we fabricated and experimentally characterized a prototype of the antenna using a probe test structure. The measurement results closely align with the simulation results, affirming the suitability of the designed antenna for radar sensing applications in future global industrial scenarios.

Keywords: slot array; horn antenna; millimeter-wave; wide bandwidth coverage; high gain

Citation: Ma, B.; Li, J.; Chen, Y.; Si, Y.; Gao, H.; Wu, Q.; Li, M. A 60 GHz Slotted Array Horn Antenna for Radar Sensing Applications in Future Global Industrial Scenarios. *Micromachines* **2024**, *15*, 728. https://doi.org/10.3390/mi15060728

Academic Editor: Haejun Chung

Received: 11 May 2024
Revised: 29 May 2024
Accepted: 29 May 2024
Published: 30 May 2024

Copyright: © 2024 by the authors. Licensee MDPI, Basel, Switzerland. This article is an open access article distributed under the terms and conditions of the Creative Commons Attribution (CC BY) license (https://creativecommons.org/licenses/by/4.0/).

1. Introduction

In recent times, there has been a notable surge in demand for radar sensors in the realms of automotive, industrial, and medical applications [1–3]. However, conventional radar sensors [4,5] are plagued by challenges including low operational frequencies, elevated costs, restricted bandwidth, and substantial form factors. To address these challenges, the 60 GHz frequency band emerges as a compelling solution due to its inherent broadband characteristics. Moreover, sensors operating in the 60 GHz band have the capability to capture rich and exceptionally accurate point cloud data [6,7], a critical factor for sustaining precision in measurements across a spectrum of applications within the global industrial radar sensing domain. Therefore, antennas characterized by both high gain and wide bandwidth [8,9] in this frequency band are poised as promising alternatives for advancing global industrial radar [10,11] sensing applications in the future. Currently, SIW technology [12–14] is gaining maturity within the high-frequency antenna domain [15]. An SIW represents a dielectric-filled metallic waveguide with a planar structure. In contrast to conventional metallic waveguides [16,17], an SIW offers distinct advantages such as a reduced profile, minimal signal loss, and seamless integration with planar circuits. As a result, SIW structures have found extensive utility in both circuit and antenna designs within the MMW frequency bands [18,19].

In 2015, S. Ramesh and his team conducted pivotal research in indoor radio link characterization for millimeter-wave wireless communications, utilizing dielectric-loaded index-tapered slot antennas [20]. This design achieved a 3.33 GHz bandwidth, 6.45 dBi gain, and a maximum reflection coefficient of 21.31 dB. However, with the evolution of high-speed communication technology [21,22], its bandwidth and gain limitations became apparent, highlighting the need for more advanced designs. By 2019 [23], Nanda Kumar Mungaru and Thangavelu Shanmuganantham had addressed these limitations with a broadband H-spaced head-shaped slot antenna based on an SIW, specifically for 60 GHz wireless communication. This innovative antenna showed an impressive impedance bandwidth of 3.22 GHz, a reflection coefficient of -38.3 dB, and a gain of 6.812 dBi. In the same year, another significant advancement was a substrate-integrated waveguide-based slot antenna, also for 60 GHz applications [24]. It boasted a 3.64 GHz impedance bandwidth, a reflection coefficient of -21.31 dB, and a 6.812 dBi gain. Compared to S. Ramesh et al.'s earlier design, these newer antennas offered an improved bandwidth and reflection coefficient [25]. Moving to 2021, T. Shanmuganantham and colleagues analyzed a tree-shaped slotted impedance-matching antenna for 60 GHz femtocell applications. They achieved remarkable improvements in size, gain, and bandwidth, with a 7.46 dBi gain and a 9.7 GHz bandwidth, surpassing S. Ramesh's earlier design. Despite these advancements, the challenge of integrating these technologies into future radar sensing applications within the global industrial landscape remained formidable. In 2023, Manish Sharma introduced a Conformal ultra-compact narrowband 60.0 GHz four-port millimeter-wave MIMO antenna for wearable short-range 5G application [26]. This state-of-the-art design featured a compact size of $16 \times 16 \times 0.254$ mm^3, a 1.735 GHz bandwidth, and a maximum reflection coefficient of -35.79 dB, exhibiting superior gain. However, the trade-off for this enhanced gain was a narrower bandwidth, potentially limiting its electrical performance stability near the central frequency [27].

Addressing the critical challenge of low gain and narrow bandwidth in antenna designs for radar industry applications, this paper introduces a novel 60 GHz slot array horn antenna based on an SIW structure. This design markedly enhances overall antenna performance by meticulously crafting the radiation band structure and gap array. The resultant antenna demonstrates exceptional capabilities, notably in terms of gain and bandwidth. The advanced performance of this antenna is pivotal for radar sensors, enabling them to achieve high accuracy and generate rich point cloud data. These attributes make the proposed antenna an optimal choice for future radar sensing applications across various global industrial scenarios. By overcoming the limitations of previous designs, this antenna stands as a significant step forward in the field, offering a robust solution tailored to meet the demanding requirements of modern radar technology.

2. Design and Simulation

In this study, we have developed an antenna substrate using Arlon AD255C (Chandler, AZ, USA), a high-frequency dielectric material, featuring copper layers on both the top and bottom surfaces to facilitate electromagnetic radiation. The copper layer of the metal substrate has a thickness (h_p) of 17.5 μm and a conductivity (δ) of 5.8×10^7 s/m; its relative permittivity is 2.55, with a dielectric loss tangent of 0.0014, optimizing its performance for high-frequency applications. Additionally, our study employed Ansys Electronics Desktop 2022 R1 software for structural simulation. Comprehensive simulations and optimizations were conducted to refine the antenna's performance. The result is an SIW slot array [28,29] horn antenna, which notably achieves a bandwidth of 14.96 GHz, a gain of 10.01 dBi, and is tuned to a resonant frequency of 60 GHz. The detailed design and specifications of this antenna are illustrated in Figure 1.

Our study aims to iteratively refine and optimize the design of the antenna described above. Next, we will begin describing the following sections:

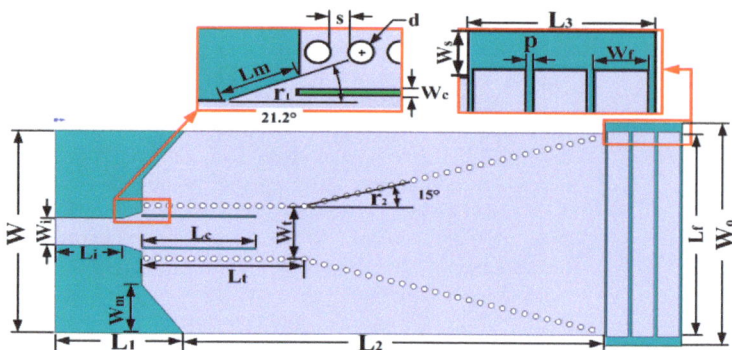

Figure 1. Antenna design parameters.

2.1. SIW Structure Design

The antenna designed in this paper adopts the traditional rectangular waveguide [30,31] and the primary mode in the rectangular waveguide TE_{10} mode [32]. It is also the lowest mode under the mode of TE_{mn}; the wavelength of λ_g is an important parameter of the rectangular waveguide mode of TE_{10} and can be expressed as below:

$$\lambda_g = \frac{\lambda}{\sqrt{\varepsilon_r - (\lambda/\lambda_c)^2}} \quad (1)$$

The column width of the metal through the hole of W_t can be obtained by using the traditional rectangular waveguide calculation formula. The width of the SIW is obtained by the following Formula (2):

$$W_t = W_i + \frac{d^2}{0.95S} \quad (2)$$

The diameter (d) to a width of waveguide (W_t) is not mentioned in Formula (3) and sometimes it will give error values, and the more appropriate equation is represented as:

$$W_i = W_t + 1.08 \times \frac{d^2}{S} - 0.1 \times \frac{d^2}{W_i} \quad (3)$$

Here, W_i represents the equivalent width of the SIW. Upon obtaining the equivalent width W_i of the SIW, we can determine the cutoff frequency of the SIW in the TE_{10} mode, as shown in Formula (4):

$$f_{TE10} = \frac{1}{2W_i\sqrt{\mu\varepsilon}} = \frac{c}{2W_i\sqrt{\mu_r\varepsilon_r}} \quad (4)$$

Here, c represents the speed of light in a vacuum, and μ_r and ε_r denote the relative permeability and relative permittivity of the substrate material.

Based on the above information, the dimension design of the SIW structure can be completed. First, it is necessary to choose an appropriate width W_i for the SIW based on the operating frequency. Taking into account the radiation leakage problem and the difficulty of machining, the size and spacing of the metallized through-holes [25] should satisfy conditions $d \leq \frac{\lambda_g}{5}$, $S \leq 2d$, and $d < 0.1 \times W_t$; S and d were chosen to be 0.6 mm and 0.3 mm, respectively. The microstrip [33] part is combined with the conical corner of $\frac{\lambda}{4}$ proposed in Ref. [24] to better radiate electromagnetic waves into the SIW rectangular waveguide, where the inclination (r_1) of the conical corner of the antenna transmission line is 21.2°. The horn aperture (r_2) is selected to be 30° to better radiate electromagnetic waves outward.

The antenna structure is shown in Figure 2. The SIW column width of the antenna is 3.383 mm, and the substrate thickness is 0.508 mm. Through simulation, the bandwidth performance of the antenna is shown in Figure 3, and the antenna's bandwidth as designed in step 1 is 0.19 GHz.

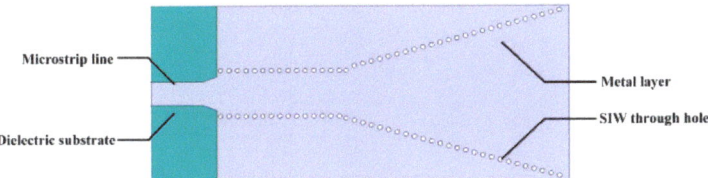

Figure 2. SIW horn antenna structure.

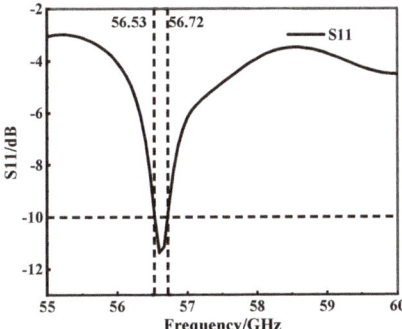

Figure 3. SIW horn antenna bandwidth performance.

2.2. Antenna Radiation Belt Design

The initial iteration of the SIW horn antenna, as developed through the aforementioned steps, exhibited a bandwidth of only 0.19 GHz. This was significantly narrower than the design requirements, with the center frequency and other parameters not aligning with our goals. To address this issue, we conducted an in-depth analysis of the antenna's electromagnetic radiation characteristics [34,35]. We discovered that the return loss at the transmitting end of a standard SIW horn antenna was notably high, negatively impacting its electromagnetic radiation, as depicted in Figure 4a. To optimize the electromagnetic performance of the antenna, we introduced a novel design element: the radiation belt, as illustrated in Figure 4b.

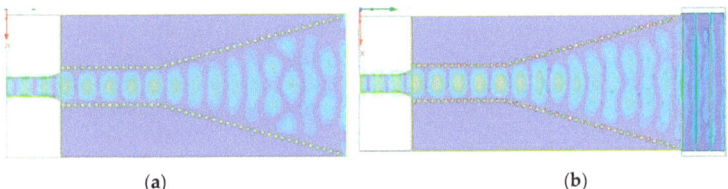

Figure 4. (**a**) Electromagnetic radiation without radiation belt; (**b**) electromagnetic radiation with radiation belt.

This strategic modification was found to not only optimize the electromagnetic radiation of the antenna but also substantially enhance its bandwidth performance. The improvements achieved through this design adjustment are clearly demonstrated in the parametric metrics shown in Figure 5. Most notably, the introduction of the radiation strip significantly expanded the antenna's bandwidth from a mere 0.19 GHz to an impressive 9.56 GHz. This dramatic increase marks a considerable advancement in the bandwidth performance of the antenna, aligning it more closely with the stringent requirements of high-frequency applications.

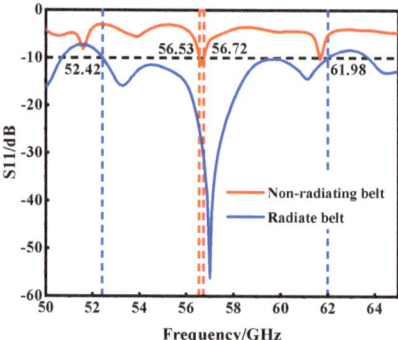

Figure 5. Results with or without radiation belt S_{11}.

2.3. Antenna Slot Array

Under the TE_{10} mode of a rectangular waveguide, opening narrow slots along the direction of current flow generally does not significantly affect the original mode. However, when etching slots on the metal surface, if the slots disrupt the original surface conduction current, displacement currents (i.e., time-varying tangential electric fields) will be induced on the aperture formed by the slots to maintain current continuity. In this case, the open slots are excited, leading to the radiation of electromagnetic energy into space.

According to the principle of electromagnetic equivalence, the radiation of an ideal slot with a length of L_c and a width of W_c ($L_c = \frac{\lambda}{2}, W_c \leq L_c$) is equivalent to a magnetic current source J_m. Following Babinet's principle, it complements a strip dipole with the same shape. Because the rectangular substrate-integrated waveguide (SIW) has transmission characteristics similar to those of traditional rectangular metal waveguides filled with dielectrics, the design methods for SIW slot antennas and traditional metal waveguide slot antennas are essentially the same. However, since the SIW is fabricated using PCB technology, its thickness is relatively thin, and the narrow sidewalls of the traditional metal waveguide are equivalent to the through-hole metallization of the SIW. Therefore, radiation can only be achieved by opening slots on the upper and lower metal surfaces of the SIW.

For the SIW operating in the TE_{10} mode, it exhibits three common types of slots based on their positions, as illustrated in Figure 6c. Slot 1 represents a longitudinal slot on the wide side, parallel to the propagation direction of the SIW. To cut off the current distribution, it needs to be offset from the centerline of the SIW by a certain distance. Slot 2 is a diagonal slot on the wide side, with its center located on the symmetrical centerline of the SIW and at an angle with respect to the propagation direction. Slot 3 is a transverse slot on the wide side, perpendicular to the propagation direction. Among these, the longitudinal slot array on the wide side is the most common type of SIW slot array. When the current is interrupted to open slots, energy leaks into space, thereby radiating electromagnetic waves. Therefore, by establishing a slot array, electromagnetic wave radiation can be controlled to achieve an increase in bandwidth.

In the quest to further augment the performance of the antenna, the concept of a slot array mode, as indicated in Reference [36], was incorporated in step 3 of our process. The SIW, fabricated using the PCB process [37], possesses a thin structure. This thinness is analogous to the narrow sidewalls of a traditional metal waveguide, achieved through metalized vias or holes. In the PCB design, electromagnetic radiation is primarily confined to the top and bottom wide edges of the SIW, where the metal layers are present. A critical aspect of this design is the presence of gaps within the SIW structure. These gaps disrupt the continuity of the transverse surface currents, leading to a distinct behavior. Specifically, the transverse surface currents are redirected towards the ends of these gaps. This redirection results in abrupt changes in the longitudinal currents along the propagation direction of the SIW, as detailed in References [38,39]. Consequently, the longitudinal slot on the wide side of the SIW can be effectively modeled as an admittance in parallel with the transmission line. This concept is instrumental in establishing the equivalent circuit model, depicted in Figure 6a.

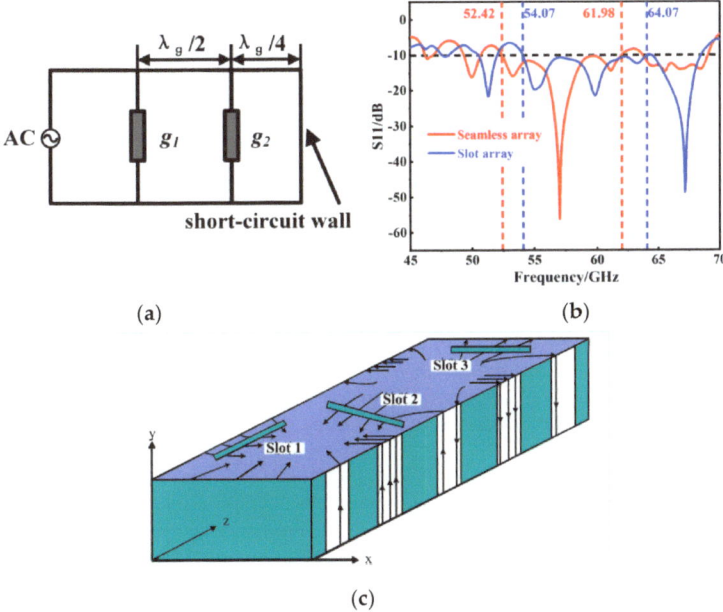

Figure 6. (a) Three slot types of the SIW TE$_{10}$ mold; (b) results with or without slot and S$_{11}$; (c) three types of slots under TE$_{10}$ mode in SIW.

When the wide-side longitudinal slot of the SIW is in resonance, Stevenson provided its normalized conductance as:

$$g = 2.09 \frac{W_i}{h} \frac{\lambda_g}{\lambda} cos^2\left(\frac{\pi \lambda}{2\lambda_g}\right) sin^2\left(\frac{\pi \lambda}{W_i}\right) \quad (5)$$

Figure 6b illustrates the significant impact of implementing a slot array on the bandwidth of the antenna. This comparison clearly shows that introducing the slot array not only extends the bandwidth from 9.56 GHz to a remarkable 10 GHz, but also beneficially shifts the center frequency towards the higher frequency spectrum, moving from 57.2 GHz to 59.07 GHz. This shift brings the antenna's performance closer to the targeted center frequency of 60 GHz, aligning more closely with our design objectives.

In addition to the slot array modification, another critical strategy for optimizing the antenna's electromagnetic performance involves reducing the area of the metal layer. This is achieved by cutting the metal at specific angles, as discussed in Reference [40]. Reducing the metal layer area minimizes metal radiation and coupling effects, thereby enhancing the electromagnetic properties of the antenna. The simulations and comparisons of the S$_{11}$ results provide insightful evidence of this improvement. Notably, the bandwidth's left and right boundaries at the metal layer's tangent angle are expanded outwards, thereby broadening the overall bandwidth from the initial 10 GHz to an impressive 14.48 GHz. The culmination of these design modifications and optimizations is the finalized SIW slot array horn antenna, as showcased in Figure 7a. This design represents the integration of innovative techniques to significantly improve the antenna's bandwidth and electromagnetic performance, marking a substantial advancement in antenna technology for high-frequency applications.

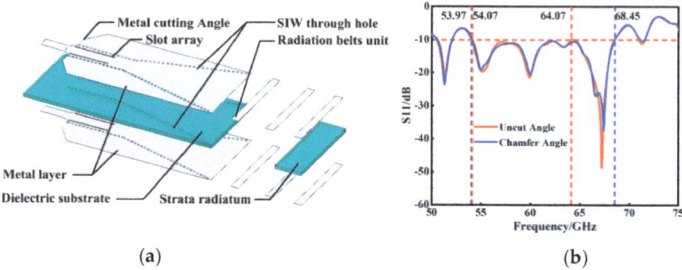

Figure 7. (a) Three-dimensional explosion diagram of the overall antenna structure; (b) results of cutting angle S_{11} with or without metal layer.

3. Analysis, Results, and Discussions

After the above three stages of the design process, the overall structure of the antenna is finally obtained. The effects of the geometric parameters are investigated in this part. The following parameters are selected for this study: the length (L_c) and width (W_c) of the slot array, the distance (P), length (L_f), and width (W_f) of the radiation belt unit, and the width (W_m) of the metal cutting angle. However, the parameters described above will exhibit a hierarchical relationship. The initial design parameters of the antenna are as follows (Table 1).

Table 1. Initial values of optimization parameters.

Symbol	Value (mm)	Description
W_m	2.735	Width of metal cut corners
P	0.150	Radiation belt unit spacing
W_f	1.000	Radiation band unit width
L_f	11.000	Radiation belt unit length
W_c	0.092	Width of slot array element
L_c	5.750	Length of slot array element

The scan optimization and discussion of the above parameters are as follows (Figure 8).

This part of the study focuses on the value of S_{11} under different W_m values. A step scanning optimization method was employed, incrementing W_m by 0.02 mm steps. The simulation results revealed that as W_m increased, the optimum value was identified at 2.835 mm. At this width, the antenna achieved its widest bandwidth, spanning 14.96 GHz (from 54.06 GHz to 69.02 GHz), with the maximum reflection coefficient reaching −39.27 dB. After careful analysis, a W_m of 2.835 mm was determined to be the most effective for achieving optimal bandwidth and reflection characteristics.

The study then analyzed the impact of varying the W_f on the antenna's performance, specifically looking at the S_{11} parameters. It was observed that the antenna's bandwidth shifted at both ends as W_f increased. The most favorable bandwidth expansion occurred at a W_f of 1.15 mm. Meanwhile, when W_f is set to 1.15 mm, the resonance frequency aligns well with the W_m at 2.835 mm, yielding optimal bandwidth. Consequently, an optimal antenna performance was achieved when the width of the W_f was set to 1.15 mm.

This part of the study examines the S_{11} results for different L_f values. The simulations indicate that the S_{11} reflection coefficient peaks at an L_f of 13 mm. However, the optimal length for L_f is determined to be 12 mm. At L_f = 12 mm, the length of the radiation belt closely aligns with the aperture of the SIW horn antenna, facilitating more effective radiation of electromagnetic waves. To minimize electromagnetic wave scattering, the width of the radiation belt is designed to be larger than the distance between the two metallized holes in the SIW horn antenna layer. The study further investigates how the variation in the P affects the antenna's reflection coefficients. It was observed that the antenna's bandwidth decreases as P increases, particularly when the band spacing exceeds 0.15 mm. The antenna achieves its optimal impedance bandwidth of 14.96 GHz at $P = 0.15$ mm, marking this as the best value for antenna bandwidth performance.

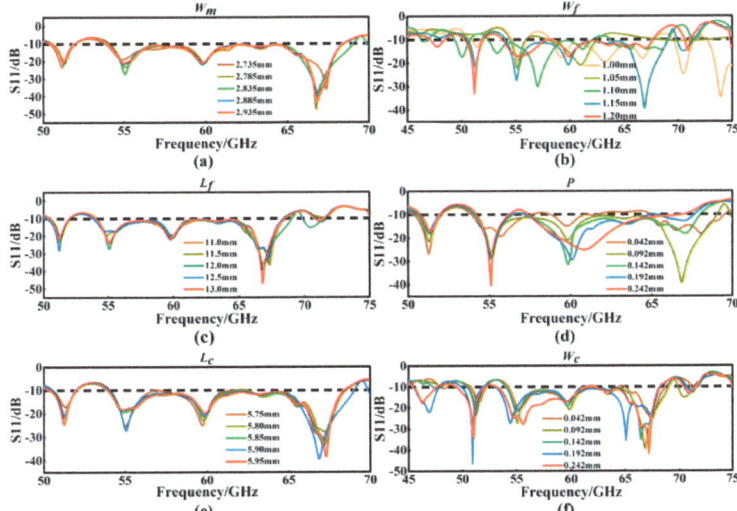

Figure 8. Antenna parameter optimization diagram. (**a**) the impact of the metal cutting angle width (W_m) on antenna performance. (**b**) the impact of the radiation belt unit width (W_f) on antenna performance. (**c**) the impact of the radiation belt unit length (L_f) on antenna performance. (**d**) the impact of the radiation belt unit distance (P) on antenna performance. (**e**) the impact of the slot array length (L_c) on antenna performance. (**f**) the impact of the slot array width (W_c) on antenna performance.

This segment of the study focuses on the influence of the L_c on the antenna's bandwidth. It was observed that the antenna's bandwidth reaches its maximum of 14.96 GHz when L_c is set to 5.90 mm. Consequently, an L_c of 5.90 mm was identified as the optimal length for maximizing the antenna bandwidth. The study then examines how varying the W_c affects the antenna's impedance bandwidth. The results show a gradual widening of the antenna's impedance bandwidth with an increase in W_c, with the maximum bandwidth of 15.15 GHz achieved at a W_c of 0.192 mm. Therefore, the optimal width for the slot array is determined to be 0.192 mm, as it yields the highest bandwidth performance.

Following the meticulous optimization of the core parameters of the antenna, we achieved notable results as depicted in Figure 9, which displays the gain exponent. The extensive simulation process culminated in the antenna achieving a significant bandwidth of 14.96 GHz along with a substantial gain of 10.01 dBi. These figures mark a considerable advancement in antenna performance, particularly in terms of bandwidth and gain. The detailed bandwidth simulation, showcasing these results, is presented in Figure 10b. This diagram illustrates the effective bandwidth coverage of the antenna, confirming its suitability for high-frequency applications. The substantial bandwidth, coupled with the high gain, indicates that the antenna is well-equipped to meet the demanding requirements of contemporary wireless communication systems. The specific size parameters of the designed 60 GHz SIW slot array horn antenna are given in Table 2:

The final design of the SIW slot array horn antenna, as depicted in Figure 10a, was meticulously crafted and manufactured based on the optimized parameters previously outlined. The antenna boasts a compact size of $32 \times 13 \times 0.543$ mm^3, highlighting its small footprint and potential for seamless integration into various applications. The comprehensive simulation and validation results of this antenna are presented in Figure 10b. These results confirm its performance metrics and suitability for the intended high-frequency applications. The combination of its small size and the results from the simulation and validation processes underscore the antenna's innovative design and its potential impact in the field of wireless communication.

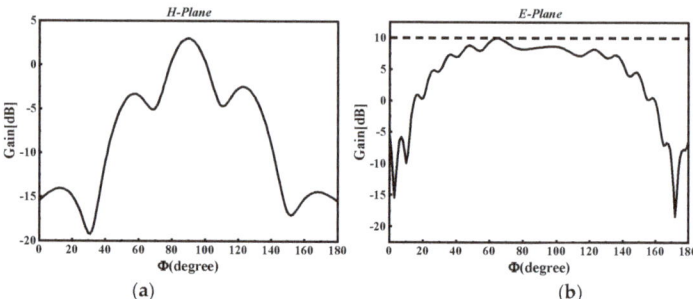

Figure 9. Radiation patterns of E and H planes of SIW slot array horn antenna; (**a**) H-plane radiation pattern; (**b**) E-plane radiation pattern.

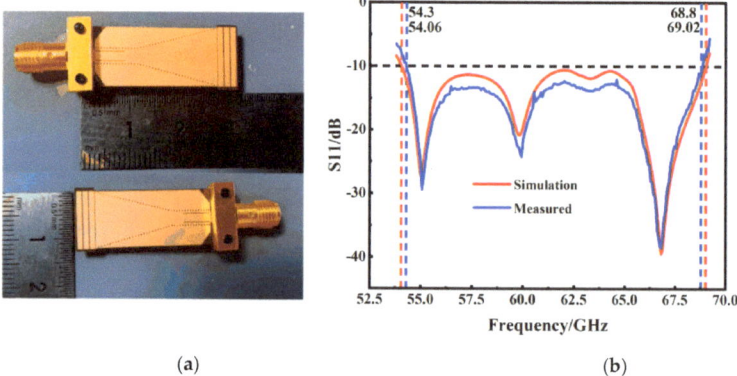

Figure 10. Antenna diagram and performance verification; (**a**) picture of real products; (**b**) analog and measurement bandwidth.

Table 2. Specific values of antenna design parameters.

Symbol	Value (mm)	Description
W	11.800	Antenna width
L_1	6.705	Transmission line and metal cutting length
L_2	21.795	Horn length
L_3	3.980	Radiation layer length
W_i	1.575	50 Ω feeder width
L_i	3.500	50 Ω feeder length
W_t	3.383	Metallized through hole row wide
L_t	8.405	Rectangular waveguide length
W_m	2.835	Width of metal cut corners
L_m	1.064	Conical corner length
W_f	1.150	Radiation band unit width
L_f	12.000	Radiation belt unit length
d	0.300	Metal through hole diameter
S	0.600	Metallized through hole aperture
P	0.150	Radiation belt unit spacing
W_c	0.192	Width of slot array element
L_c	5.900	Length of slot array element
W_o	13.000	Width of horn antenna output

The measurement results of the SIW slot array horn antenna revealed a bandwidth of 14.5 GHz and a reflection coefficient of −37.51 dB. These figures are largely in alignment with the theoretical predictions, affirming the success of the antenna design. However, a minor shift in frequency was observed, which warrants consideration.

This deviation from the theoretical values is likely attributable to a few key factors encountered during the manufacturing and testing phases. Firstly, errors in the positioning and sizing of the through-hole apertures during the manufacturing process could have contributed to these discrepancies. Secondly, the loss of joints in the measurement process is another potential source of variation. Lastly, environmental conditions [41] during testing may have had an impact on the results. These factors underscore the complexities involved in translating theoretical designs into practical applications. They highlight the need for precision in manufacturing and the influence of external conditions on the performance of high-frequency antennas.

4. Comparison

Table 3 presents a comprehensive comparison of the antenna's performance parameters across the three distinct design phases discussed in Section 2. This comparative analysis clearly illustrates that the antenna exhibits superior performance in the third phase compared to the first two phases. The enhanced performance in the final phase can be attributed to the strategic additions made to the antenna design, namely the incorporation of radiation bands and slot arrays. These modifications have proven to be effective in significantly improving both the bandwidth and gain of the antenna. This improvement highlights the importance of iterative design and optimization in antenna engineering, demonstrating how each phase of development contributes to the overall enhancement of antenna performance.

Table 3. Compares the performance of the three phases.

	Bandwidth (GHz)	Gain (dBi)	Maximum Reflection Coefficient (dB)
Phase 1	0.19	5.43	−11.35
Phase 2	9.56	10.56	−56.08
Phase 3	14.96	10.01	−39.47

Table 4 effectively compares the performance of the antenna developed in this paper with other antennas previously reported in the literature. The comparison is based on key parameters such as bandwidth, gain, size, and reflection coefficient. The findings indicate that the antenna designed in this study holds a distinct advantage in terms of bandwidth performance, surpassing most of the other antennas. Additionally, it demonstrates superior performance in terms of reflection coefficient, and its gain and size are also comparatively better than most of the antennas in the study.

The excellence in these performance metrics is particularly significant for industrial radar sensing systems. The wide bandwidth of the antenna contributes to a higher distance resolution, which is crucial in accurately determining the position of objects. Moreover, the center frequency of 60 GHz enhances the velocity resolution, enabling more stable detection of moving objects. The combination of a wide bandwidth and an optimal center frequency is vital for collecting rich and highly accurate point cloud data. This, in turn, facilitates the generation of high-resolution radar images, crucial for the efficacy of industrial radar systems.

Table 4. Comparison between the published SIW structural antennas at 60 GHz and our work.

Reference	Size (mm)	Gain (dBi)	Bandwidth (GHz)	Return Loss (dB)
[20]	29.5 × 8 × 0.787	10	3.33	−12.23
[23]	14 × 8 × 0.381	/	4.18	−33.95
[24]	14 × 8.4 × 0.381	/	3.5	−21.98
[25]	7 × 12 × 0.508	6.54	6.8	−36.35
[26]	16 × 16 × 0.254	10.56	1.73	<−50.0
[42] (simulation)	17.5 × 14.5 × 0.42	11.8	10	/
[43]	13 × 9 × 0.381	/	5.3	−29.6
This work	32 × 13 × 0.543	10.01	14.5	−37.51

5. Conclusions

This study focused on the development of a 60 GHz slot array horn antenna, leveraging an SIW structure. A key aspect of this design was the strategic control of electromagnetic radiation direction through the arrangement of apertures. Additionally, the antenna's bandwidth performance was optimized by incorporating a radiation band at the radiation front. The use of a slot array further expanded the working bandwidth of the antenna.

Experimental evaluations of the antenna demonstrated its exceptional performance, with a strong correlation between simulation predictions and actual measurement results. The antenna's impressive performance characteristics are pivotal for providing high-resolution radar imaging, crucial for accurate target detection and tracking in industrial applications. Moreover, the antenna's compact size contributes to cost reductions in the deployment of radar sensing systems in industrial settings.

In conclusion, the designed antenna, with its combination of excellent performance, high resolution, and cost-effectiveness, emerges as a promising alternative for future industrial radar sensing applications. Its innovative design and successful implementation mark a significant advancement in the field of radar technology.

Author Contributions: Conceptualization, B.M.; methodology, B.M. and J.L.; software, B.M.; validation, Y.S., H.G. and Y.C.; formal analysis, B.M. and Y.C.; investigation, B.M. and J.L.; resources, B.M.; data curation, B.M.; writing—original draft preparation, B.M. and J.L.; writing—review and editing, Y.C. and Y.S.; visualization, H.G. and Y.S.; supervision, Q.W.; project administration, Q.W. and M.L.; funding acquisition, M.L. All authors have read and agreed to the published version of the manuscript.

Funding: This research was funded by Double First-Class Talent Plan Construction (11012315); Double First-Class disciplines National first-class curriculum construction (11013168); Double First-Class Disciplines Construction (11013351); and the National Future Technical College Construction Project (11013169).

Data Availability Statement: The data that support the findings of this study are available from the corresponding author upon reasonable request.

Acknowledgments: Thank you very much for the support of the School of Instrument and Intelligent Future Technology of North University of China.

Conflicts of Interest: The authors declare no conflicts of interest.

References

1. Ruan, X.; Chan, C.H. An Endfire Circularly Polarized Complementary Antenna Array for 5G Applications. *IEEE Trans. Antennas Propag.* **2020**, *1*, 266–274. [CrossRef]
2. Vettikalladi, H.; Sethi, W.; Tariq, H.; Mohammed, A.M. 60 GHz beam-tilting coplanar slotted SIW antenna array. *Frequenz* **2022**, *1–2*, 29–36. [CrossRef]
3. Kao, T.Y.J.; Yan, Y.; Shen, T.M.; Chen, A.Y.K.; Lin, J. Design and Analysis of a 60-GHz CMOS Doppler Micro-Radar System-in-Package for Vital-Sign and Vibration Detection. *IEEE Trans. Microw. Theory Tech.* **2013**, *61*, 1649–1659. [CrossRef]

4. Shen, T.M.; Kao, T.Y.J.; Huang, T.Y.; Tu, J.; Lin, J.; Wu, R.B. Antenna Design of 60-GHz Micro-Radar System-In-Package for Noncontact Vital Sign Detection. *IEEE Antennas Wirel. Propag. Lett.* **2012**, *11*, 1702–1705. [CrossRef]
5. Salarpour, M.; Farzaneh, F.; Staszewski, R.B. A low cost-low loss broadband integration of a CMOS transmitter and its antenna for mm-wave FMCW radar applications. *AEU-Int. J. Electron. Commun.* **2018**, *95*, 313. [CrossRef]
6. Ye, S.; Liang, X.; Wang, W.; Jin, R.; Geng, J.; Bird, T.S.; Guo, Y.J. High-gain planar antenna arrays for mobile satellite communications. *IEEE Antennas Propag. Mag.* **2012**, *54*, 256–268.
7. Li, C.; Cummings, J.; Lam, J.; Graves, E.; Wu, W. Radar remote monitoring of vital signs. *IEEE Microw Mag.* **2009**, *1*, 47–56. [CrossRef]
8. Olk, A.E.; Macchi, P.E.M.; Powell, D.A. Powell, High-Efficiency Refracting Millimeter-Wave Metasurfaces. *IEEE Trans. Antennas Propag.* **2020**, *7*, 5453–5462. [CrossRef]
9. Liu, D.; Gu, X.; Baks, C.W.; Valdes-Garcia, A. Antenna-in-Package Design Considerations for Ka-Band 5G Communication Applications. *IEEE Trans. Antennas Propag.* **2017**, *12*, 6372–6379. [CrossRef]
10. Rabbani, M.S.; Ghafouri-Shiraz, H. Ultra-Wide Patch Antenna Array Design at 60 GHz Band for Remote Vital Sign Monitoring with Doppler Radar Principle. *J Infrared Milli. Terahz. Waves.* **2017**, *38*, 548–566. [CrossRef]
11. Zhang, Z.-Y.; Zuo, S.; Zhao, Y.; Ji, L.-Y.; Fu, G. Broadband Circularly Polarized Bowtie Antenna Array Using Sequentially Rotated Technique. *IEEE Access* **2018**, *6*, 12769–12774. [CrossRef]
12. Abdel-Wahab, W.M.; Safavi-Naeini, S. Wide-Bandwidth 60-GHz Aperture-Coupled Microstrip Patch Antennas (MPAs) Fed by Substrate Integrated Waveguide (SIW). *IEEE Antennas Wirel. Propag. Lett.* **2011**, *10*, 1003–1005. [CrossRef]
13. Guan, D.F.; Ding, C.; Qian, Z.P.; Zhang, Y.S.; Guo, Y.J.; Gong, K. Gong, Broadband high-gain SIW cavity-backed circular-polarized array antenna. *IEEE Trans. Antennas Propag.* **2016**, *64*, 1493–1497. [CrossRef]
14. Li, Y. Circularly Polarized SIW Slot LTCC Antennas at 60 GHz. In *Substrate-Integrated Millimeter-Wave Antennas for Next-Generation Communication and Radar Systems*; John Wiley & Sons: Hoboken, NJ, USA, 2021; pp. 177–195.
15. Statnikov, K.; Sarmah, N.; Grzyb, J.; Malz, S.; Heinemann, B.; Pfeiffer, U.R. A 240 GHz circular polarized FMCW radar based on a SiGe transceiver with a lens-integrated on-chip antenna. In Proceedings of the European Radar Conference 2014, Rome, Italy, 8–10 October 2014; pp. 447–450.
16. Xu, F.; Wu, K. Guided-wave and leakage characteristics of substrate integrated waveguide. *IEEE Trans. Microw. Theory Tech.* **2005**, *53*, 66–73.
17. Wu, P.; Yu, Z.; Liu, K.; Zhan, F.; Wang, S.; Wang, K.; Hao, C. Low-Profile and High-Integration Phased Array Antenna Technology. In Proceedings of the IEEE Conference on Antenna Measurements and Applications (CAMA), Guangzhou, China, 14–17 December 2022; pp. 1–6.
18. Chuang, H.; Kuo, H.; Lin, F.; Huang, T.; Kuo, C.; Ou, Y. 60-GHz millimeter-wave life detection system (MLDS) for noncontact human vital-signal monitoring. *IEEE Sens. J.* **2012**, *3*, 602–609. [CrossRef]
19. Zhu, H.R.; Li, K.; Lu, J.G.; Mao, J.F. Millimeter-Wave Active Integrated Semielliptic CPW Slot Antenna With Ultrawideband Compensation of Ball Grid Array Interconnection. *IEEE Trans. Compon. Packag. Technol.* **2022**, *1*, 111–120. [CrossRef]
20. Ramesh, S.; Rao, T.R. Indoor radio link characterization studies for millimeter wave wireless communications utilizing dielectric-loaded exponentially tapered slot antenna. *J. Electromagn. Waves Appl.* **2015**, *29*, 551–564. [CrossRef]
21. Singh, A.; Rehman, S.U.; Yongchareon, S.; Chong, P.H.J. Multi-Resident Non-Contact Vital Sign Monitoring Using Radar: A Review. *IEEE Sens. J.* **2021**, *4*, 4061–4084. [CrossRef]
22. Kathuria, N.; Seet, B.C. 24 GHz Flexible LCP Antenna Array for Radar-Based Noncontact Vital Sign Monitoring. In Proceedings of the Asia-Pacific Signal and Information Processing Association Annual Summit and Conference (APSIPA ASC), Auckland, New Zealand, 7–10 December 2020; pp. 1472–1476.
23. Mungaru, N.K.; Shanmuganantham, T. Broad-band H-spaced head-shaped slot with SIW-based antenna for 60 GHz wireless communication applications. *Microw. Opt. Technol. Lett.* **2019**, *61*, 1911–1916. [CrossRef]
24. Mungaru, N.K.; Shanmuganantham, T. Substrate-integrated waveguide-based slot antenna for 60 GHz wireless applications. *Microw. Opt. Technol. Lett.* **2019**, *8*, 1945–1951. [CrossRef]
25. Shanmuganantham, T.; Kumar, K.B.; Kumar, S.A. *Analysis of Tree-Shaped Slotted Impedance Matching Antenna for 60 GHz Femtocell Applications*; The Korean Institute of Communications and Information Sciences (KICS): Seoul, Republic of Korean, 2021; pp. 426–431.
26. Manish, S.; Ashwni, K. Conformal ultra-compact narrowband 60.0 GHz four-port millimeter wave MIMO antenna for wearable short-range 5G application. *Wirel. Netw.* **2024**, *6*, 1572–8196.
27. Tsai, C.C.; Cheng, Y.S.; Huang, T.Y.; Hsu, Y.P.A.; Wu, R.B. Design of microstrip-to-microstrip via transition in multilayered LTCC for frequencies up to 67 GHz. *IEEE Trans. Compon. Packag. Manuf. Technol.* **2011**, *4*, 595–601. [CrossRef]
28. Zhang, Z.; Cao, X.; Gao, J.; Li, S.; Han, J. Broadband SIW Cavity-Backed Slot Antenna for Endfire Applications. *IEEE Antennas Wirel. Propag. Lett.* **2018**, *7*, 1271–1275. [CrossRef]
29. Hong, Y.; Bang, J.; Choi, J. Gain-enhanced 60-GHz SIW cavity-backed slot array antenna using metallic grooves. *Microw. Opt. Technol. Lett.* **2020**, *62*, 2429–2433. [CrossRef]
30. Liu, P.; Pedersen, G.F.; Hong, W.; Zhang, S. Dual-Polarized Wideband Low-Sidelobe Slot Array Antenna for V-Band Wireless Communications. *IEEE Trans. Antennas Propag.* **2023**, *8*, 6667–6677. [CrossRef]

31. Gouveia, C.; Loss, C.; Raida, Z.; Lacik, J.; Pinho, P.; Vieira, J. Textile Antenna Array for Bio-Radar Applications. In Proceedings of the International Microwave and Radar Conference (MIKON), Warsaw, Poland, 5–8 October 2020; pp. 315–319.
32. Yang, T.; Zhao, Z.; Yang, D.; Nie, Z. A single-layer circularly polarized antenna with improved gain based on quarter-mode substrate integrated waveguide cavities array. *IEEE Antennas Wirel. Propag. Lett.* **2020**, *19*, 2388–2392. [CrossRef]
33. Lin, S.K.; Lin, Y.C. A compact sequential-phase feed using uniform transmission lines for circularly polarized sequential-rotation arrays. *IEEE Trans. Antennas Propag.* **2011**, *7*, 2721–2724. [CrossRef]
34. Abdelaziz, A.; Hamad, E.K.I. Design of a compact high gain microstrip patch antenna for tri-band 5G wireless communication. *Frequenz* **2019**, *1*, 45–52. [CrossRef]
35. Grzyb, J.; Statnikov, K.; Sarmah, N.; Heinemann, B.; Pfeiffer, U.R. A 210–270-GHz Circularly Polarized FMCW Radar With a Single-Lens-Coupled SiGe HBT Chip. *IEEE Trans. Terahertz Sci. Technol.* **2016**, *6*, 771–783. [CrossRef]
36. Zhu, Y.; Chen, K.; Tang, S.Y.; Yu, C.; Hong, W. Ultrawideband Strip-Loaded Slotted Circular Patch Antenna Array for Millimeter-Wave Applications. *IEEE Antennas Wirel. Propag. Lett.* **2023**, *9*, 2230–2234. [CrossRef]
37. Zheng, B.; Shen, Z. Effect of a finite ground plane on microstrip-fed cavity-backed slot antennas. *IEEE Trans. Antennas Propag.* **2005**, *53*, 862–865. [CrossRef]
38. Feng, W.; Ni, X.; Shen, R.; Wang, H.; Qian, Z.; Shi, Y. High-Gain 100 GHz Antenna Array Based on Mixed PCB and Machining Technique. *IEEE Trans. Antennas Propag.* **2022**, *8*, 7246–7251. [CrossRef]
39. Chang, L.; Li, Y.; Zhang, Z.; Li, X.; Wang, S.; Feng, Z. Low-Sidelobe Air-Filled Slot Array Fabricated Using Silicon Micromachining Technology for Millimeter-Wave Application. *IEEE Trans. Antennas Propag.* **2017**, *8*, 4067–4074. [CrossRef]
40. Agrawal, M.; Kumar, T.A. Wideband Circularly Polarized Antenna Using Substrate Integrated Waveguide and Truncated-Corner Patch Loaded Slot. *Wirel. Pers. Commun.* **2023**, *131*, 399–413. [CrossRef]
41. Statnikov, K.; Grzyb, J.; Sarmah, N.; Heinemann, B.; Pfeiffer, U.R. A lens-coupled 210–270 GHz circularly polarized FMCW radar transceiver module in SiGe technology. In Proceedings of the European Microwave Conference (EuMC), Paris, France, 7–10 September 2015; pp. 550–553.
42. Zhu, H.; Li, X.P.; Feng, W.W.; Yao, L.; Xiao, J.Y.; Wang, T.H. Surface micro-machined beam-steering antenna array for 60 GHz radios. *Infrared Millim. Waves* **2017**, *36*, 563–569.
43. Mungaru, N.K.; Thangavelu, S. Broadband substrate-integrated waveguide venus-shaped slot antenna for V-band applications. *Microw Opt. Technol. Lett.* **2019**, *61*, 2342–2347. [CrossRef]

Disclaimer/Publisher's Note: The statements, opinions and data contained in all publications are solely those of the individual author(s) and contributor(s) and not of MDPI and/or the editor(s). MDPI and/or the editor(s) disclaim responsibility for any injury to people or property resulting from any ideas, methods, instructions or products referred to in the content.

Article

Design and Analysis of 5-DOF Compact Electromagnetic Levitation Actuator for Lens Control of Laser Cutting Machine

Chuan Zhao [1], Qinwei Zhang [1], Wenzhe Pei [1], Junjie Jin [1,*], Feng Sun [1], Hongkui Zhang [2], Ran Zhou [1], Dongning Liu [1], Fangchao Xu [1], Xiaoyou Zhang [3] and Lijian Yang [4]

1. School of Mechanical Engineering, Shenyang University of Technology, Shenyang 110870, China; zhaochuan@sut.edu.cn (C.Z.); zhangqinwei@smail.sut.edu.cn (Q.Z.); peiwenzhe@smail.sut.edu.cn (W.P.); sunfeng@sut.edu.cn (F.S.); zhouran@sut.edu.cn (R.Z.); dongningliu@sut.edu.cn (D.L.); xufangchao@sut.edu.cn (F.X.)
2. China Coal Technology & Engineering Group Shenyang Research Institute, Fushun 113122, China
3. Department of Mechanical Engineering, Nippon Institute of Technology, Saitama 345-8501, Japan; zhang.xiaoyou@nit.ac.jp
4. School of Information Science and Engineering, Shenyang University of Technology, Shenyang 110870, China
* Correspondence: jinjunjie@sut.edu.cn

Abstract: In laser beam processing, the angle or offset between the auxiliary gas and the laser beam axis have been proved to be two new process optimization parameters for improving cutting speed and quality. However, a traditional electromechanical actuator cannot achieve high-speed and high-precision motion control with a compact structure. This paper proposes a magnetic levitation actuator which could realize the 5-DOF motion control of a lens using six groups of differential electromagnets. At first, the nonlinear characteristic of a magnetic driving force was analyzed by establishing an analytical model and finite element calculation. Then, the dynamic model of the magnetic levitation actuator was established using the Taylor series. And the mathematical relationship between the detected distance and five-degree-of-freedom was determined. Next, the centralized control system based on PID control was designed. Finally, a driving test was carried out to verify the five-degrees-of-freedom motion of the proposed electromagnetic levitation actuator. The results show it can achieve a stable levitation and precision positioning with a desired command motion. It also proves that the proposed magnetic levitation actuator has the potential application in an off-axis laser cutting machine tool.

Keywords: electromagnetic levitation actuator; laser cutting machine; PID control; coordinate transform; magnetic field analysis

1. Introduction

With the rapid development of the aerospace, automobile and semiconductor industries, the processing difficulty of key parts with complex structures and high-hardness materials is gradually increasing. Improving efficiency under the premise of ensuring machining accuracy and indicating quality has become an important challenge for advanced manufacturing technology. Laser cutting has been widely applied due to its characteristics of high precision, high efficiency, small thermal deformation, low noise, strong flexibility and small thermal deformation.

In laser cutting processing, the processing method of coaxial laser beam and auxiliary gas nozzle has basically been adopted, and the research on the laser cutting process is mostly focused on the analysis of the nozzle auxiliary gas flow field and structural improvement. The research on traditional laser cutting technology mostly focuses on the improvement of the nozzle structure and the analysis of the auxiliary gas flow field [1–3]. Its processing quality and efficiency are affected by laser power, pulse frequency, gas pressure, feed rate, defocusing amount, sheet properties and thickness. [4] proposed a new process optimization scheme, which can effectively improve the processing efficiency of the laser cutting machine and the processing quality of the kerf by making the blowing direction of the auxiliary gas and the laser beam form a certain angle. However, this different axis will also lead to a different processing efficiency and quality of the laser cutting machine in each processing feed direction, which will seriously affect the processing accuracy. In order to solve the influence of this inconsistency, a high-speed, high-precision and compact driver must be used to control the lens to ensure the position relationship between the laser beam and the auxiliary airflow in real time in different feed directions.

The electromechanical actuators cannot achieve a multi-degree of freedom driving motion with a compact structure due to the existence of the contact transmission mechanism. The magnetic levitation technology can realize force regulation without contact. In a multi-DOF driving system, the magnetic levitation technology is highly favorable for the following characteristics: (i) a compact structure as the driving force and torque are generated by an integrated maglev actuator; (ii) a high motion precision with active control and positioning errors compensation; (iii) the elimination of vibration noise owing to its non-contact advantage [5,6]

The maglev actuators can be categorized into Lorentz actuators and reluctance actuators. Dyck [7] developed a 6-DOF magnetically levitated rotary table for micro-positioning. This stage uses a combination of four Lorentz-force magnetically levitated linear motors to achieve an unlimited rotation motion range about the vertical axis. Heyman [8] designed a Lorentz force-based magnetically levitated stage which can achieve a 10 mm stroke in all XYZ directions. Gloess [9] presented a magnetically levitated hub actuator. This stage prototype can generate thrust forces in the X and Y directions of up to 200 N. Zhang [10] proposed a MagTable which consists of a planar array of square coils and a permanent magnet type carrier. The maximum levitation height of the carrier is 30 mm within a 400 mm × 200 mm horizontal translation range. Huang [11,12] proposed a min–max model predictive control (MPC) method of planar motors, which can achieve robust precision position tracking. The Lorentz actuators can achieve a long stroke with nanometer positioning. However, it is difficult to achieve laser lens driving by Lorentz force due to a large volume caused by the PM array or armature winding.

The magnetic bearing is a typical application of reluctance levitation technology [13]. In order to support the rotor of rotating machinery, the five-degree-of-freedom magnetic bearing system often adopts a distributed scheme [14]. Two radial magnetic bearings realize the three-degree-of-freedom control on both sides of the rotor, and a thrust magnetic bearing is used to move the rotor axially [15,16]. Masahiro [17] designed a maglev motor with a 5-DOF active control. The movable ranges of the rotor in the axial and radial direction are restricted to ±0.3 mm and ±0.5 mm, respectively. Luan [18] and Zhang [19] designed a controllable magnetic levitation actuator for an EDM machine tool to improve the stability of the inter pole voltage, hence the machining speed increases to 3.925 µs. Dongjue He [20,21] designed a novel air core coil type electro-magnetic driving unit to actuate the lens holder, which can achieve a range of ±5 mm with a tracking error of less than 12 µm and a bandwidth of more than 100 Hz in the axial direction. However, the above magnetic levitation driver has a large volume, a large mover mass, and a large motion inertia, resulting in slow control accuracy and response speed.

Therefore, this project proposes a five-degree-of-freedom magnetic levitation driver with a compact structure, which adopts six sets of differential electromagnets to achieve five-degree-of-freedom motion. The dynamic model of the five-degree-of-freedom magnetic levitation drive device is established. The characteristics of the electromagnetic force of the linearized model are analyzed, and the mathematical model between the sensor and the actual displacement of the suspension platform is derived. The PID control is used to verify the five-degrees-of-freedom motion of the system, and the displacement response and position control characteristics of the system are analyzed. In the control of each degree of freedom motion, it can achieve a stable suspension and good tracking effect of the desired command, and has a certain robustness.

2. 5-DOF Magnetic Levitation Driver

2.1. Magnetic Levitation Driver Function

Traditional laser cutting technology requires the laser beam to be coaxial with the auxiliary gas in order to ensure the consistency of the cutting quality in the processing feed direction. A review of the literature reveals that the eccentricity of the laser beam with the auxiliary gas improves the quality and efficiency of the process. This paper proposes a magnetic levitation drive for a five-degree-of-freedom laser light path. It realizes high cutting quality and efficiency, non-contact 5-DOF motion, reduced friction and improved system response characteristics. The ranges of the magnetic levitation actuator in five-degrees-of-freedom are, respectively, 0.05 mm in axial range, 0.1 mm in radial range, 0.001 rad in α direction, and 0.001 rad in β direction.

The magnetic levitation drive is shown in Figure 1. The drive as a whole consists of a top cover, a bottom cover and a connecting ring in the center. There is an aluminum ring in the middle of the suspended platform to place the laser lens, and the control of the suspended platform realizes the control of the laser lens to achieve the purpose of controlling the laser light path and the off-axis effect. Among them, four sets of axial differential electromagnets are evenly distributed in the upper and lower covers, which can make the floating platform realize the z-axis direction and α, β direction movement; the connecting ring in the middle part is likewise evenly distributed with two sets of radial differential solenoids in 45° relation to the axial solenoids, which can realize the movement of the floating platform in the direction of the X and Y axes.

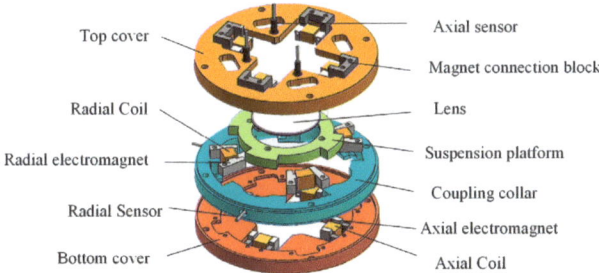

Figure 1. Structural diagram of a 5-DOF magnetic levitation actuator.

2.2. 5-DOF Magnetic Levitation Actuator Principle of Operation

The 5-DOF motion of the magnetic levitation actuator is controlled by the electromagnetic force of a differential electromagnet. In the experiments, the translations along the x and y directions are similar, as are the α and β direction rotations. Therefore, in this paper, only the principles of translation in the z and x directions and the rotation in the α direction are presented. As shown in Figure 2, there are four sets of differential electromagnets labeled 1 (1'), 2 (2'), 3 (3'), and 4 (4') in the vertical direction; and there are two sets of differential electromagnets 5 (5') and 6 (6') in the horizontal direction in a diagonal arrangement. In Figure 2a, the combined force generated by the four sets of differential electromagnets in the vertical direction is in the same direction as the axis, thereby driving the suspended platform in that direction. In Figure 2b, the electromagnets 5 (5') and electromagnets 6 (6') generate a combined force pointing in the positive direction of the X-axis, thereby driving the suspended platform along that direction. In Figure 2c, the torque generated by electromagnet 1 (1') is in the opposite direction to the movement generated by electromagnet 3 (3'), thereby driving the levitated platform to rotate in that direction. The experimental setup is shown in Figure 3.

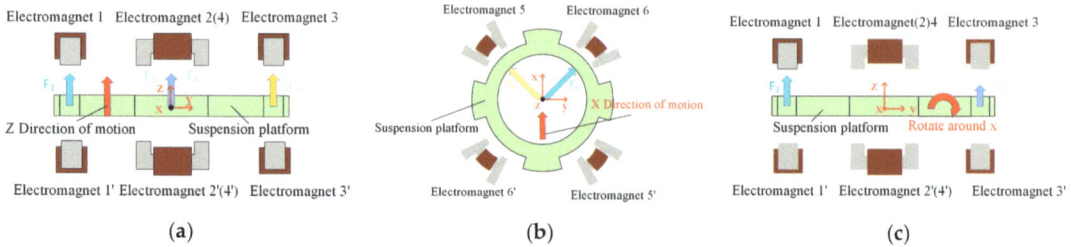

Figure 2. Motion control principle of suspended platform: (**a**) z direction, (**b**) x direction, (**c**) α direction.

Figure 3. The 5-DOF magnetic levitation drive.

3. Analysis of Magnetic Field Characteristics of Magnetic Drive Platform

3.1. Axial Single Electromagnetic Force Analysis

First, the magnetic field finite element software is used to simulate the five-degree-of-freedom magnetic levitation drive model. The core magnetic material is set as silicon steel, the levitation platform material is set as Q235, the coil material is set as copper, and other materials are set as aluminum. In order to ensure the accuracy of the calculation results, the model adopts adaptive mesh and refines the key parts such as arc air gap and levitation air gap. The levitation platform is set as the force object, the variable is set as the control current, and the range of the parameterized scan is 0–1.5 A, with a step size of 0.1 A.

Then, the simulated electromagnetic force is compared with the theoretical value calculated by the simplified model of differential electromagnetic force Equation (1) using the simulated electromagnetic force, and the change in the electromagnetic force with the excitation current at the axial equilibrium position is compared and analyzed, as shown in Figure 4.

$$F = F_1 - F_2 = \frac{\mu_0 N^2 A (i + i_0)^2}{4(d_0 - z)^2} - \frac{\mu_0 N^2 A (i - i_0)^2}{4(d_0 + z)^2} \tag{1}$$

In the above equation, N is the number of turns of coil required to wind the solenoid, A is the cross-sectional area of the magnetic circuit air gap, i is the current in the coil of the solenoid, i_0 is the bias current, and d_0 is the balance air gap.

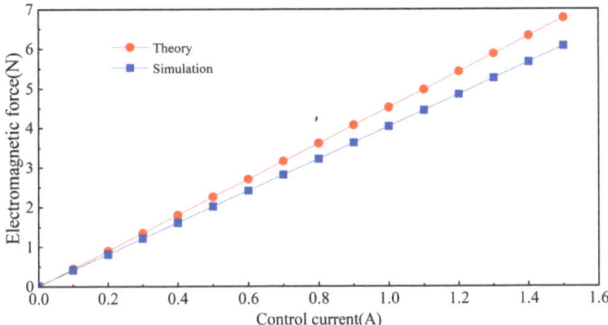

Figure 4. Comparison of theoretical and simulated electromagnetic forces.

The comparison between simulation and theoretical calculation shows that the simulated electromagnetic force has the same trend as the theoretical electromagnetic force. At the maximum control current of 1.5 A, the maximum error between theory and simulation is 10%, which meets the design requirements and indicates that the structural design is reasonable. The results also show that the magnetic force of 3.64 N at a control current of 0.9 A satisfies the experimental requirements for a levitated platform under current differential control. The structural parameters and solenoid parameters of which are simulated are shown in Table 1.

Table 1. Theoretical calculation and simulation parameters.

Parameter Name	Symbol	Value
number of turns	N	100
Balance air gap	d_0 (mm)	1
permeability in vacuum	μ_0	$4\pi \times 10^{-7}$
Air gap cross-sectional area	A (mm^2)	60
Suspended platform inner diameter	d_1 (mm)	67
Suspended platform outer diameter	d_3 (mm)	114
Electromagnet cross section width	b (mm)	10
Electromagnet cross section length	L_4 (mm)	6
Coupling collar inner diameter	d_2 (mm)	126

3.2. Equilibrium Position Electromagnetic Force Analysis

The variation in the axial electromagnetic force with the bias current under different control currents is analyzed using the finite element method and the simulation results are shown in Figure 5.

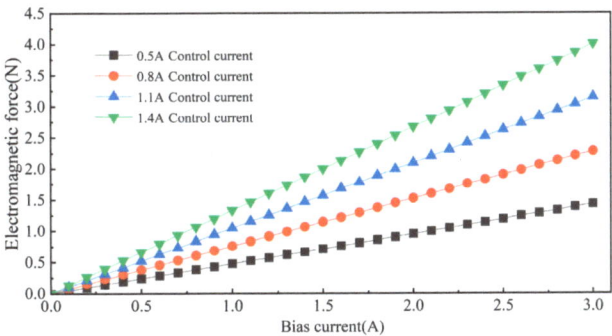

Figure 5. Variation in electromagnetic force with different bias currents.

The analysis results show the change in electromagnetic force at the equilibrium position with different bias currents; different control currents are selected to observe the change in electromagnetic force and it can be seen from the simulation that the electromagnetic force has a good linear relationship at the equilibrium position.

Furthermore, the impact of axial displacement at the equilibrium position and the selection of the control current on the performance of the axial electromagnetic force was analyzed. The equilibrium position was set at 0 mm, and the variation in the axial electromagnetic force with the control current was examined as the axial displacement was varied from 0.1 mm to 0.4 mm. The results of this analysis are presented in Figure 6.

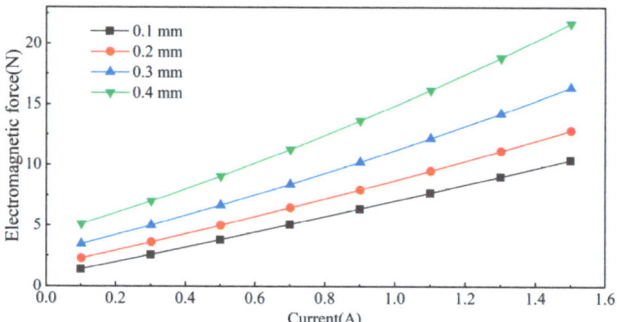

Figure 6. Variation in electromagnetic force with control current under different axial displacements.

The results show that the slope of the curve remains basically unchanged when the displacement is less than 0.4 mm; there is a significant increase in the slope and the nonlinear characteristics of the electromagnetic force begin to appear.

In the design of the 5-DOF magnetic levitation actuator, in order to ensure stability during operation, the levitation platform should have a good linear workspace at the steady state operating point, which is expected by the design. So, the electromagnetic force is simulated at different bias currents and at bias current 1.5 A and the electromagnetic force has a good linear space. Additionally, the variation in the electromagnetic force with the control current in the displacement case is also simulated. The simulation results show that the electromagnetic force within the equilibrium position range exhibits good linear characteristics. Consequently, the designed 5-DOF maglev actuator demonstrates a favorable linear workspace at the equilibrium position.

3.3. Magnetic Field Analysis

Figure 7 depicts the magnetic field simulation in the axial and radial directions, respectively. To facilitate the clear observation of the axial magnetic field simulation, a set of electromagnet simulations is utilized. The five-degree-of-freedom magnetic drive platform model is simulated using simulation software, with a maximum current of 3 A applied to the axial and radial electromagnets for excitation, respectively. The simulation results indicate that the magnetic field strength and magnetic circuit of the levitated platform are consistent with theoretical expectations. Furthermore, it is demonstrated that the structural design of the five-degree-of-freedom magnetic levitation actuator avoids magnetic leakage.

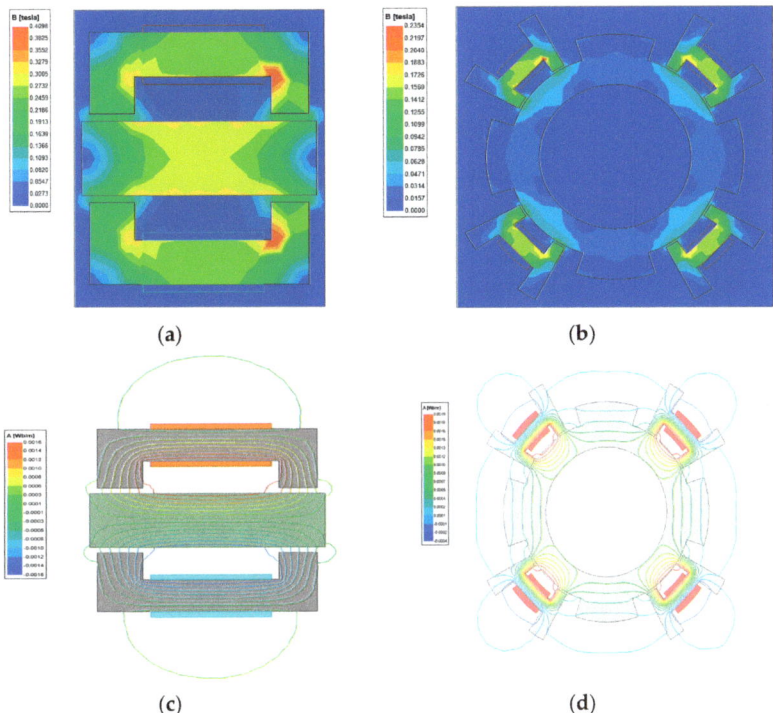

Figure 7. Magnetic field simulation of a 5-DOF magnetic levitation actuator. (**a**) Axial magnetic field intensity. (**b**) Radial magnetic field intensity. (**c**) Axial magnetic circuit. (**d**) Radial magnetic circuit.

4. Mathematical Modeling of a 5-DOF Maglev Actuator

4.1. Sensor Coordinate Transformation for Magnetic Levitation Actuators

When performing levitation experiments with a five-degree-of-freedom magnetic levitation actuator, it is necessary to transform the coordinates of the sensor and the degrees of freedom. According to the working principle of a differential electromagnet, by controlling the size of the control current of the electromagnet, the size of the electromagnetic force of the levitation platform can be changed to realize the movement of five-degrees-of-freedom. The axial sensors are C_1, C_2 and C_3, and the radial sensors are C_5 and C_6. When the platform moves in the x and y directions, it can be measured directly without solving the derivation. When the platform moves in other degrees of freedom, the offset detected using the sensors for each degree of freedom is not the actual controlled offset and needs to be solved. This discrepancy is caused by the fact that the centerline of the sensor is not in the same line as the centerline of the magnetic poles. Where the relationship between the measurement signal of the sensor and the degrees of freedom is shown in Equation (2):

$$\begin{cases} z = \frac{1}{3}(d_1 + d_2 + d_3) \\ \alpha = \frac{d_2 - d_1}{2L_1 \sin\theta} \\ \beta = \frac{d_3 - d_2}{2L_1 \sin\theta} \\ x = d_5 \\ y = d_6 \end{cases} \qquad (2)$$

Rewrite in matrix form:

$$\begin{bmatrix} z \\ \alpha \\ \beta \\ x \\ y \end{bmatrix} = N \begin{bmatrix} d_1 \\ d_2 \\ d_3 \\ d_5 \\ d_6 \end{bmatrix} \quad (3)$$

N is the coordinate transformation matrix of the sensor and the degrees of freedom:

$$N = \begin{bmatrix} \frac{1}{3} & \frac{1}{3} & \frac{1}{3} & 0 & 0 \\ -\frac{1}{2L_1 \sin\theta} & -\frac{1}{2L_1 \sin\theta} & 0 & 0 & 0 \\ 0 & -\frac{1}{2L_1 \sin\theta} & \frac{1}{2L_1 \sin\theta} & 0 & 0 \\ 0 & 0 & 0 & 1 & 0 \\ 0 & 0 & 0 & 0 & 1 \end{bmatrix}$$

4.2. Modeling of Magnetic Levitation Drive Systems

From the 5-DOF magnetic levitation actuator system, the force analysis of the levitated platform is shown in Figure 8. F is the magnetic force of the electromagnet. z, α, and β are the displacement of the suspended platform along the Z-axis, the angle of rotation around the X-axis, and the angle of rotation around the Y-axis, respectively. L_1, L_2, and L_3 are the distances between the sensor, axial electromagnet, and radial electromagnet, respectively. θ is the angle between the sensor and the X-axis. m is the mass of the suspended platform.

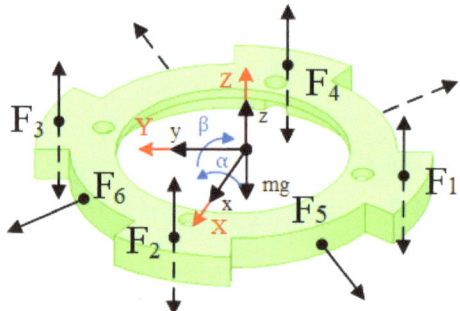

Figure 8. Force analysis of suspension platform.

The dynamics of a 5-DOF magnetic levitation actuator is modeled according to the Lagrange equations:

$$\begin{aligned} m\ddot{x} &= \frac{\sqrt{2}}{2}(F_5 + F_6) - c_x \dot{x} + f_x \\ m\ddot{y} &= \frac{\sqrt{2}}{2}(F_5 - F_6) - c_y \dot{y} + f_y \\ m\ddot{z} &= F_1 + F_2 + F_3 + F_4 - c_z \dot{z} - mg + f_z \\ J_\alpha \ddot{\alpha} &= F_1 L_2 - F_3 L_2 - c_\alpha \dot{\alpha} + T_\alpha \\ J_\beta \ddot{\beta} &= F_4 L_2 - F_2 L_2 - c_\beta \dot{\beta} + T_\beta \end{aligned} \quad (4)$$

where F_1 to F_6 are the magnetic forces of each of the six electromagnets. c_z, c_x, c_y, c_α, c_β are the damping coefficients for the Z-axis, X-axis, Y-axis, rotation around the X-axis, and rotation around the Y-axis, respectively. f_z, f_x, f_y, T_α, T_β are the perturbation forces in the Z-axis, X-axis, Y-axis, rotation around the X-axis, and rotation around the Y-axis, respectively.

$$\begin{aligned} F_n &= k_i i_n + k_d d_n, n = 1,2,3,4 \\ F_m &= k'_i i_m + k'_d d_m, m = 5,6 \end{aligned} \quad (5)$$

$$k_i = \frac{4Ki_0}{d_0^2}, k_d = \frac{4Ki_0^2}{d_0^3} \quad (6)$$

The linear differential equation of the system is obtained, as shown in Equation (7), by applying the linearization Equation (5) to Equation (4).

$$\begin{aligned}
m\ddot{x} &= \tfrac{\sqrt{2}}{2}k'_i(i_5 + i_6) + \tfrac{\sqrt{2}}{2}k'_d(d_5 + d_6) - c_x\dot{x} + f_x \\
m\ddot{y} &= \tfrac{\sqrt{2}}{2}k'_i(i_5 - i_6) + \tfrac{\sqrt{2}}{2}k'_d(d_5 - d_6) - c_y\dot{y} + f_y \\
m\ddot{z} &= k_i(i_1 + i_2 + i_3 + i_4) + k_d(d_1 + d_2 + d_3 + d_4) - c_z\dot{z} + f_z \\
J_\alpha\ddot{\alpha} &= k_i L_2(i_1 - i_3) + k_d L_2(d_1 - d_3) - c_\alpha\dot{\alpha} + T_\alpha \\
J_\beta\ddot{\beta} &= k_i L_2(i_4 - i_2) + k_d L_2(d_4 - d_2) - c_\beta\dot{\beta} + T_\beta
\end{aligned} \quad (7)$$

Organized in matrix form:

$$M\begin{bmatrix}\ddot{x}\\ \ddot{y}\\ \ddot{z}\\ \ddot{\alpha}\\ \ddot{\beta}\end{bmatrix} = K_I\begin{bmatrix}i_1\\ i_2\\ i_3\\ i_4\\ i_5\\ i_6\end{bmatrix} + K_D\begin{bmatrix}d_1\\ d_2\\ d_3\\ d_4\\ d_5\\ d_6\end{bmatrix} - C\begin{bmatrix}\dot{x}\\ \dot{y}\\ \dot{z}\\ \dot{\alpha}\\ \dot{\beta}\end{bmatrix} + \begin{bmatrix}f_x\\ f_y\\ f_z\\ T_\alpha\\ T_\beta\end{bmatrix} \quad (8)$$

M is the inertial matrix of the suspension platform, K_I is the current coefficient matrix, K_D is the displacement coefficient matrix, and C is the damping coefficient matrix.

$$M = \begin{bmatrix} m & 0 & 0 & 0 & 0 \\ 0 & m & 0 & 0 & 0 \\ 0 & 0 & m & 0 & 0 \\ 0 & 0 & 0 & J_\alpha & 0 \\ 0 & 0 & 0 & 0 & J_\beta \end{bmatrix}$$

$$K_I = \begin{bmatrix} 0 & 0 & 0 & 0 & \tfrac{\sqrt{2}}{2}k'_i & \tfrac{\sqrt{2}}{2}k'_i \\ 0 & 0 & 0 & 0 & \tfrac{\sqrt{2}}{2}k'_i & -\tfrac{\sqrt{2}}{2}k'_i \\ k_i & k_i & k_i & k_i & 0 & 0 \\ k_i L_2 & 0 & -k_i L_2 & 0 & 0 & 0 \\ 0 & -k_i L_2 & 0 & k_i L_2 & 0 & 0 \end{bmatrix}$$

$$K_D = \begin{bmatrix} 0 & 0 & 0 & 0 & \tfrac{\sqrt{2}}{2}k'_d & \tfrac{\sqrt{2}}{2}k'_d \\ 0 & 0 & 0 & 0 & \tfrac{\sqrt{2}}{2}k'_d & -\tfrac{\sqrt{2}}{2}k'_d \\ k_d & k_d & k_d & k_d & 0 & 0 \\ k_d L_2 & 0 & -k_d L_2 & 0 & 0 & 0 \\ 0 & -k_d L_2 & 0 & k_d L_2 & 0 & 0 \end{bmatrix}$$

$$C = \begin{bmatrix} c_x & 0 & 0 & 0 & 0 \\ 0 & c_y & 0 & 0 & 0 \\ 0 & 0 & c_z & 0 & 0 \\ 0 & 0 & 0 & c_\alpha & 0 \\ 0 & 0 & 0 & 0 & c_\beta \end{bmatrix}$$

The coordinate transformations of magnetic pole displacements and degrees of freedom:

$$\begin{cases} x = \tfrac{\sqrt{2}}{4}d_5 + \tfrac{\sqrt{2}}{4}d_6 \\ y = \tfrac{\sqrt{2}}{4}d_5 - \tfrac{\sqrt{2}}{4}d_6 \\ z = \tfrac{1}{4}(d_1 + d_2 + d_3 + d_4) \\ \alpha = \tfrac{1}{2L_2}(d_1 - d_3) \\ \beta = \tfrac{1}{2L_2}(d_4 - d_2) \end{cases} \quad (9)$$

Organized in matrix form:

$$\begin{bmatrix} x \\ y \\ z \\ \alpha \\ \beta \end{bmatrix} = \mathbf{N}_1 \begin{bmatrix} d_1 \\ d_2 \\ d_3 \\ d_4 \\ d_5 \\ d_6 \end{bmatrix} \tag{10}$$

$\mathbf{N}_1$ is the distribution matrix

$$\mathbf{N}_1 = \begin{bmatrix} 0 & 0 & 0 & 0 & \frac{\sqrt{2}}{4} & \frac{\sqrt{2}}{4} \\ 0 & 0 & 0 & 0 & \frac{\sqrt{2}}{4} & -\frac{\sqrt{2}}{4} \\ \frac{1}{4} & \frac{1}{4} & \frac{1}{4} & \frac{1}{4} & 0 & 0 \\ \frac{1}{2L_2} & 0 & -\frac{1}{2L_2} & 0 & 0 & 0 \\ 0 & -\frac{1}{2L_2} & 0 & \frac{1}{2L_2} & 0 & 0 \end{bmatrix}$$

Organize the matrix:

$$\begin{bmatrix} d_1 \\ d_2 \\ d_3 \\ d_4 \\ d_5 \\ d_6 \end{bmatrix} = \mathbf{N}_2 \begin{bmatrix} x \\ y \\ z \\ \alpha \\ \beta \end{bmatrix} \tag{11}$$

After organizing the matrix, a system dynamics model is obtained, with the model parameters shown in Table 2. The model can be expressed as follows.

$$M \begin{bmatrix} \ddot{x} \\ \ddot{y} \\ \ddot{z} \\ \ddot{\alpha} \\ \ddot{\beta} \end{bmatrix} = K_I \mathbf{N}_2 \begin{bmatrix} i_x \\ i_y \\ i_z \\ i_\alpha \\ i_\beta \end{bmatrix} + K_D \mathbf{N}_2 \begin{bmatrix} x \\ y \\ z \\ \alpha \\ \beta \end{bmatrix} - C \begin{bmatrix} \dot{x} \\ \dot{y} \\ \dot{z} \\ \dot{\alpha} \\ \dot{\beta} \end{bmatrix} + \begin{bmatrix} f_x \\ f_y \\ f_z \\ T_\alpha \\ T_\beta \end{bmatrix} \tag{12}$$

Table 2. Model parameter.

Parameter Name	Symbol	Value
Quality	M (kg)	0.3
Moment of inertia about the X-axis	J_α (kg·m^2)	9.7×10^{-4}
Moment of inertia about the Y-axis	J_β (kg·m^2)	9.7×10^{-4}
Sensor distance	L_2 (mm)	42.5
Axial current coefficient	K_i (N/A)	1.13
Axial displacement coefficient	K_d (N/m)	1.7×10^3
Radial current coefficient	K_i' (N/A)	1.13
Radial displacement coefficient	K_d' (N/m)	1.7×10^3
Damping coefficient in Z direction	c_z (N/(m/s))	0.3
Damping coefficient in α direction	c_β (N/(m/s))	9.7×10^{-4}

5. Levitation Experiments with a 5-DOF Magnetic Levitation Actuator

5.1. Centralized Control Strategy for Magnetic Levitation Drives

The system model of the five-degree-of-freedom magnetic levitation actuator comprises two parts: the model for the five-degree-of-freedom motion and the model for each group of electromagnets. The closed-loop control system adopts a series-level control structure. The outer loop controls the five-degrees-of-freedom of the platform and utilizes a PID control law, serving as the primary regulation loop of the control system. The inner loop controls the current of the electromagnet and adopts a PI control. A platform PID control system is established, as depicted in Figure 9. The reference inputs for the platform's five-degrees-of-freedom (x_{ref}, y_{ref}, z_{ref} α_{ref}, β_{ref}) are set, and the outer loop PID controller is adjusted based on the error. The inner loop employs a PI current loop to ensure that the output current quickly tracks the output voltage of the levitation controller within a certain frequency range, thereby enhancing the current response of the magnetic levitation drive and achieving control over the five-degrees-of-freedom motion.

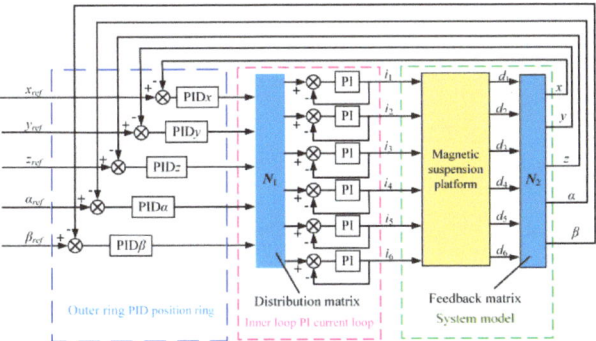

Figure 9. The block diagram of five-degree-of-freedom centralized control system.

5.2. Experimental System Composition

The experimental system for the five-degree-of-freedom magnetic levitation actuator is illustrated in Figure 10, comprising the prototype, hardware equipment, and control system. The control system is based on the DS1202 control board from dSPACE, with ControlDesk 7.6 software toolkits installed on the host computer. The hardware circuit utilizes drivers, while air gap detection employs eddy current displacement sensing technology from Zhuzhou Liulingba Technology and Science Co., Ltd. (Zhuzhou, China). The detection range is 0.65 mm to 2.65 mm and the analog output voltage range is 0 V to 10 V.

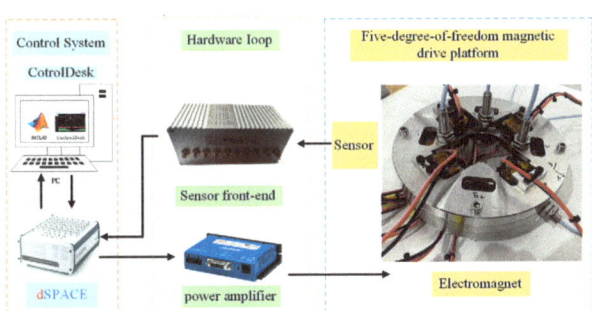

Figure 10. Experimental system.

5.3. Magnetic Levitation Drive Experiment

In this system, the sensor displacement needs to be determined. The sensor measurement is resolved for the z, α and β degrees of freedom. When the sensor measurement is 0 mm, it indicates that the levitated platform is attracted by the electromagnet; when the air gap is measured to be 1.12 mm, the levitated platform reaches its lowest point.

We set the levitated platform to float at 0.6 mm, midway within the air gap. To stabilize the levitation platform at this height, the magnitude of the current is adjusted so that the electromagnetic force on the levitation platform equals the force of gravity. The changes in the levitation displacement and current are demonstrated in Figure 11. This lays the foundation for subsequent translational and rotational experiments. Additionally, the levitation experiment can verify whether the magnetic levitation actuator can achieve a stable levitation state at the midpoint position. The results indicate that the system can achieve a stable levitation at 0.6 mm after levitation.

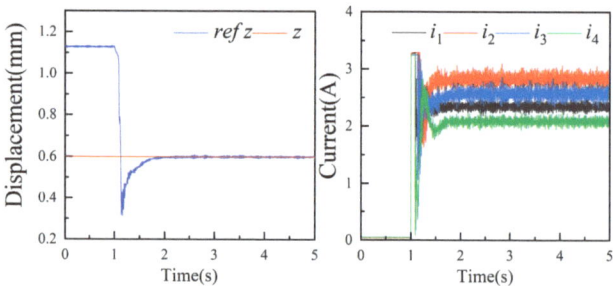

Figure 11. Floatation experiments and current changes.

When the levitation platform is stably suspended, step signals are applied to the z degree of freedom, α degree of freedom, and β degree of freedom in turn, and the experimental results are shown in Figure 12. In the initial state, the platform is in a stable suspension position, corresponding to an air gap length of 0.6 mm, and the platform deflection angles α and β are both zero radians. As shown in Figure 12a, a 0.05 mm step signal is input to the z degree of freedom at 0.5 s. The platform reaches a new levitation state after about 0.7 s, at which time the platform levitates with an air gap of 0.65 mm. Additionally, the platform deflection angles α and β remain zero radians during the adjustment process, and the z degree of freedom step input does not interfere with the platform rotational degrees of freedom. As shown in Figure 12b, a step signal of 0.001 rad is applied to the α degree of freedom at 0.5 s, and the system stabilizes after about 1.1 s. The system is then stabilized. The z degree of freedom and the B degree of freedom remain in the same state as before the step is applied; the suspension height is 0.6 mm. As shown in Figure 12c, a step signal of 0.001 rad is applied to the β degree of freedom at 0.5 s, and the system stabilizes after about 0.6 s. The system is then stabilized. At this point, the three degrees of freedom changes in the platform relative to the initial state after stabilizing the suspension are 0.6 mm, 0.001 rad, and 0.001 rad, respectively.

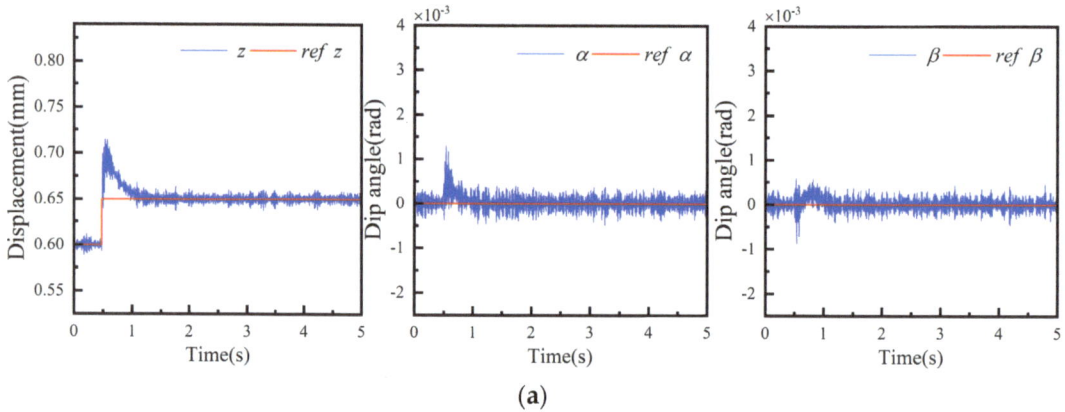

(a)

Figure 12. *Cont.*

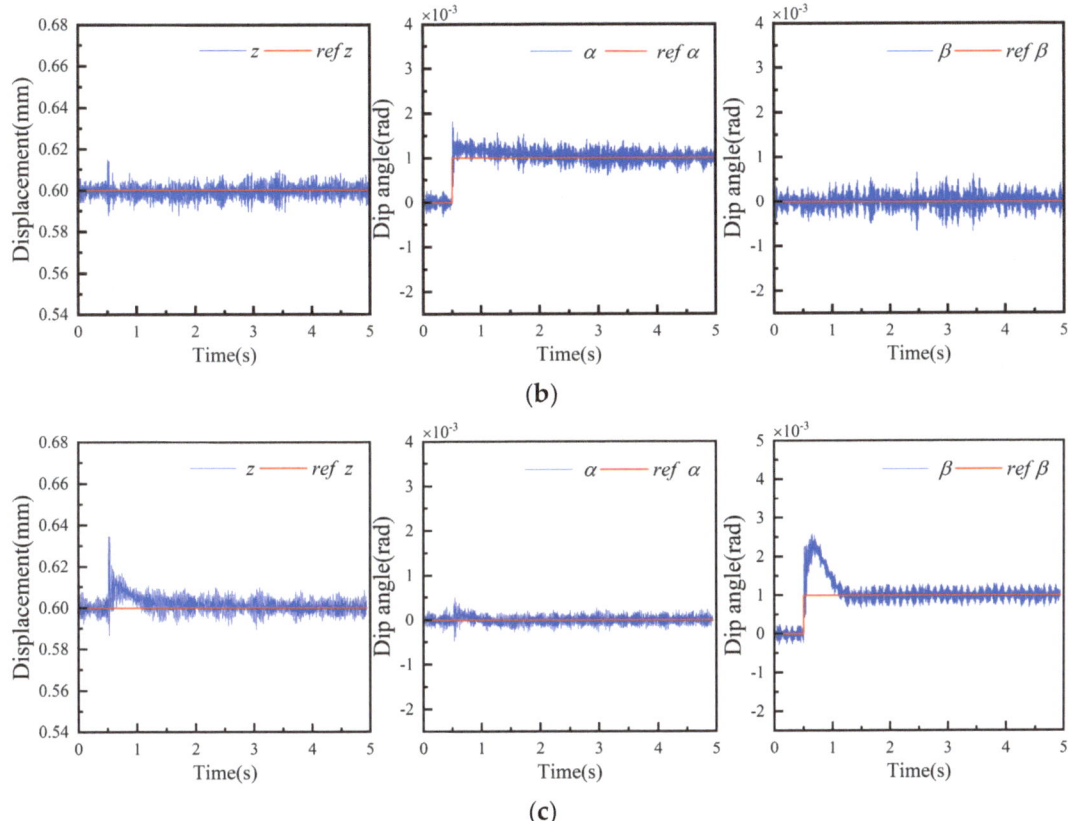

Figure 12. The z, α, β step response of PID control. (**a**) z direction; (**b**) α direction; (**c**) β direction.

Similarly, when the platform is stably levitated, the step and sinusoidal signals are applied to the x and y degrees of freedom in turn, and the experimental results are shown in Figure 13. In the initial state, the platform is in a stable levitation position, with the air gap length corresponding to z being 0.6 mm, and the platform deflection angles A and B are both zero radians. As shown in Figure 13a, a step signal of 0.1 mm is input to the x degree of freedom at 0.5 s, and the platform reaches a new levitation state after about 0.8 s, at which point the platform has an air gap of 0.6 mm. The x degree of freedom is also tracked. The tracking characteristics of the x degree of freedom are further analyzed by applying a sinusoidal signal with a frequency of 0.5 Hz. The trajectory is tracked with an amplitude ratio of 1.12 and a phase difference of 2.9°. Additionally, the other degrees of freedom of the suspended platform remain at zero during the adjustment process. As shown in Figure 13b, a step signal of 0.1 mm is applied to the y degree of freedom at 0.5 s, and the system stabilizes after about 0.8 s. A sinusoidal signal with a frequency of 0.5 Hz and an amplitude of 0.1 mm is applied, and the trajectory is tracked with an amplitude ratio of 1.17 and a phase difference of 0.2°.

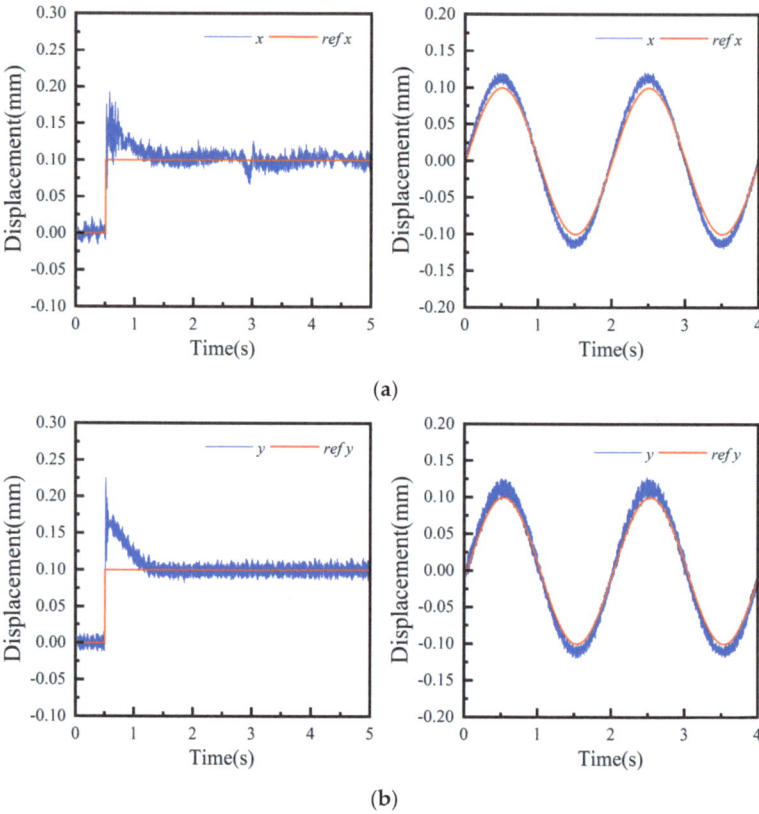

Figure 13. *y* step response of PID control. (**a**) *x* direction. (**b**) *y* direction.

During the above experiments, the levitation platform is able to maintain stable levitation after applying a step to each degree of freedom. It can be seen that the structure of the 5-DOF magnetic levitation actuator is reasonably designed.

6. Discussion and Recommendations

The experimental results of the five-degree-of-freedom magnetic levitation actuator are given in Figures 12 and 13 for the step response and trajectory tracking in different degrees of freedom. The main reason for the large overshoot of each step response is that the magnetic levitation system is a non-damped system, and the overshoot and response time for PID control are in some contradiction; a small overshoot will inevitably be sacrificed for a certain response time. Other reasons may be that the selection of PID parameters is not optimal. The comparison of the magnetic levitation drive designed in this paper with the published results shows that the experimental results are somewhat deficient. Both the overshoot and the response time are not at the expected level. At the same time, there are power loss and temperature problems. However, the five-degree-of-freedom magnetic levitation actuator with the six group differential control proposed in this paper has the advantages of a compact structure and a certain robustness in five-degree-of-freedom motion. We believe that through subsequent optimization, this design will show higher application potential.

7. Conclusions

In this paper, a 5-DOF compact electromagnetic levitation actuator for lens control was designed. The nonlinear characteristics of the magnetic driving force were analyzed by establishing an analytical model and conducting finite element calculations. Next, we established the dynamic model of the magnetic levitation actuator. A centralized control system based on the PID control was designed, and driving experiments were conducted to verify the motion in five-degrees-of-freedom. The main conclusions are as follows:

1. The five-degree-of-freedom magnetic levitation actuator exhibits a positive correlation between the electromagnetic force and the control current within the range of 0 to 1.5 A. The maximum output electromagnetic force reaches 6.1 N. Specifically, at a control current of 0.9 A, the electromagnetic force measures 3.64 N, ensuring the stability of the levitation platform.
2. When the suspended platform was in the equilibrium position, the different bias currents ranging from 0.5 A to 1.4 A were applied to observe the change in electromagnetic force. Similarly, we set the equilibrium position at 0 mm and selected four sets of control currents to observe the change in electromagnetic force as the displacement varied from 0.1 mm to 0.4 mm. It was found that the slopes of the electromagnetic force curves remained relatively consistent. However, when the displacement exceeded 0.4 mm, the slope increased significantly, indicating the onset of electromagnetic force nonlinearity. These results suggest that the electromagnetic force exhibits a strong linear relationship within the equilibrium position range.
3. In the experiments, step signals were applied to the z, α, β, x, and y degrees of freedom. The experimental results indicate that the axial range is 0.05 mm, the radial range is 0.1 mm, and the range for the α and β degrees of freedom is 0.001 rad. Furthermore, sinusoidal signals were applied to the radial actuator, and the tracking characteristics were also analyzed, achieving the desired results in both cases.

In the future, our first priority will be to optimize the controller to address issues related to overshooting and response time. We plan to explore different control methods for regulating the five-degree-of-freedom magnetic levitation drive. Alternatively, we intend to integrate the 5-DOF magnetic levitation drive into a laser cutting head to investigate its impact on processing efficiency and quality under various control methods.

Author Contributions: Conceptualization, methodology, validation, formal analysis, C.Z.; writing—original draft preparation, Q.Z.; data processing, W.P.; funding acquisition, project administration, J.J.; supervision, F.S. and H.Z.; visualization, R.Z. and D.L.; writing—review and editing, F.X. and X.Z.; resources, L.Y. All authors have read and agreed to the published version of the manuscript.

Funding: This research is supported by National Natural Science Fund of China, grant number 52375258, 52005345, and 52005344; National Key Research and Development Project grant number, 2020YFC2006701; Scientific Research Fund Project of Liaoning Provincial Department of Education grant number LJKMZ20220506, LJKMZ20220460, and JYTMS20231191; Major Project of the Ministry of Science and Technology of Liaoning Province, grant number 2022JH1/10400027; Key Project of the Ministry of Science and Technology of Liaoning Province, grant number 2022JH1/10800081).

Data Availability Statement: Data are contained within the article.

Acknowledgments: We thank S.R. (Shaojun Ren), Y.L. (Yuhang Liu) and X.W. (Xin Wang) for their suggestions and recommendations.

Conflicts of Interest: The authors declare no conflicts of interest.

References

1. Anghel, C.; Gupta, K.; Jen, T.C. Analysis and optimization of surface quality of stainless steel miniature gears manufactured by CO_2 laser cutting. *Optik* **2020**, *203*, 164049. [CrossRef]
2. Santosh, S.; Thomas, J.K.; Pavithran, M.; Nithyanandh, G.; Ashwath, J. An experimental analysis on the influence of CO_2 laser machining parameters on a copper-based shape memory alloy. *Opt. Laser Technol.* **2022**, *153*, 108210. [CrossRef]

3. Ghozali, R.G.; Pangaribawa, M.R. Analysis of the Effect of Cutting Motion Speed in CNC laser cutting on roughness and accuracy. *Eng. Proc.* **2024**, *63*, 10. [CrossRef]
4. Quintero, F.; Pou, J.; Fernandez, J.L. Optimization of an off-axis nozzle for assist gas injection in laser fusion cutting. *Opt Lasers Eng.* **2006**, *44*, 1158–1167. [CrossRef]
5. Chen, F.Y.; Wang, W.J.; Wang, S.J. Rotating Lorentz force agnetic bearings dynamics modeling and adaptive controller design. *Sensors* **2023**, *20*, 8543. [CrossRef] [PubMed]
6. Zhou, L.; Wu, J.J. Magnetic levitation technology for precision motion systems: A review and future perspectives. International Journal of Automation. *Int. J. Autom. Technol.* **2022**, *16*, 386–402. [CrossRef]
7. Dyck, M.; Lu, X.; Altintas, Y. Magnetically levitated rotary table with six degrees of freedom. *IEEE/ASME Trans.* **2016**, *22*, 530–540. [CrossRef]
8. Heyman, I.L.; Wu, J.; Zhou, L. LevCube: A six-degree-of-freedom magnetically levitated nanopositioning stage with centimeter-range XYZ motion. *Precis. Eng.* **2023**, *83*, 102–111. [CrossRef]
9. Gloess, R.; Goos, A. Magnetic levitation stages for planar and linear scan application with nanometer resolution. *Int. J. Appl. Electro.* **2020**, *63*, 105–117.
10. Zhang, X.; Trakarnchaiyo, C.; Zhang, H. MagTable: A tabletop system for 6-DOF large range and completely contactless operation using magnetic levitation. *Mechatronics* **2021**, *77*, 102600. [CrossRef]
11. Huang, S.D.; Peng, K.Y.; Cao, G.Z.; Wu, C.; Xu, J.Q.; He, J.B. Robust precision position tracking of planar motors using min-max model predictive control. *IEEE Trans. Ind. Electron.* **2022**, *69*, 13265–13276. [CrossRef]
12. Huang, S.D.; Cao, G.Z.; Xu, J.Q.; Cui, Y.K.; Wu, C.; He, J.B. Predictive position control of long-stroke planar motors for high-precision positioning applications. *IEEE Trans. Ind. Electron.* **2021**, *68*, 796–811. [CrossRef]
13. Berkelman, P.; Dzadovsky, M. Magnetic levitation over large translation and rotation ranges in all directions. *IEEE/ASME Trans. Mech.* **2013**, *18*, 44–52. [CrossRef]
14. Li, Q.; Hu, Y.F.; Wu, H.C. Structure design and optimization of the radial magnetic bearing. *Actuators* **2023**, *12*, 27. [CrossRef]
15. Song, S.W.; Kim, W.H.; Lee, J.; Jung, D.H. A study on weight reduction and high performance in separated magnetic bearings. *Energies* **2023**, *7*, 3136. [CrossRef]
16. Wu, C.; Li, S.S. Modeling, Design and suspension force analysis of a novel AC six-pole heteropolar hybrid magnetic bearing. *Appl. Sci.* **2023**, *13*, 1643. [CrossRef]
17. Masahiro, O.; MSUZAWA, T.; Saito, T.; Tatsumi, E. Magnetic levitation performance of miniaturized magnetically levitated motor with 5-DOF active control. *Mech. Eng. J.* **2017**, *5*, 17-00007.
18. Luan, B.; Xu, F.C.; Yang, G.; Jin, J.J.; Xu, C.C.; Sun, F. Experimental study of EDM characteristics using a 5-DOF controllable magnetic levitation actuator. *Int. J. Adv. Manuf. Technol.* **2023**, *10*, 3423–3437. [CrossRef]
19. Zhang, X.Y.; Tanaka, S. High speed electrical discharge machining using 3-DOF controlled magnetic drive actuator. *Univ. J. Mech. Eng.* **2015**, *6*, 215–221. [CrossRef]
20. He, D.J.; Tadahiko, S.; Takahiro, N. Development of a maglev lens driving actuator for off-axis control and adjustment of the focal point in laser beam machining. *Precis. Eng.* **2013**, *2*, 255–264. [CrossRef]
21. He, D.J.; Tadahiko, S.; Takahiro, N. Development of a Lens Driving Maglev Actuator for Laser Beam Off-Axis Cutting and Deep Piercing. *Precis. Eng.* **2012**, *523*, 774–779. [CrossRef]

Disclaimer/Publisher's Note: The statements, opinions and data contained in all publications are solely those of the individual author(s) and contributor(s) and not of MDPI and/or the editor(s). MDPI and/or the editor(s) disclaim responsibility for any injury to people or property resulting from any ideas, methods, instructions or products referred to in the content.

Review

Planar Printed Structures Based on Matryoshka Geometries: A Review

Alfredo Gomes Neto [1], Jefferson Costa e Silva [1], Joabson Nogueira de Carvalho [1] and Custódio Peixeiro [2,*]

[1] Group of Telecommunications and Applied Electromagnetism (GTEMA), Instituto Federal da Paraíba, João Pessoa 58015-435, Brazil; alfredogomes@ifpb.edu.br (A.G.N.); jefferson@ifpb.edu.br (J.C.e.S.); joabson@ifpb.edu.br (J.N.d.C.)

[2] Instituto de Telecomunicações, Instituto Superior Técnico, University of Lisbon, 1049-001 Lisbon, Portugal

* Correspondence: custodio.peixeiro@lx.it.pt

Abstract: A review on planar printed structures that are based on Matryoshka-like geometries is presented. These structures use the well-known principle of Matryoshka dolls that are successively nested inside each other. The well-known advantages of the planar printed technology and of the meandered nested Matryoshka geometries are combined to generate miniaturized, multi-resonance, and/or wideband configurations. Both metal and complementary slot structures are considered. Closed and open configurations were analyzed. The working principles were explored in order to obtain physical insight into their behavior. Low-cost and single-layer applications as frequency-selective surfaces, filters, antennas, and sensors, in the microwave frequency region, were reviewed. Potential future research perspectives and new applications are then discussed.

Keywords: printed circuits; microstrip; ring resonators; Matryoshka sets; frequency selective surfaces; filters; antennas; sensors

Citation: Neto, A.G.; Silva, J.C.e.; Carvalho, J.N.d.; Peixeiro, C. Planar Printed Structures Based on Matryoshka Geometries: A Review. *Micromachines* **2024**, *15*, 469. https://doi.org/10.3390/mi15040469

Academic Editor: Haejun Chung

Received: 16 February 2024
Revised: 21 March 2024
Accepted: 27 March 2024
Published: 29 March 2024

Copyright: © 2024 by the authors. Licensee MDPI, Basel, Switzerland. This article is an open access article distributed under the terms and conditions of the Creative Commons Attribution (CC BY) license (https:// creativecommons.org/licenses/by/ 4.0/).

1. Introduction

The name Matryoshka comes from the well-known Russian dolls, shown in Figure 1, that are successively nested inside each other. It has been used to refer to nested sets in many areas of electrical and electronics engineering, such as electronics packaging [1], implantable medical devices [2], biomedical imaging [3], computer network security [4], silicon compounds [5], software protection [6], Internet-of-Things [7], image retrieval [8] and reconstruction [9], cancer gene analysis [10], cellular biophysics [11], cloud computing [12], 5G network slicing [13], acoustic wave resonators [14], pattern recognition [15], and astronomy [16]. It has also been used associated with planar printed configurations. This combination of printed circuits with Matryoshka-like geometries benefits from the well-known advantages of printed planar technology (low profile, lightweight, compactness, low cost, easy fabrication and integration of electronic components, wide range of characteristic impedances) and the multiband (or wideband) behavior and miniaturization associated with Matryoshka configurations. The Matryoshka-like scheme was used for the first time in planar printed circuit structures at the Group of Telecommunications and Applied Electromagnetism (GTEMA) from the Instituto Federal da Paraíba in João Pessoa, Brazil. A multi-resonant frequency selective surface (FSS) was proposed in 2014 [17,18]. Since then, the work in these planar printed structures based on Matryoshka geometries has progressed steadily at GTEMA, as reported in many MSc theses [17,19–31] and associated publications [32–46]. Different applications have been envisaged, such as FSSs [17–19,21,23,24,27,32,33,36,38–40,42], filters [20,22,26,30,37,43,46], antennas [25,28], and sensors [29,31,34,35,41,44,45]. All these applications are motivated mainly by the huge importance of new telecommunication systems, particularly mobile communication networks, with emphasis on the recently deployed 5G systems and the closed associated Internet-of-Things (IoT). New ideas and perspectives are being explored to further develop these new types of structures. However, having been worked on for about ten years, the topic is already sufficiently mature to justify the publication of a review paper. Therefore, the goal of this paper was to precisely present a review on the work conducted in the field of planar printed structures based on Matryoshka geometries. The paper is organized into seven sections. After this introductory section, Section 2 deals with the printed planar Matryoshka geometry. The Matryoshka cell is described, and the corresponding working principles are analyzed. Section 3, Section 4, Section 5, and Section 6 are dedicated to the description of the use of this type of cell in FSSs, filters, antennas, and sensors in already-reported applications, respectively. At the end, Section 7 contains the main conclusions and the perspectives of present and future developments.

Figure 1. Example of a Russian Matryoshka with nine nested dolls.

2. The Matryoshka Geometry

The Matryoshka geometry is based on concentric rings. As shown in Figure 2, Matryoshka geometries have been conceived as an evolution of the split ring resonators (rings). Starting with a set of rings (so far homothetic), a gap is introduced in each one and then the consecutive rings are connected near the gaps. However, differently from the SRR [47], the rings are connected. As in an SRR, the rings may take different shapes, from simple ones (as square or circular) to other, more complex, canonical or non-canonical geometries. Matryoshka geometries have been implemented in printed circuit board (PCB) technology, both with and without a ground plane. As the SRR, they can be formed by metal strips or by slots in the metal (complimentary configurations). As complimentary configurations, Matryoshka geometries have been used in defected ground structures (DGSs) [30,48] and FSSs [24].

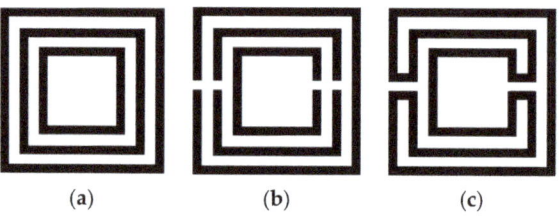

(a) (b) (c)

Figure 2. Example of the evolution from concentric rings to a Matryoshka geometry. (**a**) Concentric rings without gaps. (**b**) SRR. (**c**) Matryoshka geometry.

For a specific type of Matryoshka geometry, there are two sub-types: the open and the closed. It is open when there is a gap in the smaller inner ring (Figure 3). As is detailed in the next sections, this gap has a remarkable effect on the structure's characteristics, namely, on its resonance frequencies. Due to the metal continuity, in the closed configuration, for the first resonance

$$L_{ef} \approx \lambda_{refclose}, \qquad (1)$$

whereas, for the open one,

$$L_{ef} \approx \lambda_{refopen}/2, \qquad (2)$$

where L_{ef} is the effective length of the structure and λ_{ref} is the effective wavelength [19] for the first resonance. Naturally, there are other (higher-order) resonances. This difference in the behavior of the closed and the open structure can be explained by the continuity required by the closed structure and the interference standing wave pattern imposed by the reflection at the gap of the open structure's inner ring. This means that, for structures with the same dimensions, the open structure has a first resonance frequency that is approximately half the one of the closed structure. In other words, for the same first resonance frequency, the open structure has an equivalent electrical length that is approximately half of the one of the closed structure, meaning a much more effective miniaturization capability.

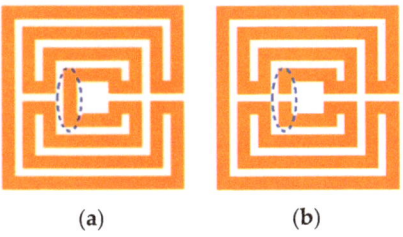

Figure 3. Example of Matryoshka geometries. (**a**) Closed. (**b**) Open.

The Matryoshka configurations are highly meandered, and the total area occupied is defined by the external ring. The physical parameters of an open square Matryoshka configuration, with four rings, are indicated in Figure 4. For the closed Matryoshka geometry, there is no gap at the inner ring (s = 0). When a Matryoshka geometry is used in an FSS, it is also necessary to specify the unit-cell size.

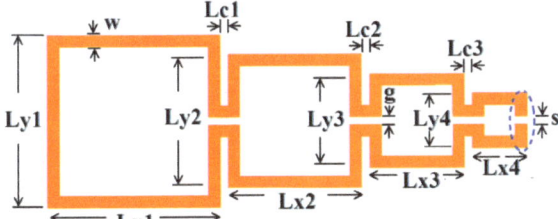

Figure 4. Open Matryoshka geometry, expanded, with the definition of its physical parameters.

The average perimeter of the closed geometry (P_N), corresponding to the physical length defined at the middle of each segment, can be obtained using Equation (3). For the open geometry, s must be subtracted from P_N.

$$P_N = 2\left[\sum_{n=1}^{N}(L_{xn} + L_{yn}) + \sum_{n=1}^{N-1} L_{cn} - 2Nw - (N-1)g\right] \quad (3)$$

In printed planar structures, it is also necessary to specify the substrate characteristics (ε_r—relative electric permittivity, h—thickness, and tanδ—loss tangent) and the presence or absence of the ground plane. In both cases, the structures are transversally non-homogeneous, and an equivalent homogenous medium can be conceived. For the commonly used substrates, with normal magnetic behavior, an effective relative electric permittivity (ε_{ref}) and an effective wavelength (λ_{ef}) can be defined.

$$\lambda_{ef} = \frac{\lambda_0}{\sqrt{\varepsilon_{ref}}} \quad (4)$$

where λ_0 is the wavelength in a vacuum. The procedure used to calculate ε_{ref} depends on the type of configuration used, which is associated with the envisaged application. For filters, a microstrip structure has been used, whereas for FSSs, a simple substrate without a ground plane has been selected. For antenna applications, so far, only microstrip structures with DGSs have been employed.

There are some features of the Matryoshka geometries that depend on the specific type of structure and application. These specific features will be detailed in the next sections, where applications as FSSs, filters, antennas, and sensors are analyzed. However, there are some features that are intrinsic of the Matryoshka geometries and therefore are common to all type of applications. These common features re analyzed here using microstrip filters as application examples.

There are different formulas to obtain the ε_{ref} of a microstrip line. A simple non-dispersive model, valid for low frequency, is given in Equations (5)–(7) [49].

$$\varepsilon_{ref} = \frac{\varepsilon_r + 1}{2} + \frac{\varepsilon_r - 1}{2}\left(1 + 10\frac{h}{w}\right)^{-ab} \quad (5)$$

$$a = 1 + \frac{1}{49}\ln\left[\frac{\left(\frac{w}{h}\right)^4 + \left(\frac{w}{52h}\right)^2}{\left(\frac{w}{h}\right)^4 + 0.432}\right] + \frac{1}{18.7}\ln\left[1 + \left(\frac{w}{18.1h}\right)^3\right] \quad (6)$$

$$b = 0.564\left(\frac{\varepsilon_r - 0.9}{\varepsilon_r + 3}\right)^{0.053} \quad (7)$$

For a 2.0 mm wide microstrip line printed on a FR4 substrate with ε_r = 4.4 and h = 1.5 mm, Equation (5) leads to ε_{ref} = 3.23.

The outline of a microstrip filter, with an open Matryoshka geometry of two rings, is shown in Figure 5. The input and output microstrip lines are 2.8 mm wide (50 Ohm characteristic impedance). W = 2.0 mm and g = s = 1.0 mm are used.

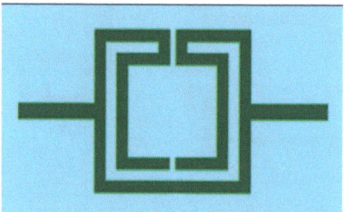

Figure 5. Example of microstrip filter based on an open Matryoshka geometry with two rings.

The four configurations, indicated in Table 1, were numerically simulated in Ansoft HFSS [50]. Square rings were used ($L_n = L_{xn} = L_{yn}$). All the four configurations had the same average perimeter (P_N = 178.00 mm).

Table 1. Physical characterization of Matryoshka geometries with two rings.

Configuration	L_1 (mm)	L_2 (mm)	L_{c1} (mm)
Config1	27.25	21.25	1.00
Config2	28.00	20.00	2.00
Config3	31.00	15.00	6.00
Config4	34.00	10.00	10.00

The simulated transmission coefficient of the four configurations is shown is Figure 6 for the open configuration and Figure 7 for the closed one.

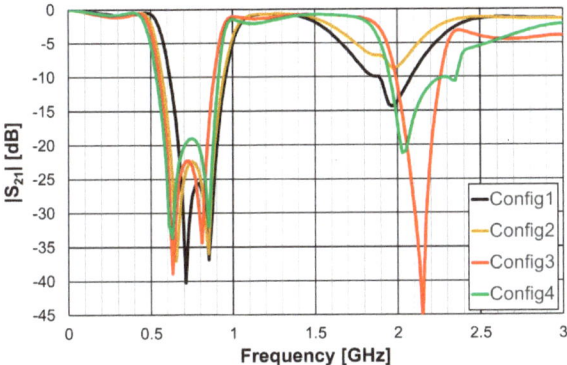

Figure 6. Simulated transmission coefficient of the open Matryoshka filters with constant length.

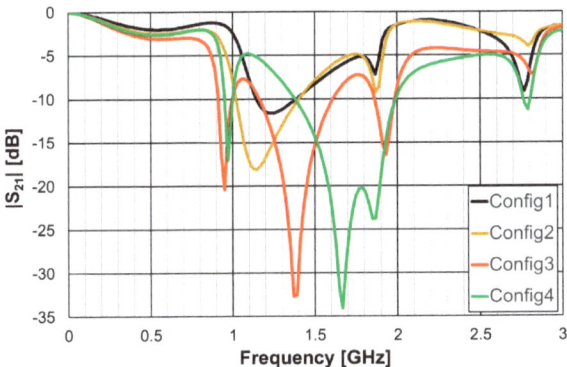

Figure 7. Simulated transmission coefficient of the closed Matryoshka filters with constant length.

As can be concluded from Figure 6, the open configurations present adequate characteristics for a stopband filter, that is, high attenuation in the stopband, low attenuation in the passband, steep slope transition from passband to stopband (and vice-versa), and large bandwidth. However, that is not the case for the closed configurations (Figure 7). The open configuration provides a higher order filter because it offers two different resonance paths and the closed configurations just one. Moreover, as predicted, the open configurations have much lower first resonance frequencies. As is verified in the next sections, this conclusion was also obtained for the Matryoshka configurations used for other envisage applications (FSSs, antennas, sensors). Table 2 summarizes the characteristics of the open Matryoshka filter configurations.

Table 2. Main characteristics of open Matryoshka filter configurations.

Configuration	fr_1 (GHz)		fr_2 (GHz)		f_0 (GHz)	BW * (%)
	Equation (2)	Simulation	Equation (2)	Simulation		
Config1	0.681	0.71	0.800	0.85	0.785	43.4
Config2	0.684	0.65	0.802	0.85	0.756	49.6
Config3	0.695	0.63	0.810	0.81	0.717	46.4
Config4	0.707	0.63	0.818	0.85	0.720	52.0

* Defined for a -10 dB reference level.

For a single ring configuration, the effective length can be simply calculated as the average perimeter. However, for multiring Matryoshka configurations, there is coupling between the rings, and there is not a simple physical interpretation of the effective length. To pre-design two rings' configurations, curve fitting was used to obtain L_{ef} associated with the first two resonances [20].

$$L_{ef1} = 2[3(L_1 + L_2 - 4w) + L_c] \quad (8)$$

$$L_{ef2} = 2(3L_1 + 2L_2 - 10w) \quad (9)$$

Equation (2) tends to provide a better estimation of the first resonance frequency for intensive coupling (small Lc). Although the relative error can reach about 12% (for the first resonance of configuration 4), Equation (2) is still very useful at the pre-design stage of these filters. These configurations provide miniaturized filters with very large bandwidth. The four configurations have different sizes for the rings and separation between them, but, because they have the same perimeter, the stopband characteristics of the open configurations are very similar.

Microstrip filters based on open Matryoshka geometries with two, three, and four rings were also simulated. The corresponding dimensions are indicated in Table 3. The previously indicated FR4 substrate, w = 2.0 mm and g = s = 1.0 mm, were used, again.

Table 3. Physical characterization of open Matryoshka geometries with 2, 3, and 4 rings.

Configuration	N	L_1 (mm)	L_2 (mm)	L_3 (mm)	L_4 (mm)	L_{c1} (mm)	L_{c2} (mm)	L_{c3} (mm)	P_N (mm)
Config5	2			NA	NA		NA	NA	210.00
Config6	3	32.00	24.00		NA	2.00		NA	268.00
Config7	4			16.00	8.00		2.00	2.00	294.00

The obtained |S21| results are shown in Figure 8.

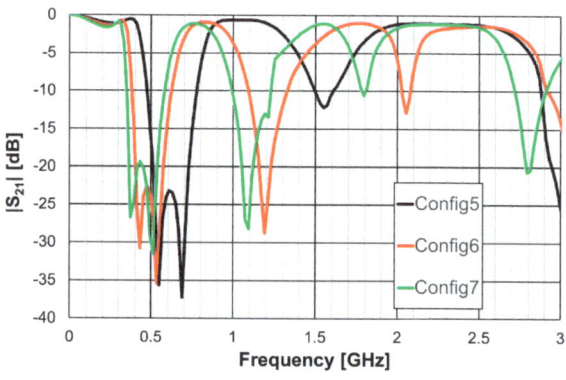

Figure 8. Simulated transmission coefficient of open Matryoshka filters with 2, 3, and 4 rings.

Table 4 contains the main simulation results associated with the first two resonances shown in Figure 8.

Table 4. Main characteristics of open Matryoshka filter configurations with 2, 3 and 4 rings.

Configuration	N	P_N (mm)	f_{r1} (GHz)	f_{r2} (GHz)	f_0 (GHz)	BW * (%)
Config5	2	210.0	0.55	0.69	0.627	48.7
Config6	3	268.0	0.43	0.53	0.501	48.4
Config7	4	294.0	0.37	0.51	0.462	50.3

* Defined for a −10 dB reference level.

The use of more rings leads to the appearance of more resonances and, if the average perimeter increases, to a decrease in the frequency associated with the first two resonances.

The surface current density, on configuration 6, at the resonance frequencies and for frequencies between them, is shown in Figure 9.

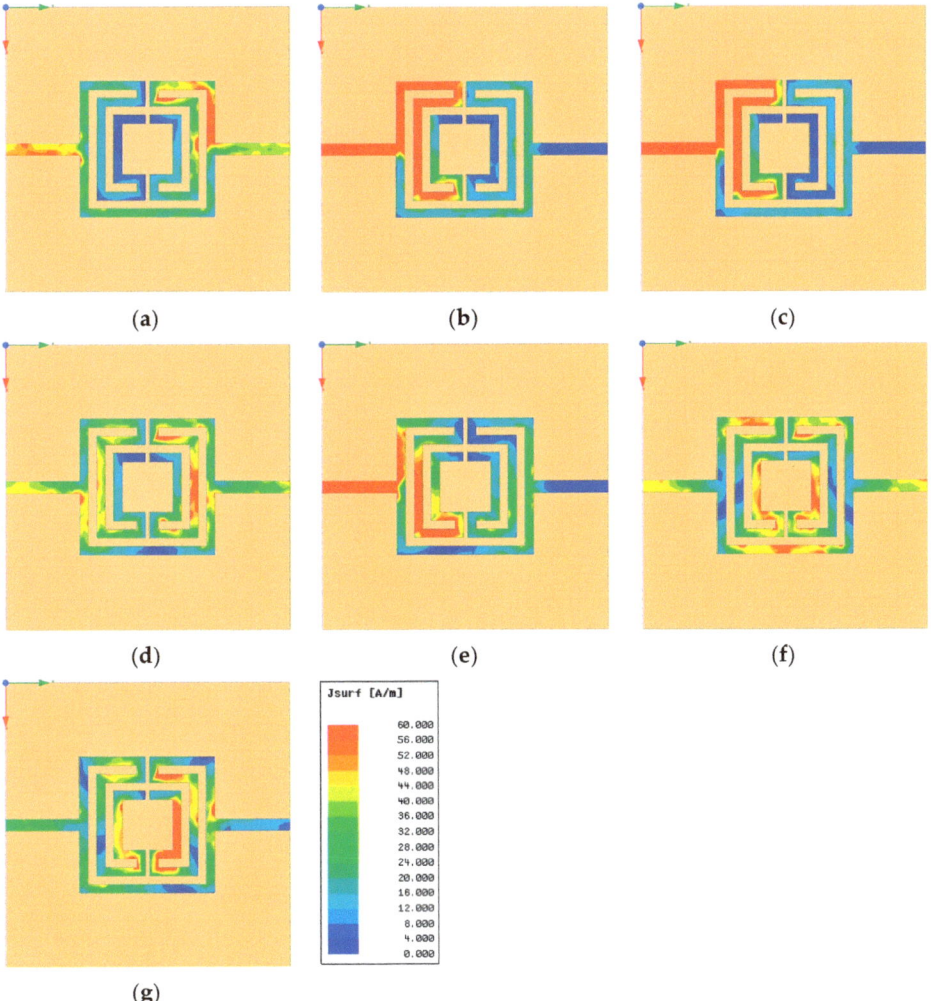

Figure 9. Simulated surface current density of the open Matryoshka filter with three rings (configuration 6). (**a**) f = 0.33 GHz, (**b**) f = f_{r1} = 0.43 GHz, (**c**) f = f_{r2} = 0.53 GHz, (**d**) f = 0.83 GHz, (**e**) f = f_{r3} = 1.19 GHz, (**f**) f = 1.79 GHz, (**g**) f = f_{r4} = 2.05 GHz.

There was a common pattern of the surface current distribution at the resonance frequencies. Being a stopband filter, there was no transmission at the resonance frequencies. In fact, for such frequencies (Figure 9b,c,e,g), the current at the output port was negligible, and the current at the input port was very strong due to a positive interference of the incident wave and the waves reflected at the two parallel paths, mostly if there was a good input impedance matching. For the frequencies between resonance frequencies (Figure 9a,d,f), there was almost perfect transmission. It was also noticeable that the current magnitude on the inner rings increased as frequency went up.

3. Examples of Application as FSS

The first application suggested for the Matryoshka geometry was as FSS [17,18]. This is quite logical since, at the time, there was already a strong and continued research activity in the topic of FSS at GTEMA, and FSS was already widely used in microwave, millimeter wave, and Terahertz frequency bands. There are many specific applications of FSSs, such as RFID, absorbers, rasorbers, reconfigurable intelligent surfaces (RIS), RF energy harvesting, polarizers, dichroic sub-reflector and reflector antennas, reflectarray antennas, beam-switching antennas, lens antennas, and radio astronomy [51–53]. A very important initial choice in the design of an FSS is the geometry of the unit-cell. Despite the variety of available geometries, with the rapid growing of wireless technologies, telecommunication system requirements impose a continuing challenge to meet characteristics such as miniaturization, multiband operation, and polarization independence.

3.1. Closed Matryoshka FSSs

In [17,18], the closed square Matryoshka geometry with two rings, shown in Figure 10, was introduced. A 0.97 mm thick FR4 substrate with $\varepsilon_r = 4.4$ and $\tan\delta = 0.02$ was used. The dimensions (in mm) of two configurations are indicated in Table 5, and Wx = Wy = 24.0 mm.

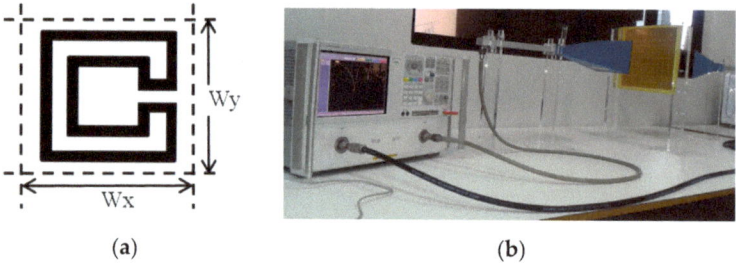

Figure 10. Closed Matryoshka FSS. (**a**) Unit-cell geometry. (**b**) Photo of the prototype with 10 × 10 unit-cells and measurement setup.

Table 5. Dimensions of the closed square Matryoshka geometry with two rings.

Configuration	L_1	L_2	L_{c1}	w	g	P_2 (mm)
Config1	22.0	12.0	3.5	1.5	1.0	129.0
Config2	22.0	7.0	6.0	1.5	1.0	114.0

Using Equations (1), (4), and (5) and the procedure proposed in [17,18] to estimate L_{ef} and ε_{ref}, the initial dimensions of the Matryoshka unit-cell, fulfilling the specifications, can be obtained. L_{ef} depends on the polarization considered. Usually the two orthogonal linear polarization (vertical-V and horizontal-H) are employed. The use of a numerical simulator can then provide the necessary optimization.

The simulation and experimental transmission coefficient results, obtained for configuration 1, are shown in Figure 11. The simulation results correspond to an infinite

FSS (Floquet boundary conditions [51,52]). The horn antennas, available at the time, only allowed measurements above 4.5 GHz.

A general good agreement was obtained between the simulation and experimental results. The ripple in the experimental results was caused by the reflections on the objects present in the non-anechoic environment of the laboratory. The transmission coefficient results depended on the polarization (horizontal-H or vertical-V) of the incident electric field. However, the first resonance frequency was the same for both polarizations (1.75 GHz), but the −10 dB bandwidth was larger for the vertical polarization (19.8% and 10.0%).

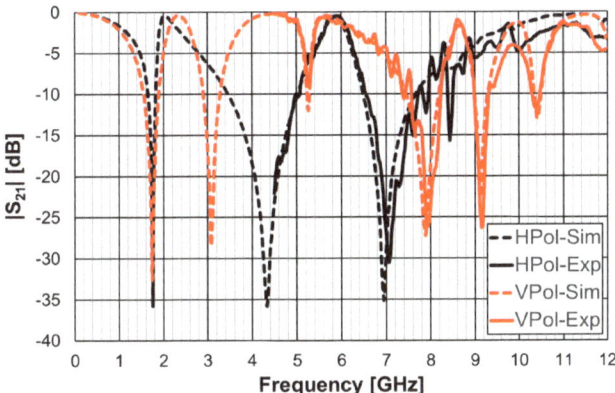

Figure 11. Simulated and experimental |S21| results for configuration 1.

The simulation results obtained for the two configurations are compared in Figure 12.

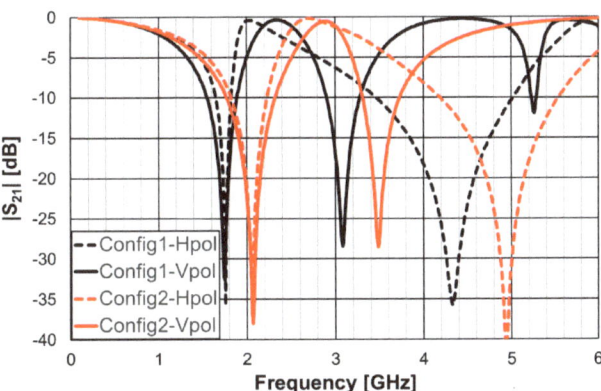

Figure 12. Comparison of the simulated |S21| results for configurations 1 and 2.

Although both configurations had the same external ring dimensions, configuration 1 had a larger L_{ef}, and consequently its resonance frequencies were lower. For instance, f_{res11} = 1.75 GHz and f_{res12} = 2.06 GHz. Configuration 2 had a much larger bandwidth for both polarizations. For the vertical polarization, BW_1 = 19.8%, whereas BW_2 = 26.7%. The dimensions of the internal ring can be used to fine tune the FSS and to control the bandwidth.

3.2. Open Matryoshka FSSs

In [19,32], the open Matryoshka geometry was introduced, reducing the first resonance frequency to approximately half, when compared to the closed Matryoshka geometry FSS, previously analyzed. It is interesting to compare, now for application in FSSs, the simple rings, with closed and open Matryoshka geometries, as shown in Figure 13. Again, a 0.97 mm thick FR4 substrate with $\varepsilon_r = 4.4$ and $\tan\delta = 0.02$ was used. A unit-cell with size Wx = Wy = 24.0 mm, L_1 = 22.0 mm, L_2 = 12 mm, L_c = 3.5 mm, w = 1.5 mm, and g = s = 1.0 mm was chosen.

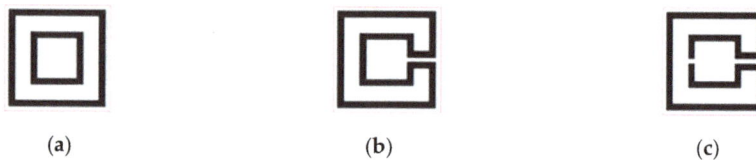

Figure 13. FSS square unit-cells with two square rings. (**a**) Simple rings. (**b**) Closed Matryoshka geometry. (**c**) Open Matryoshka geometry.

The simulated |S21| results are shown in Figure 14 for horizontal polarization, and in Figure 15 for vertical polarization.

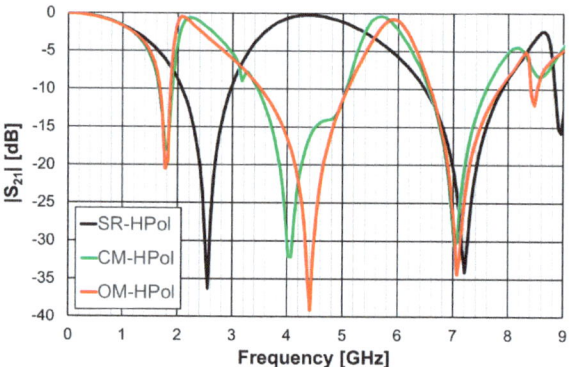

Figure 14. Simulated |S21| results of the simple rings (SR), closed Matryoshka (CM) and open Matryoshka (OM), for horizontal polarization (HPol).

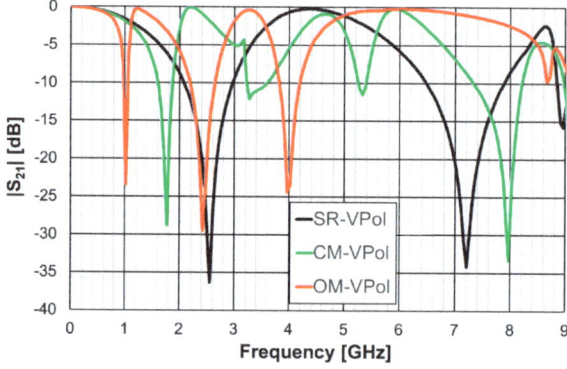

Figure 15. Simulated |S21| results of the simple rings (SR), closed Matryoshka (CM) and open Matryoshka (OM), for vertical polarization (VPol).

The simple rings configuration was physically symmetric and therefore its response was the same for the horizontal and vertical polarizations. This was also the case for the closed Matryoshka configuration, but only for the first resonance. However, the open Matryoshka configuration had completely different responses for the two polarizations. Although the three configurations occupied the same area, the two unit-cell with Matryoshka geometry provided much lower first resonance frequencies, especially for vertical polarization. The results associated with the first resonance are summarized in Table 6.

Table 6. Summary of the first resonance results for the square ring (SR) unit-cell FSSs with simple rings and closed (CM) and open (OM) Matryoshka geometries.

Unit-Cell Geometry	First Resonance Frequency (GHz)		Bandwidth * (%)	
	HPol	VPol	HPol	VPol
SR	2.56	2.56	35.4	35.4
CM	1.78	1.78	13.6	15.3
OM	1.78	1.01	10.9	8.1

* Defined for a -10 dB reference level.

The FSS with open Matryoshka geometry provided a remarkable reduction of the first resonance frequency, especially for vertical polarization, when compared with the closed configuration (43%) and with the simple rings (61%). However, there was a substantial reduction of the bandwidth.

The simulation results for the open Matryoshka configuration were validated with experimental results, as shown in Figure 16.

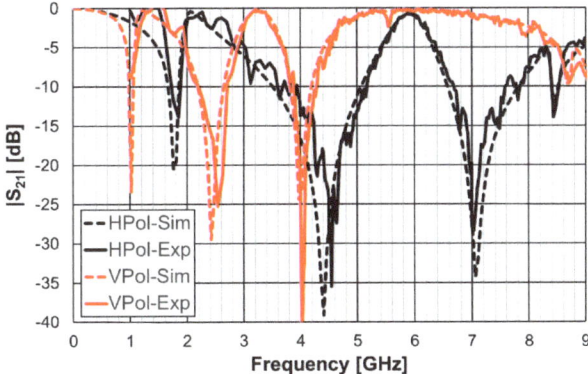

Figure 16. Comparison of simulated and experimental |S21| results of the open Matryoshka for horizontal and vertical polarizations.

More rings can be used to increase the effective length and therefore further reduce the first resonance frequency, increase the number of resonances, and provide a fine-tuned control of the resonances and of the bandwidth [19,32]. To confirm these conclusions, an FSS with an open Matryoshka unit-cell with three rings was designed, fabricated, and tested. Again, a 0.97 mm thick FR4 substrate with $\varepsilon_r = 4.4$ and $\tan\delta = 0.02$ was used. A unit-cell with size $W_x = W_y = 24.0$ mm, $L_1 = 22.0$ mm, $L_{c1} = L_{c2} = 2.25$ mm, $L_2 = 14.5$ mm, $L_3 = 7.0$ mm, $w = 1.5$ mm, and $g = s = 1.0$ mm was chosen. Photos of the prototype and of the experimental setup are shown in Figure 17.

The simulation and experimental results of the three rings of open Matryoshka geometry for vertical and horizontal polarizations are shown in Figure 18. The experimental results were only able to be measured starting at 1 GHz. There was a general good agreement between the simulation and experimental results. As mentioned before, and as can be verified in Figure 17b, the ripple in the experimental results was caused by the reflections on the objects present in the non-anechoic environment of the laboratory.

The simulation results of the FSS with open Matryoshka unit-cells with two and three rings are summarized in Table 7.

It was confirmed that, for the same dimension of the external ring, the increase in the number of rings (two to three) provided a reduction of the first resonance frequency, an increase in the number of resonances, and fine tune control of the resonances. However, a reduction of the bandwidth was verified.

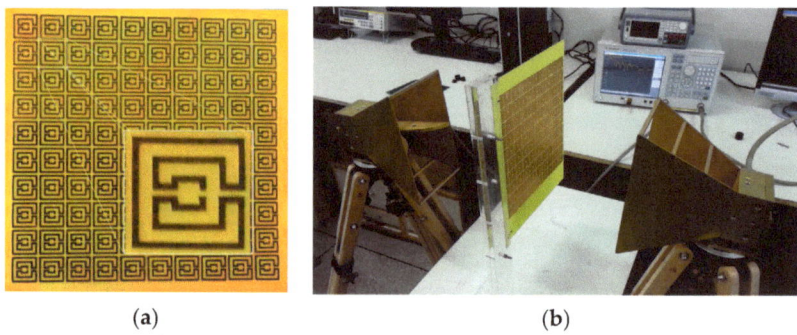

Figure 17. Photo of the FSS test procedure. (**a**) Prototype. (**b**) Experimental setup.

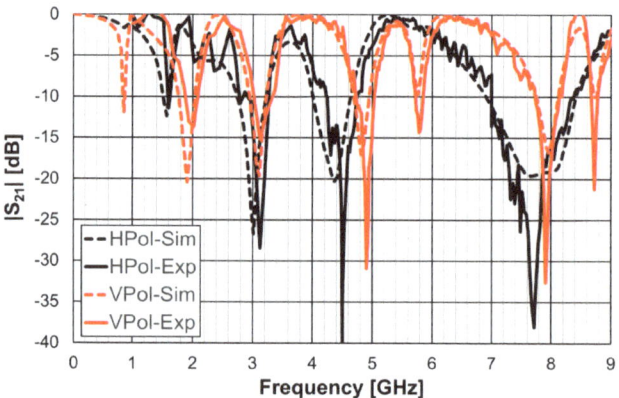

Figure 18. Comparison of simulated and experimental |S21| results of the open Matryoshka with 3 rings for horizontal and vertical polarizations.

Table 7. Summary of the simulation results of the FSS with open Matryoshka unit-cells with 2 and 3 rings.

N	Area (mm^2)	f_{r1} (GHz)		Bandwidth * (%)		f_{r2} (GHz)		f_{r3} (GHz)	
		HPol	VPol	HPol	VPol	HPol	VPol	HPol	VPol
2	22 × 22	1.78	1.01	10.9	8.1	4.36	2.41	7.66	3.96
3		1.56	0.86	5.9	2.9	3.01	1.91	4.36	3.11

* Defined for a −10 dB reference level.

3.3. Polarization of Independent FSSs

To overcome the inconvenient polarization dependence, analyzed in the previous section, a new configuration of the Matryoshka geometry was proposed in [21,36]. As shown in Figure 19, this new type of configuration has been conceived as an evolution from the simple circular ring keeping the main Matryoshka characteristics, that is, the area occupied is defined by the external ring only, and more rings can be added internally, maintaining electrical continuity. The physical characterization of the circular Matryoshka geometry is defined in Figure 20.

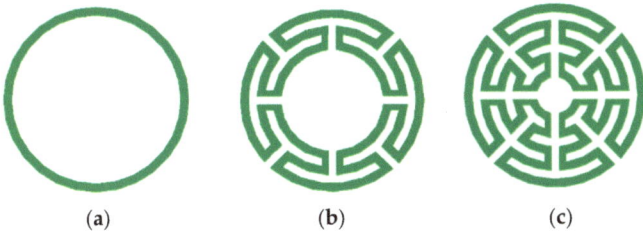

Figure 19. Evolution from a simple circular ring to a circular multiring Matryoshka geometry. (a) Simple circular ring. (b) Circular Matryoshka geometry with three rings. (c) Circular Matryoshka geometry with five rings.

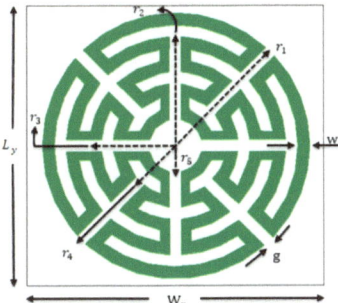

Figure 20. Definition of the physical parameters of an FSS unit-cell with circular multiring Matryoshka geometry.

The FSS unit-cells with the three geometries shown in Figure 19 were designed, fabricated, and tested [21,36]. A 0.762 mm thick FR4 substrate with ε_r = 4.4 and $\tan\delta$ = 0.02 was used. An FSS with 10 × 10 unit-cells with size Wx = Wy = 20.0 mm and w = g = 0.8 mm was chosen. The radius of the unit-cells' rings are indicated in Table 8. The radius reduction rate was maintained from ring to ring.

Table 8. Radius of the circular Matryoshka unit-cells.

Configuration	Number of Rings	r_1 (mm)	r_2 (mm)	r_3 (mm)	r_4 (mm)	r_5 (mm)
FSS1	1			NA		
FSS2	3	9.0	7.4	5.8	NA	
FSS3	5				4.2	2.6

As the FSSs were horizontally and vertically symmetric, they were polarization independent. Therefore, for these three cases, only results obtained for vertical polarization are shown. An important characteristic of an FSS is its sensitivity to the angle of the incident electromagnetic wave. Four angles of incidence were considered, from normal incidence (θ = 0) to θ = 45°, with a 15° interval.

In [21,36], the formulas indicated in Equations (10) to (12) are proposed to estimate the first resonance frequency of the FSS with one, three, and five rings, respectively. These formulas were used to specify the radius of the three configurations at the initial stage of the design procedure.

$$f_{rFSS1} = \frac{3 \times 10^8}{2\pi r_1 \sqrt{\varepsilon_{ref}}} \quad (10)$$

$$f_{rFSS2} = \frac{3 \times 10^8}{2\pi (r_1 + r_3) \sqrt{\varepsilon_{ref}}} \quad (11)$$

$$f_{rFSS3} = \frac{3 \times 10^8}{2\pi (r_1 + r_3 + r_5) \sqrt{\varepsilon_{ref}}} \quad (12)$$

where ε_{ref} is the effective relative permittivity of the equivalent homogeneous structure [21,36].

The prototypes, shown in Figure 21, were fabricated, and they were measured using the setup shown in Figure 22.

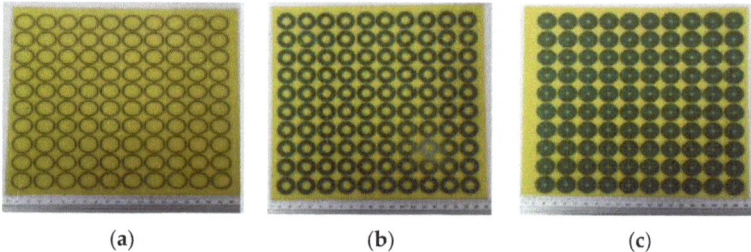

Figure 21. Prototypes of the FSSs with circular Matryoshka unit-cells. (**a**) FSS1; (**b**) FSS2; (**c**) FSS3.

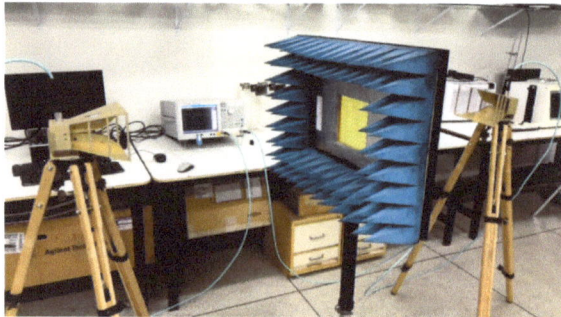

Figure 22. Setup for the measurement of the FSS prototypes with circular Matryoshka unit-cells.

The simulation and experimental results obtained for the transmission coefficient of the three prototypes are shown in Figures 23–25. The simulation results for the four different incident angles were very similar and, therefore, for the sake of clarity, only one curve is represented.

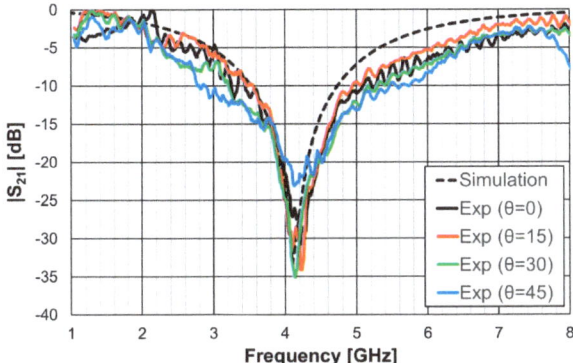

Figure 23. Comparison of the simulated and experimental |S21| results of the FSS1 prototype.

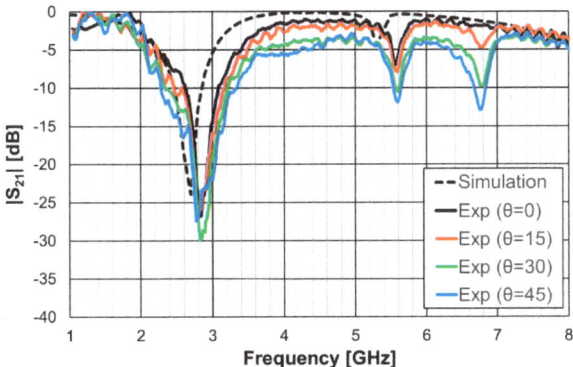

Figure 24. Comparison of the simulated and experimental |S21| results of the FSS2 prototype.

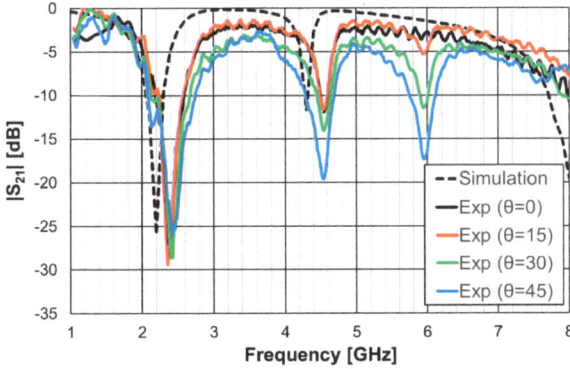

Figure 25. Comparison of the simulated and experimental |S21| results of the FSS3 prototype.

There was a good agreement between the numerical simulations and experimental results for the three prototypes. In general, the measured results were below the simulation ones. This difference was about 5 dB, on average, and tended to increase as the oblique angle of incidence increased. This effect may have been caused by the finite size of the window used in the measurement setup (Figure 22).

As the perimeter of the FSS unit-cell increased with the number of rings, more resonances appeared. The results, for the first resonance frequency, are compared in Table 9.

Table 9. Comparison of first resonance frequencies of the FSSs prototypes.

Configuration	Number of Rings	First Resonance Frequency (GHz)					
		Estimation	Simulation	Experimental			
				$\Theta = 0$	$\Theta = 15°$	$\Theta = 30°$	$\Theta = 45°$
FSS1	1	4.45 [1]	4.10	4.224	4.211	4.133	4.120
FSS2	3	2.71 [2]	2.70	2.846	2.833	2.833	2.768
FSS3	5	2.30 [3]	2.20	2.378	2.352	2.404	2.417

[1] Equation (10); [2] Equation (11); [3] Equation (12).

There was a good agreement between the numerical simulation and experimental results. Moreover, also the estimation provided by Equations (10)–(12) was accurate enough for the initial stage of the design (relative error below 5%). It is clear that the three prototypes had low sensitivity from the inclination angle and that there was a remarkable reduction in the first resonance frequency as the number of rings increased (44% from FSS1 to FSS3).

Recently a polarization-insensitive miniaturized multiband FSS with Matryoshka geometry elements was proposed [42]. From an initial polarization-sensitive unit-cell with a single element, there was an evolution to a combination of four orthogonals of such unit-cells (Figure 26).

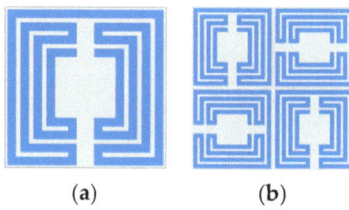

(a) (b)

Figure 26. Matryoshka unit-cells proposed in [42]. (a) Single element. (b) Combination of four orthogonal elements.

The simulation results for the transmission coefficient of the two Matryoshka unit-cells, for normal incidence ($\theta = 0$), are shown in Figures 27 and 28.

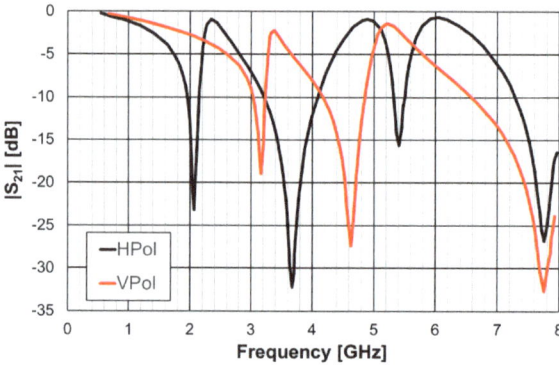

Figure 27. |S21| simulation results of the FSS with the single-element Matryoshka unit-cell.

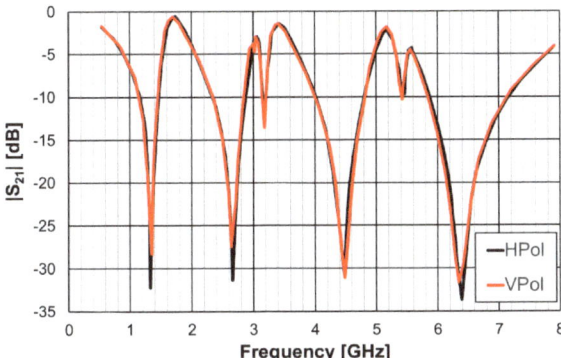

Figure 28. |S21| simulation results of the FSS with the four orthogonal Matryoshka elements unit-cell.

The single-element unit-cell had a strong polarization dependence, but the four orthogonal elements unit-cell was almost perfectly polarization independent. Moreover, from the results presented in Figures 29 and 30, it can be concluded that the new four elements arrangement also provided low sensitivity to the angle of incidence. For horizontal polarization, the curves for θ = 0, θ = 20°, and θ = 40° were almost coincident; only the curve for θ = 60° was slightly different. For vertical polarization, the situation was almost the same, but both the curves for θ = 40° and for θ = 60° were slightly different from the other two.

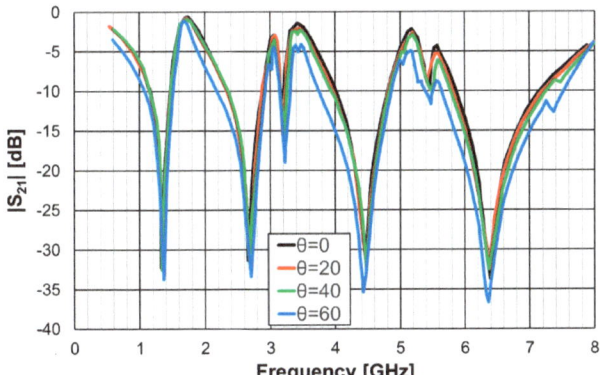

Figure 29. |S21| simulation results of the FSS with the four orthogonal Matryoshka elements unit-cell for horizontal polarization.

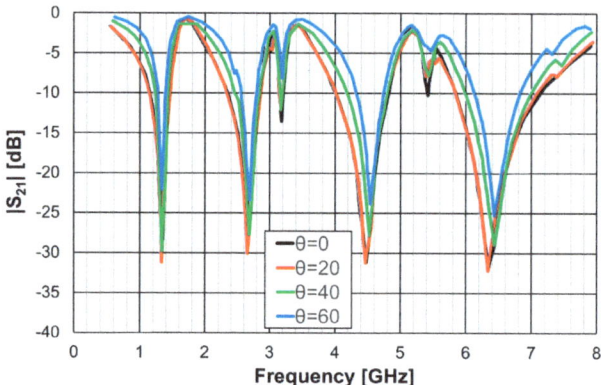

Figure 30. |S21| simulation results of the FSS with the four orthogonal Matryoshka elements unit-cell for vertical polarization.

3.4. Combination of an FSS with Dipoles

One of the advantages of the Matryoshka geometry is that it can be combined with other geometries in order to obtain an FSS with low coupling between the fields of each geometry, allowing for the control of the respective frequency responses. This is particularly interesting for the design of multiband FSS. In [23,27], cross-dipoles and Matryoshka geometries were combined to achieve a polarization-independent, triple-band FSS. The combined geometries are shown in Figure 31. The prototype, shown in Figure 32, was fabricated and characterized.

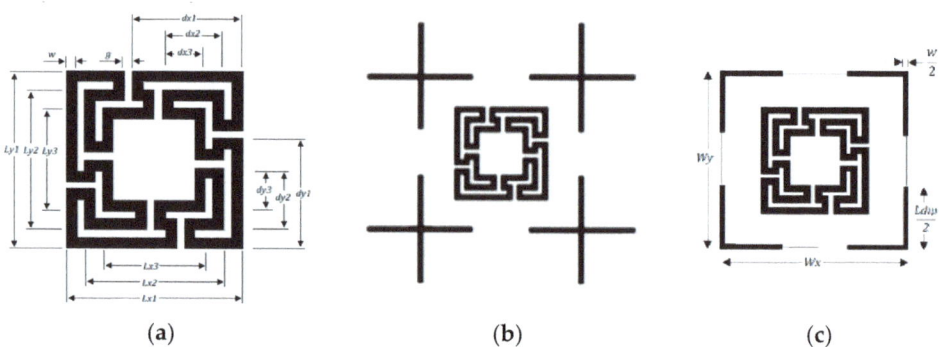

Figure 31. Combination of a Matryoshka geometry with cross-dipoles to form an FSS. (**a**) Matryoshka geometry. (**b**) Combination of Matryoshka with cross-dipoles. (**c**) FSS unit-cell.

 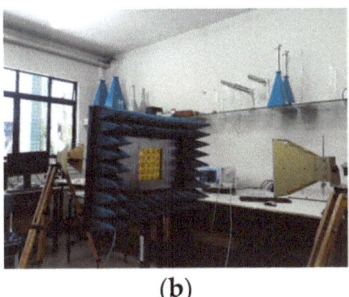

Figure 32. Photos of the FSS with the combination of a Matryoshka geometry with cross-dipoles. (**a**) Prototype. (**b**) Experimental setup.

In [23], a 1.6 mm thick FR4 substrate with ε_r = 4.4 and tanδ = 0.02 was used. The simulation results shown in Figure 33 correspond to an FSS with 5 × 5 unit-cells with size Wx = Wy = 40.0 mm, w = 1.5 mm, and g = 1.0 mm. Moreover, $L_{x1} = L_{y1}$ = 24.0 mm, $L_{x2} = L_{y2}$ = 19.0 mm, $L_{x3} = L_{y3}$ = 14.0 mm, $d_{x1} = d_{y1}$ = 15.0 mm, $d_{x2} = d_{y2}$ = 8.5 mm, $d_{x3} = d_{y3}$ = 6.0 mm, and L_{dip} = 39.0 mm. Three resonances can be observed (f_{r1} = 1.81 GHz, f_{r2} = 2.43 GHz, and f_{r3} = 3.19 GHz). They corresponded to the superposition of the first and second resonances of the Matryoshka geometry (f_{r1} = 1.82 GHz and f_{r2} = 3.18 GHz) with the first resonance of the cross-dipoles (f_{r1} = 2.46 GHz). The three resonances can be controlled separately, which is a very interesting feature that adds flexibility in the design for different potential applications. For instance, it is possible to design the dipole so that its first resonance frequency is close to one of the resonance frequencies of the Matryoshka geometry (first or second). By doing this, an increase in the bandwidth of the combined resonances can be obtained. Numerical simulation results are compared with experimental results, obtained for different angles of incidence, in Figure 34.

The simulation results for the four different incident angles were very similar and, therefore, for the sake of clarity, only one curve is represented. There was a general good agreement between the simulation and experimental results, which validates the design procedure. Moreover, the angular stability was confirmed. The discrepancies observed may have been caused by the finite size of the window used in the measurement setup (Figure 32b).

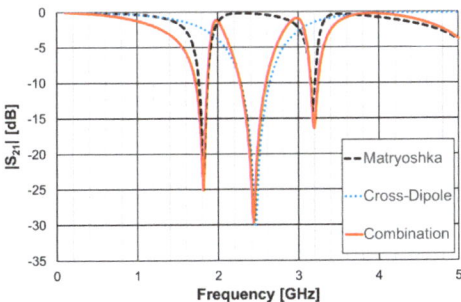

Figure 33. |S21| response of an FSS for the Matryoshka geometry, the cross-dipoles, and the combination of the two.

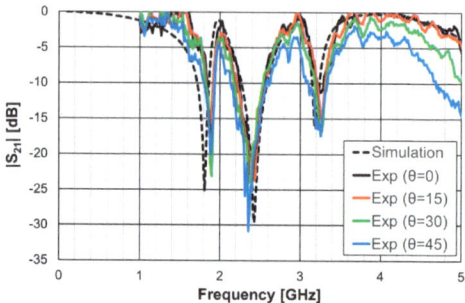

Figure 34. Comparison of the |S21| simulation and experimental results for the FSS combination of the Matryoshka geometry with the cross-dipoles.

3.5. Complimentary FSSs

The Matryoshka geometry, described in the previous section, is also used in its complementary form [24,39]. The FSS unit-cell is obtained as described in Figure 35.

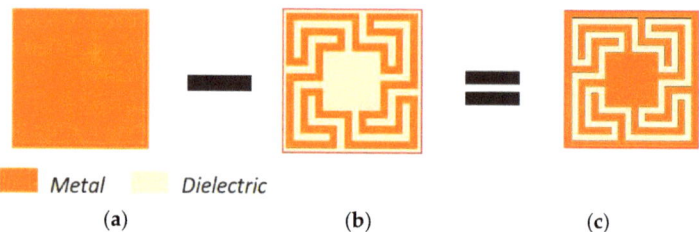

Figure 35. Complementary form of FSS Matryoshka geometry unit-cell. (a) Metal patch. (b) Metal Matryoshka geometry. (c) Complementary Matryoshka geometry.

It was shown, in the previous section, that the Matryoshka geometry with metal strips had a stopband response associated with its resonances. The complimentary Matryoshka geometry had a passband behavior. Two prototypes of these complimentary Matryoshka configurations with 9 × 9 unit-cells, each cell with 22.4 × 22.4 mm², were designed, fabricated, and tested [24,39]. A 1.6 mm thick FR4 substrate with ε_r = 4.4 and tanδ = 0.02 and w = g = 1.0 mm was used. The simulation results shown in Figure 36 correspond to FSS1 ($L_{x1} = L_{y1}$ = 20.4 mm, $L_{x2} = L_{y2}$ = 16.4 mm, $L_{x3} = L_{y3}$ = 12.4 mm, $d_{x1} = d_{y1}$ = 11.4 mm, $d_{x2} = d_{y2}$ = 7.4 mm, $d_{x3} = d_{y3}$ = 5.5 mm) and FSS2 ($L_{x1} = L_{y1}$ = 15.4 mm, $L_{x2} = L_{y2}$ = 11.4 mm, $L_{x3} = L_{y3}$ = 7.4 mm, $d_{x1} = d_{y1}$ = 9.0 mm, $d_{x2} = d_{y2}$ = 5.0 mm, $d_{x3} = d_{y3}$ = 3.0 mm). These dimensions were calculated using the design formulas proposed in [24] to provide passbands centered at 1.5 GHz and 3.5 GHz for FSS1 and 2.5 GHz and 5.1 GHz for FSS2, as well as a stopband centered at 2.45 GHz for FSS1 and 3.5 GHz for FSS2. Photos of these complimentary FSS prototype are shown in Figure 37.

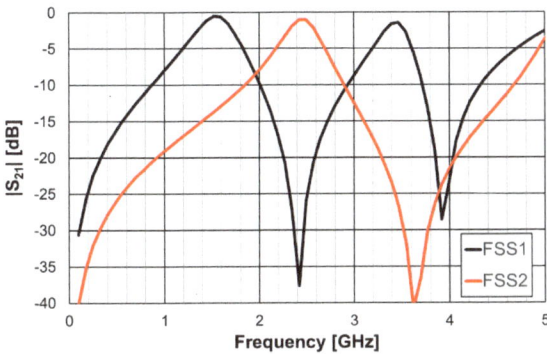

Figure 36. |S21| simulation results of the complimentary Matryoshka configurations FSS1 and FSS2.

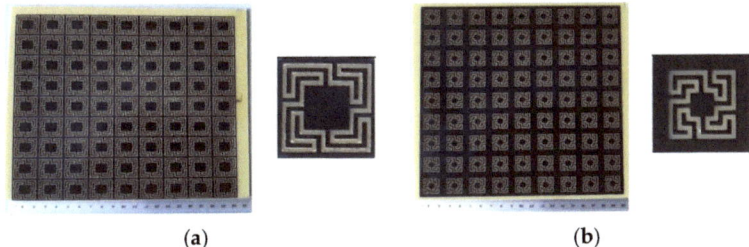

(a)   (b)

Figure 37. Photos of the prototypes of the complimentary Matryoshka configurations FSS1 and FSS2. (a) FSS1. (b) FSS2.

The results shown in Figure 36 demonstrate that it is possible to adjust the range of the stopbands and passbands according to the specifications.

As shown in the previous section, this Matryoshka configuration has very good angular stability. From the results presented in Figures 38 and 39, it can also be concluded that the complimentary Matryoshka configuration presents the same good angular stability.

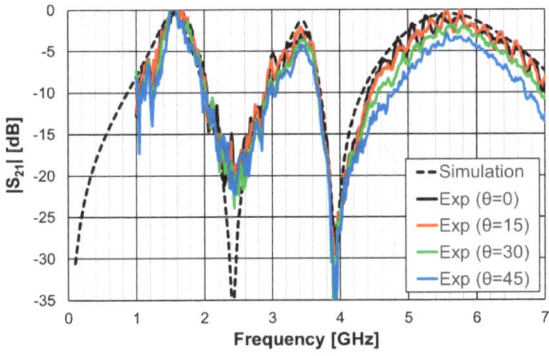

Figure 38. Comparison of the |S21| simulation and experimental results for the FSS1.

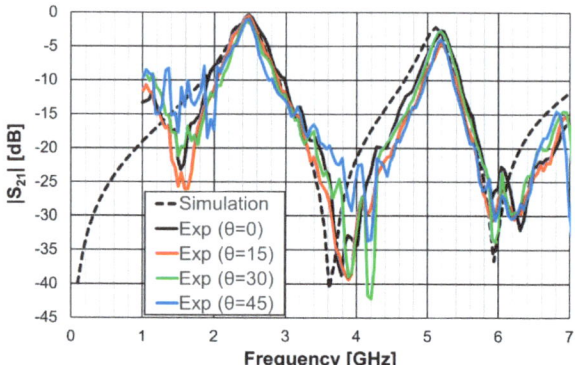

Figure 39. Comparison of the |S21| simulation and experimental results for the FSS2.

The simulation results for the four different incident angles were very similar and, therefore, for the sake of clarity, only one curve is represented. Taking into account that the experimental results were obtained in a simple non-anechoic room, there was a good agreement between the numerical simulation and experimental results. The resonance and antiresonance frequencies were confirmed experimentally with relative error differences below 4.5% [24].

3.6. Reconfigurable FSSs

For some applications, reconfigurability is a very attractive feature for an FSS. In this case, combined geometries can be useful. Adding a PIN diode between the vertical dipoles of the FSS presented in [23], and analyzed in Section 3.4, a reconfigurable FSS (RFSS) was obtained, as described in [27]. Additionally, as shown in Figure 40, a RF inductor was added between the horizontal dipoles to act as an RF choke [49]. The RFSS prototype, shown in Figure 40c, was fabricated and characterized [27].

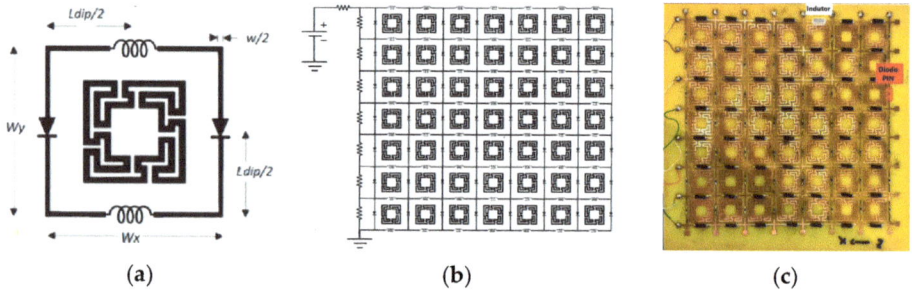

Figure 40. RFSS with Matryoshka geometry. (**a**) Unit-cell. (**b**) 7 × 7 configuration. (**c**) Prototype.

A 1.6 mm thick FR4 substrate with ε_r = 4.4 and $\tan\delta$ = 0.02 was used. The prototype shown in Figure 40c corresponds to an FSS with 7 × 7 unit-cells with size $W_x = W_y$ = 30.0 mm, w = 1.5 mm, and g = 1.0 mm. Moreover, $L_{x1} = L_{y1}$ = 24.0 mm, $L_{x2} = L_{y2}$ = 19.0 mm, $L_{x3} = L_{y3}$ = 14.0 mm, $d_{x1} = d_{y1}$ = 15.0 mm, $d_{x2} = d_{y2}$ = 8.5 mm, $d_{x3} = d_{y3}$ = 6.0 mm, and L_{dip} = 29.0 mm. The numerical simulation and measured results are presented in Figures 41–45. Figure 41 corresponds to unit-cells without PIN diodes and without inductors. These results serve as a reference. Figure 42 corresponds to unit-cells with PIN diodes but without inductors. Figures 43–45 correspond to unit-cells with both PIN diodes and inductors.

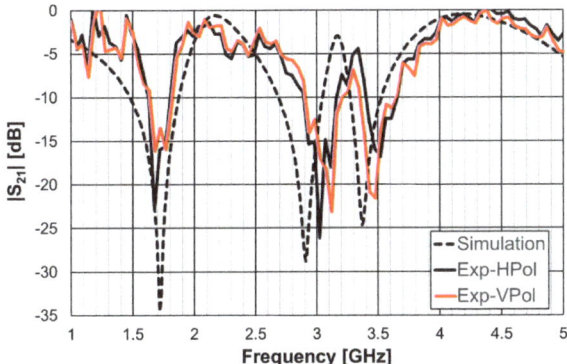

Figure 41. |S21| response of the FSS without PIN diodes and without inductors.

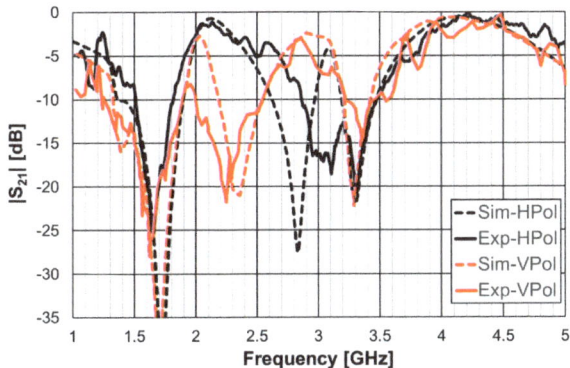

Figure 42. |S21| response of the FSS with PIN diodes but without inductors.

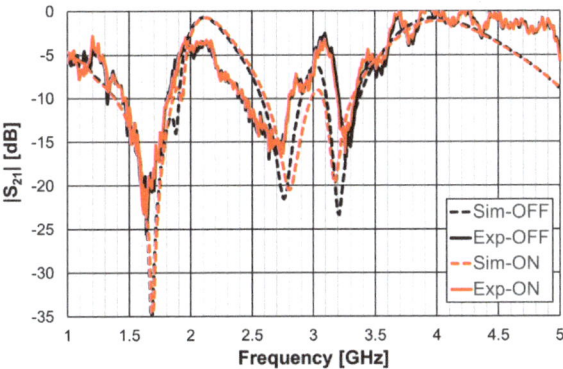

Figure 43. |S21| response of the FSS with PIN diodes and inductors for horizontal polarization.

In Figure 41, the simulation results for the horizontal and vertical polarization are identical; for the sake of clarity, only one curve is represented. There was a reasonable agreement between simulation and experimental results. The experimental second and third resonances were substantially deviated from the simulations. The difference was able to reach 6.6% and was mainly caused by the non-anechoic environment of the laboratory and eventual fabrication inaccuracies.

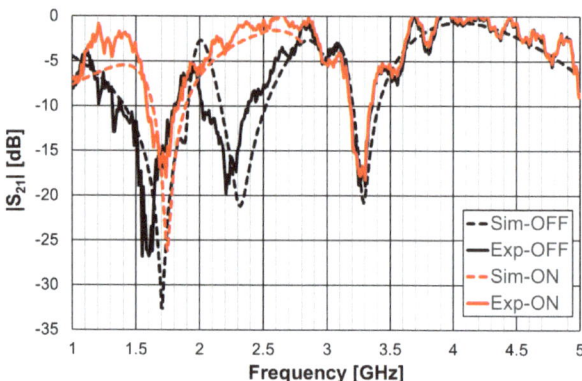

Figure 44. |S21| response of the FSS with PIN diodes and inductors for vertical polarization.

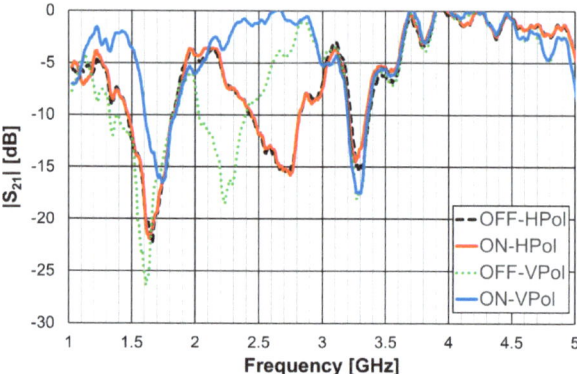

Figure 45. |S21| response of the FSS with PIN diodes and inductors for ON and OFF PIN states.

As shown in Figure 42, the agreement between numerical simulation and experimental results was good, except for the second resonance and horizontal polarization (7% relative error). In addition to the already mentioned general error causes (non-anechoic environment and fabrication inaccuracies), there must be some other problem not detected. However, as this configuration is just an intermediate step, and new prototypes would be fabricated for the next steps, the work was continued.

The results shown in Figure 43 indicate that, as expected, the state of the diode did not affect the horizontal polarization. Moreover, the problem associated with the second resonance frequency, detected in Figure 42, disappeared. The relative error for the second resonance frequency was then only about 4%. It was, therefore, verified that there is the need to use the inductors. As shown in Figure 44, a good agreement between the simulation and experimental results was verified for both OFF and ON states.

As can be concluded from the results presented in Figure 45, for vertical polarization, the reconfiguration of the FSS was effective, with a reconfigurable bandwidth of 0.37 GHz (17%), from 2.03 GHz to 2.40 GHz, with at least 10 dB of difference between OFF and ON bias states of the diodes.

4. Examples of Application as Filter

Providing access to telecommunication networks in the most diverse locations, with quality of service and without losing mobility, poses major challenges for manufacturers of mobile equipment and infrastructures. In both cases, filters play a fundamental role, separating the desired signals from the unwanted ones. Telecommunication systems, namely, 5G wireless communications, require filters with operating conditions that are increasingly challenging in terms of frequency response, in addition to low cost, weight, and volume (miniaturization). In this sense, new microwave filter configurations have been developed [54,55]. To meet these requirements, planar filters are widely used. Planar filters can be viewed as resonators, lumped or quasi-lumped, for which the resonance frequency is determined by the geometry [49,56]. There are other important applications of filters, such as Wi-Fi, global satellite navigation systems, test, and measurement equipment (spectrum and network analyzers) [53]. Aiming to take advantage of the characteristics observed for the Matryoshka geometry when used in FSSs (miniaturization and multiband operation), filters based on Matryoshka geometries were introduced in [20,26].

4.1. Filters with a Square Matryoshka Geometry

Printed planar microstrip filters based on open Matryoshka square geometries were presented in [20,37]. Filters with two and three rings are shown (Figure 46).

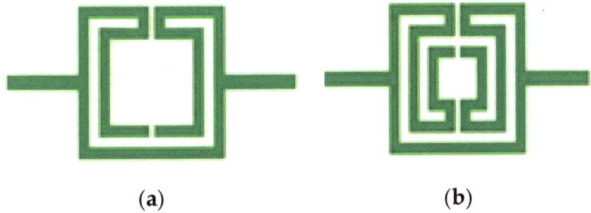

Figure 46. Open Matryoshka square geometry filters. (**a**) With two rings. (**b**) With three rings.

The physical characteristics of the five configurations chosen are described in Table 10. The definition of these physical characteristics is provided in Figure 4. A 1.5 mm thick FR4 substrate with $\varepsilon_r = 4.4$ and $\tan\delta = 0.02$ was used. The input and output microstrip lines were 2.8 mm wide (50 Ohm characteristic impedance); W = 2.0 mm and g = s = 1.0 mm were used. For the three ring configurations, $L_c = L_{c1} = L_{c2}$.

Table 10. Physical characterization of open Matryoshka geometries with two, three, and four rings.

Configuration	N	L_1 (mm)	L_2 (mm)	L_3 (mm)	L_c (mm)
Config1			20.0		2.0
Config2	2	28.0	12.0	NA	6.0
Config3			8.0		8.0
Config4	3	36.0	28.0	20.0	2.0
Config5		28.0	20.0	12.0	2.0

Photos of the fabricated prototypes are shown in Figure 47.

As explained in Section 2, a design procedure was developed to estimate the initial dimensions of the filter that fulfill the specifications. Comparisons of the numerical simulation and experimental results are shown in Figure 48 (for config1, config2, and config3) and Figure 49 (for config4 and config5).

Figure 48 shows a very good agreement between the simulation and experimental results. Table 11 summarizes the experimental characteristics of the three initial configurations.

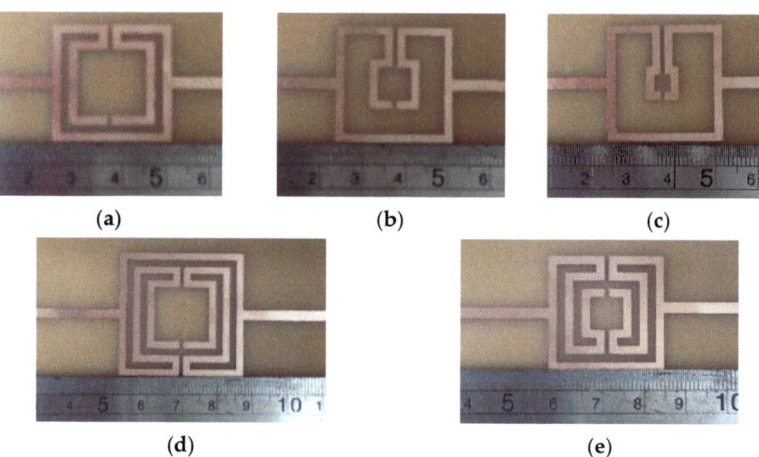

Figure 47. Photos of the prototypes of Matryoshka filters with square rings. (**a**) Config1. (**b**) Config2. (**c**) Config3. (**d**) Config4. (**e**) Config5.

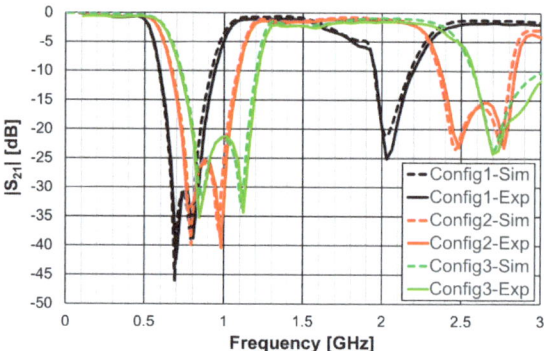

Figure 48. Comparison of the |S21| simulation and experimental results for square Matryoshka filters config1, config2, and config3.

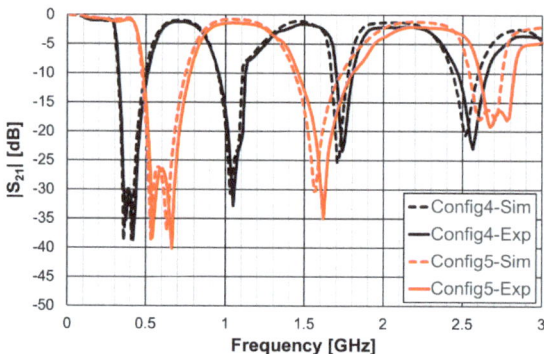

Figure 49. Comparison of the |S21| simulation and experimental results for square Matryoshka filters config4 and config5.

Table 11. Main experimental characteristics of the five square filter configurations.

Configuration	fr$_1$ (GHz)	fr$_2$ (GHz)	f$_0$ (GHz)	BW * (%)
Config1	0.700	0.805	0.769	45.4
Config2	0.770	0.980	0.876	49.0
Config3	0.840	1.120	0.964	51.4
Config4	0.375	0.420	0.421	43.8
Config5	0.540	0.660	0.626	46.4

* Defined for a −10 dB reference level.

As shown in Figure 49, a very good agreement between the simulation and experimental results was also obtained. Table 11 also summarizes the experimental characteristics of the two remaining configurations.

It can be concluded that the first two resonance frequencies (f$_{r1}$ and f$_{r2}$) depended on the perimeter of the rings. For the same external dimensions (L$_1$), the inner rings can be used to have a fine control of the stopband frequency range and of the bandwidth.

4.2. Filters with a Circular Matryoshka Geometry

Filters based on open Matryoshka circular ring configurations were studied in [26,57]. The physical characteristics of a one-ring configuration are defined in Figure 50.

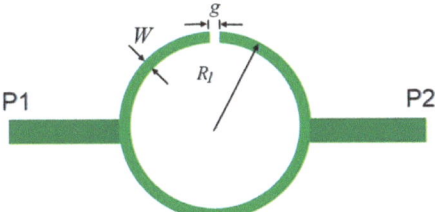

Figure 50. Stopband filter based on an open Matryoshka circular ring geometry.

Five configurations of this type of stopband filter were studied in [26] (Table 12). A 1.52 mm thick Rogers RO3003 substrate with ε_r = 3.0 and tanδ = 0.001 was used. The input and output microstrip lines (P1 and P2) were 3.8 mm wide (50 Ohm characteristic impedance). W = 1.0 mm and g = s = 1.0 mm were used.

Table 12. Physical characterization of open Matryoshka circular geometries with two and three rings.

Configuration	N	R1 (mm)	R2 (mm)	R3 (mm)
Config1			12.0	
Config2			10.0	
Config3	2	14.0	8.0	NA
Config4			6.0	
Config5	3		12	10.0

Photos of the fabricated prototypes are shown in Figure 51.

Similarly to the square Matryoshka configurations, a design procedure was developed to estimate the initial dimensions of the filter that fulfill the specifications. A comparison of the numerical simulation and experimental results is shown in Figure 52 (for config1, config2, config3, and config4) and Figure 53 (for config1 and config5).

There was a very good agreement between the numerical simulation and the experimental results, as shown in Figure 52. Table 13 summarizes the experimental characteristics of the five configurations. The first resonance frequency was almost independent of the second ring, but the second resonance frequency and the bandwidth increased substantially as the radius of the second ring decreased.

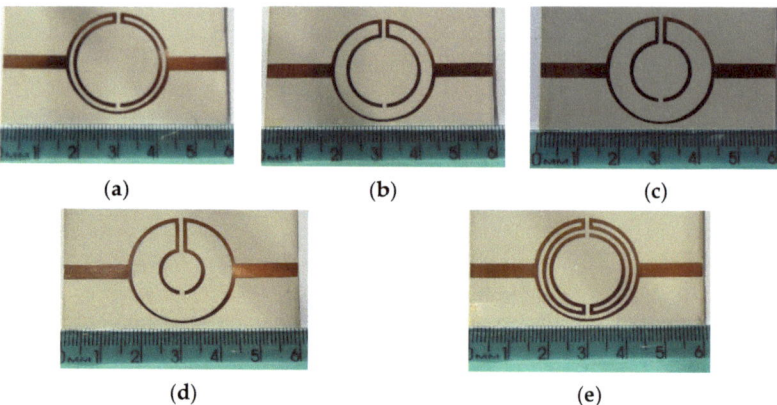

Figure 51. Photos of the prototypes of Matryoshka filters with circular rings. (**a**) Config1. (**b**) Config2. (**c**) Config3. (**d**) Config4. (**e**) Config5.

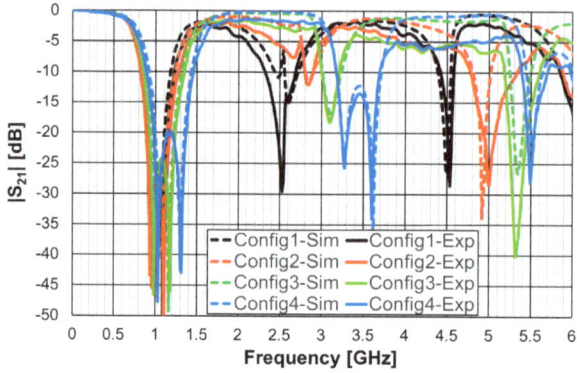

Figure 52. Comparison of the |S21| simulation and experimental results for circular Matryoshka filters config1, config2, config3, and config4.

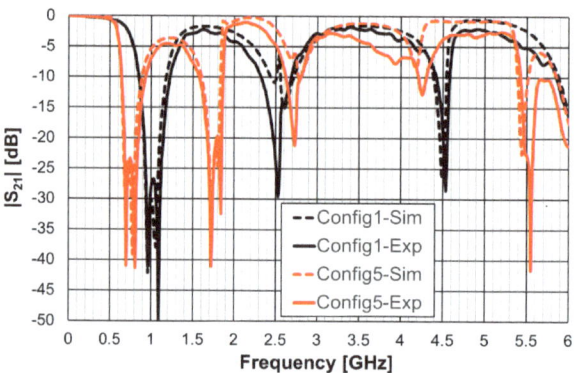

Figure 53. Comparison of the |S21| simulation and experimental results for circular Matryoshka filters config1 and config5.

Table 13. Main experimental characteristics of the five circular filter configurations.

Configuration	fr$_1$ (GHz)	fr$_2$ (GHz)	f$_0$ (GHz)	BW * (%)
Config1	0.961	1.091	1.039	35.7
Config2	0.941	1.101	1.047	43.1
Config3	0.981	1.181	1.099	45.2
Config4	1.031	1.311	1.172	48.0
Config5	0.701	0.811	0.790	36.7

* Defined for a −10 dB reference level.

As shown in Figure 53, a very good agreement between the simulation and experimental results was also obtained. A general conclusion, in line with the analysis carried out on the filters (and also the FSSs) with square Matryoshka geometry, is that, keeping the external dimension of the structure, the resonance frequency decreased (higher miniaturization) when more rings were used (higher meandering), but the bandwidth decreased.

4.3. Filters with a DGS

A DGS was formed by removing a small part from the metallic ground plane in planar printed circuit boards, most frequently in microstrip lines, as shown in Figure 54.

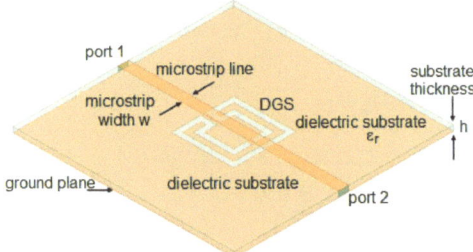

Figure 54. Example of DGS with Matryoshka geometry.

Due to ease of integration, design flexibility, and compactness, DGSs have found several applications such as in planar antennas [58,59], filters [60,61], power dividers [62,63], sensors [64,65], and wireless power transfer [66,67]. A DGS based on a Matryoshka geometry, as shown in Figure 54, was introduced in [30,43,68]. In [30], a method to design this type of DGS based on simple formulas is proposed. Four configurations were designed, fabricated, and tested [30]. A 1.6 mm thick FR4 substrate (ε_r = 4.4 and tanδ = 0.02) was used. The corresponding dimensions are indicated in Table 14. The definition of the dimensions is provided in Figure 4.

Table 14. Physical characterization of open Matryoshka geometry DGS configurations.

Configuration	L$_1$ (mm)	L$_2$ (mm)	L$_c$ (mm)	w (mm)	g = s (mm)
Config1	17.0	11.0			
Config2	15.5	9.5	1.5	1.5	1.0
Config3	14.0	8.0			
Config4	12.5	6.5			

Photos of the front and back sides of the prototypes are shown in Figure 55. Figure 56 provides the comparison of simulation and experimental |S21| results of these prototypes.

Good agreement is shown in Figure 56 between the numerical simulation and experimental results. The tendency of the resonance frequency was the same as the metal Matryoshka configuration, that is, as the area of the structure decreased, the resonance frequency increased.

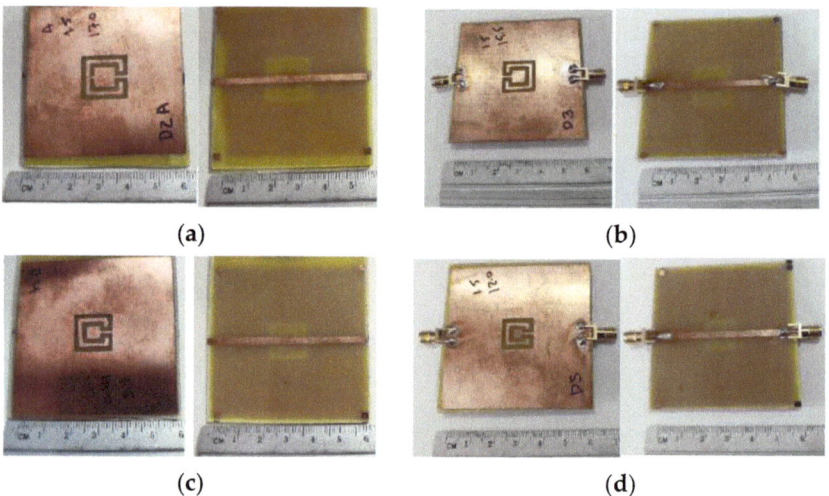

Figure 55. Photographs of the microstrip line with square Matryoshka geometry DGS configurations. (a) Config1. (b) Config2. (c) Config3. (d) Config4.

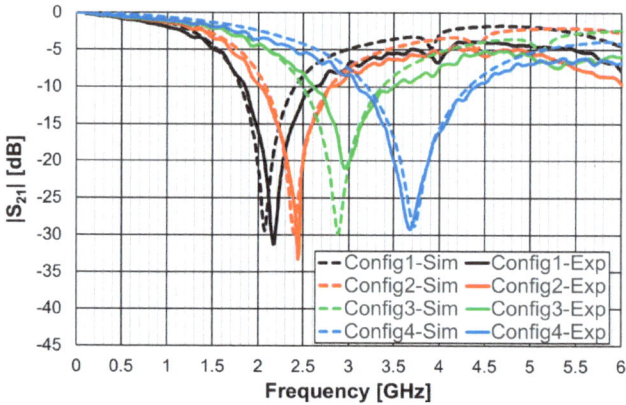

Figure 56. Comparison of the |S21| simulation and experimental responses of a microstrip line with a square Matryoshka geometry DGS.

To assess the capabilities of the Matryoshka geometry to perform as a DGS, a comparison with a DGS of the common dumbbell geometry is presented in [30]. To have a fair comparison, the square dumbbell geometry had the same area as the Matryoshka geometry. The simulation results for the four configurations indicated in Table 14 are presented in Figure 57.

A summary of the results shown in Figure 57 is provided in Table 15. The resonance frequency of the Matryoshka geometry was much lower than the resonance frequency of the dumbbell geometry (larger miniaturization), but the bandwidth was much narrower.

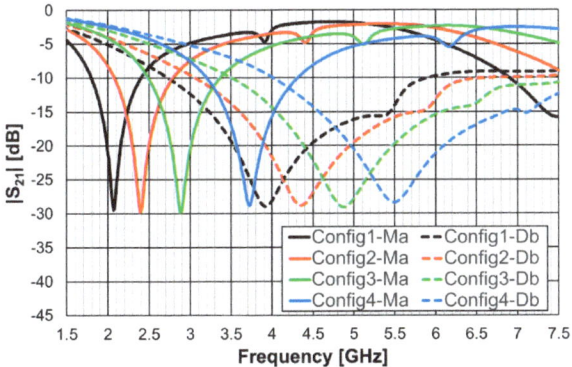

Figure 57. Comparison of the |S21| simulation results of a microstrip line with dumbbell and Matryoshka square geometry DGS.

Table 15. Comparison of first resonance characteristics of open Matryoshka and dumbbell DGS configurations.

Configuration	Matryoshka		Dumbbell	
	f_{r1} * (GHz)	BW (%)	f_{r1Ma}/f_{r1Db} (%)	BW_{Ma}/BW_{Db} (%)
Config1	2.07	28.8	52.8	19.0
Config2	2.39	29.3	55.1	20.0
Config3	2.88	29.8	59.1	20.0
Config4	3.72	29.6	67.9	22.3

* Defined for a −10 dB reference level.

4.4. Filters with a DGS and a Dielectric Resonator

Very recently, a compact filter combining a Matryoshka geometry DGS with a high-permittivity dielectric resonator was proposed [43]. The purpose was to improve the frequency response characteristics, mainly selectivity, and miniaturization. A prototype was designed, fabricated, and tested. A 1.6 mm thick FR4 substrate (ε_r = 4.4 and tan δ = 0.02) was used. As shown in Figure 58, a calcium cobaltite disk (ε_r = 90) with a diameter of 10.0 mm and a thickness of 1.9 mm was inserted into config3 of the previous section, below the ground plane, in contact with the DGS. A photograph, with a bottom view of the prototype, is shown in Figure 59. The filter transmission coefficient was simulated and measured for different positions of the dielectric disk. The corresponding results, for the disk centered on the DGS Matryoshka square geometry center, and on the DGS Matryoshka square geometry corner, are shown in Figure 60. The results obtained for the filter without a dielectric disk are also shown, for reference.

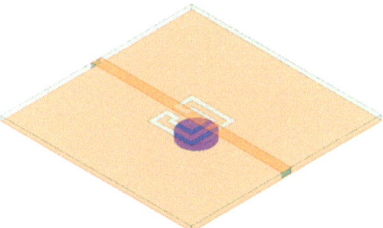

Figure 58. Filter configuration with Matryoshka square geometry DGS and dielectric resonator.

Figure 59. Photo of the prototype of the filter with Matryoshka square geometry DGS and dielectric resonator; bottom view.

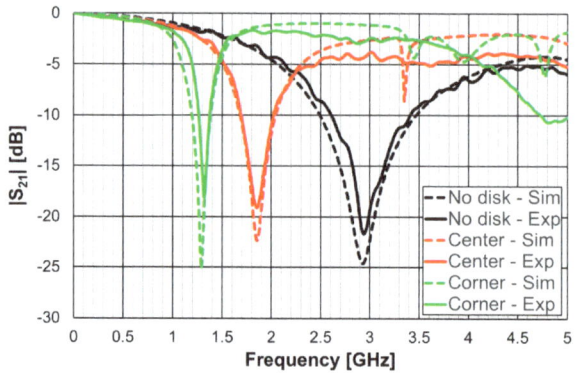

Figure 60. Comparison of the |S21| simulation and experimental results for the square Matryoshka geometry DGS with a dielectric resonator.

A very good agreement between the numerical simulation and experimental results was obtained. The use of a dielectric resonator can provide further miniaturization of the structure. The experimental resonance frequency moved from 2.939 GHz (no disk) to 1.849 GHz (center) and to 1.322 GHz (corner), which corresponded to 36.8% and 55.0% reductions, respectively. Again, the price to pay was the reduction of bandwidth, which was 58.5% (center) and 86.2% (corner). The position of the dielectric disk can be used to fine-tune the central frequency of the filter's response.

5. Examples of Application as Antenna

Due to their inherent multiresonant characteristics, Matryoshka geometries are suitable for multiband and/or wideband antenna configurations [25,28]. Moreover, because of the meandering of the nested rings, they have also been used to provide antenna miniaturization [25,28]. These features can be advantageously combined with printed antennas in general and microstrip patch antennas in particular [25,28] to be used in small mobile communication terminals and mass production electronic gadgets. Microstrip is one of the most successful antenna technologies. Such success stems from well-known advantageous and unique properties, such as a low profile, light weight, planar structure (but also conformal to non-planar geometries), mechanical strength, easy and low-cost fabrication, easy integration of passive and active components, easy combination to form arrays, and outstanding versatility in terms of electromagnetic characteristics (resonance frequency, input impedance, radiation pattern, gain, polarization). Microstrip patch antennas can be used in a very wide frequency range, extending roughly from about 1 GHz to 100 GHz [69]. So far, the Matryoshka geometries have been used in microstrip patch antennas to modify the ground plane and implement it as a DGS [25,28]. DGSs have been used in microstrip antenna implementations to provide multiband and/or wideband behavior, improve gain and cross-polarization, and suppress higher order modes and mutual coupling (in arrays) [70–74]. Many different shapes of the DGS slots have been used, ranging from canonical geometries (rectangular, triangular, circular) to non-canonical (H-shaped, dog bone-shaped) [75]. Recently, such variety was enhanced with the Matryoshka geometry [25,28].

In [25], a comparison of the performance of a microstrip patch antenna with a DGS ground plane with circular SRRs [76] and Matryoshka geometries is presented. The emphasis was on the open Matryoshka configuration. In [28], a detailed comparative analysis of the performance of open and closed Matryoshka DGS geometries was carried out. In all the cases, a cheap FR4 substrate with relative electric permittivity of 4.4, thickness of 1.6 mm, and loss tangent of 0.02 was used.

5.1. Reference Microstrip Patches

In [25], as an application example, the dimensions of a rectangular patch were chosen to provide the first resonance at 2.5 GHz. The initial dimensions obtained with the transmission line method [77] were optimized using the ANSYS Electronics Desktop 2018.1.0, release 19.1.0 [78]. A patch width (W) of 37.0 mm, length of (L) 27.8 mm, and a square ground plane with a 53.0 mm side were chosen. The patch was fed with a 2.8 mm wide microstrip transmission line (50 Ohm characteristic impedance) and inset that is 1 mm wide and 8 mm long. The corresponding simulated input reflection coefficient is shown in Figure 61, for reference.

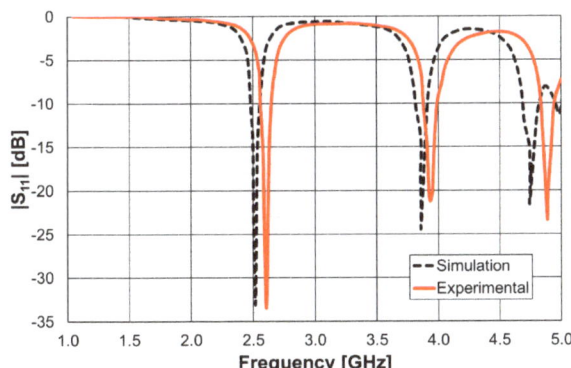

Figure 61. Simulation and experimental input reflection coefficient of the simple rectangular patch.

The first resonance (2.52 GHz), the second resonance (3.86 GHz), and the third resonance (4.74 GHz) were well matched to the 50 Ohm feed microstrip transmission line. To validate the design procedure used, a protype of the microstrip patch antenna was fabricated using a conventional photolithography technique. The amplitude of the experimental input reflection coefficient ($|S_{11}|$), also shown in Figure 61, was measured with an Agilent E5071C (Agilent, Santa Clara, CA, USA) vector network analyzer. Taking into account that the FR4 substrate used is low cost, and its characteristics are only generically known, there was a good agreement between the numerical simulations and experimental results. For the frequency of interest (first resonance), the difference in the frequency was only 3.4% (88 MHz), and the $|S_{11}|$ level was almost the same (-33 dB). This antenna presented the usual almost hemispherical broadside radiation pattern [77] with a gain of 6.18 dBi at 2.52 GHz.

Another microstrip patch was designed so that using a DGS ground plane, the same first resonance (2.5 GHz) of the simple patch, described above, could be obtained [25]. In this case, the patch was designed to have alone the first resonance frequency at 3.5 GHz.

The patch and ground plane sizes were 28.0 mm (W), 20.0 mm (L), and 38.0 mm, 45.0 mm, respectively. The corresponding simulation and experimental input reflection coefficient are shown in Figure 62.

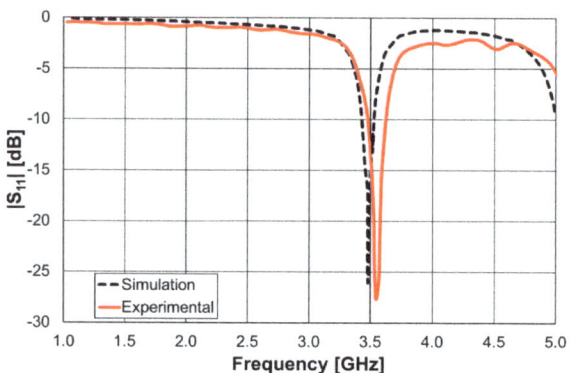

Figure 62. Simulation and experimental amplitude of the input reflection coefficient of the simple rectangular patch.

The difference in the simulation and experimental resonance frequencies was only 2.0% (69 MHz), and the $|S_{11}|$ level was below -26 dB, for both curves.

5.2. DGS Uni-Cell

From initial exploratory simulations [25], it is concluded that the patch with a DGS would present the first resonance frequency at 2.5 GHz when the DGS unit-cell alone had the first resonance at about 2.6 GHz. Therefore, both the complementary open Matryoshka and circular SRR configurations were designed to provide such a 2.6 GHz first-resonance frequency. To take into account the intrinsic characteristics of the unit-cells, their analysis was performed by considering an infinite FSS with 20×20 mm^2 unit-cells. The two configurations are shown in Figure 63. Ansys HFSS [78] was used for the simulations.

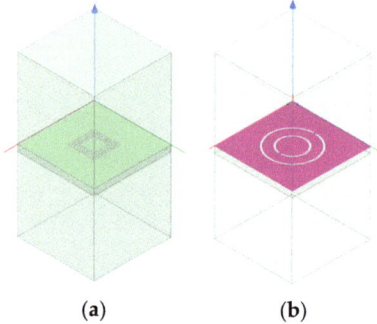

Figure 63. FSS unit-cells. (**a**) Complementary open Matryoshka geometry. (**b**) Complementary circular SRR geometry.

The complementary open square Matryoshka configuration had dimensions of $L_1 = 6.8$ mm and $L_2 = 4.8$ mm. For the circular complementary SRR, $r_1 = 5.4$ mm and $r_2 = 3.5$ mm was used. In both cases, the trace and slot widths were 0.50 mm and 0.25 mm, respectively. The simulated $|S21|$ results are shown in Figure 64.

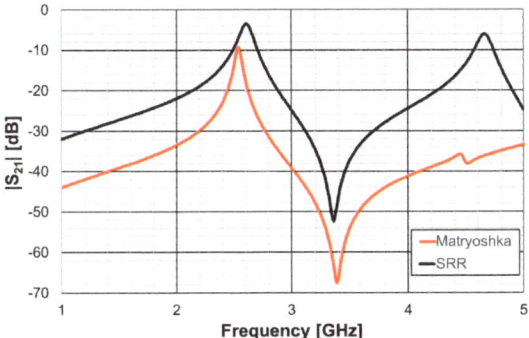

Figure 64. Simulated |S21| results of the open square Matryoshka and SRR FSS configurations.

It can be concluded that, as required, both unit-cells provided the first resonance frequency at about 2.6 GHz.

5.3. Patch with DGS

The patch's ground plane was changed by the introduction of a DGS with an open square Matryoshka and a circular SRR [25], as shown in Figure 65.

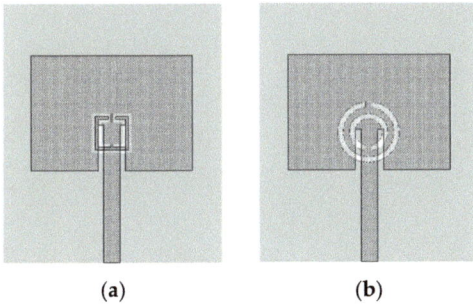

Figure 65. Microstrip patch with DGS. (**a**) Square Matryoshka geometry. (**b**) Circular SRR geometry.

The simulation and experimental results for the amplitude of the input reflection coefficient, for both DGS unit-cell geometries, are shown in Figure 66.

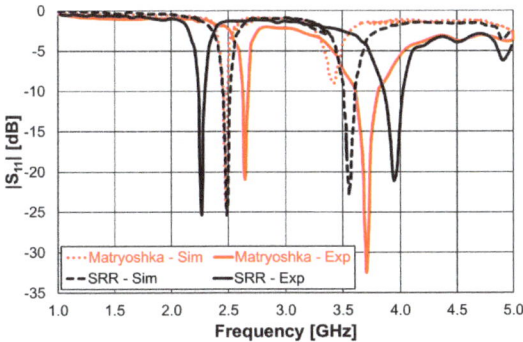

Figure 66. Simulation and experimental |S_{11}| results of the patch with DGS ground plane.

There were some differences between the simulation and experimental results. For the first resonance frequency, the experimental result for the SRR geometry was 8.8% (218 MHz) below the simulation one, whereas for the Matryoshka geometry, the experimental result was 6.7% (165 MHz) above the simulation. These differences were mainly caused by the inaccuracy of the fabrication process, mostly related with the narrow (0.25 mm) slots. However, these unwanted differences did not jeopardize the envisage proof of concept, that is, both the DGS configurations provided a remarkable miniaturization of about 46% (in the area of the microstrip patch).

The farfield radiation patterns of the patch with the two DGS configurations are shown in Figures 67–69.

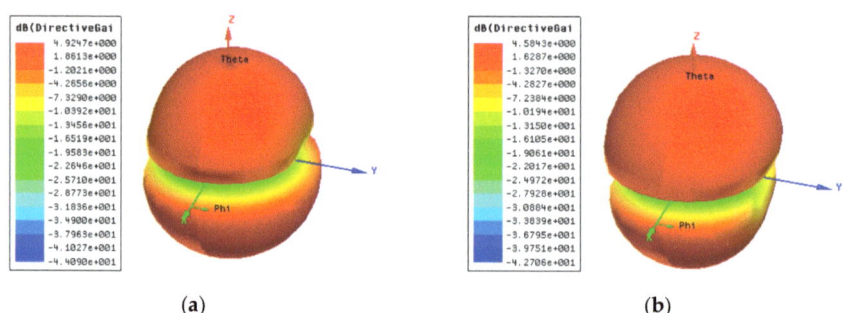

(a)   (b)

Figure 67. Simulation 3D radiation pattern (gain scale) of the patch with DGS at the first resonance frequency. (**a**) Matryoshka geometry. (**b**) SRR geometry.

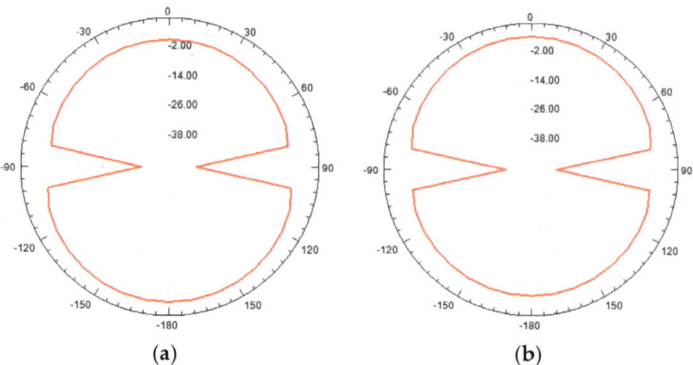

(a)   (b)

Figure 68. Simulation H-plane radiation pattern (gain scale) of the patch with DGS at the first resonance frequency. (**a**) Matryoshka geometry. (**b**) SRR geometry.

When compared with the radiation pattern of the common patch, the main difference was the high radiation level below the ground plane. In contrast with the typical hemispherical type of radiation pattern observed for the common microstrip patch [77], a bi-hemispherical type of radiation pattern was caused by the DGS. This was expected, first due to the introduction of slots in the ground plane and second because the slots were near resonance and therefore with enhanced radiation. This type of radiation pattern may be interesting for application where a more uniform spatial radiation power distribution is required. For the DGS with Matryoshka geometry, the maximum gain (4.9 dBi) was obtained in the back hemisphere ($\theta \approx 180°$). For the DGS with SRR geometry, the direction of maximum radiation was kept on the front hemisphere ($\theta \approx 0$), with 4.6 dBi gain, but the front-to-back (FBR) ratio was low (2.8 dB). The drop of about 1.4 dB in the gain was related to the more uniform distribution of radiated power in space.

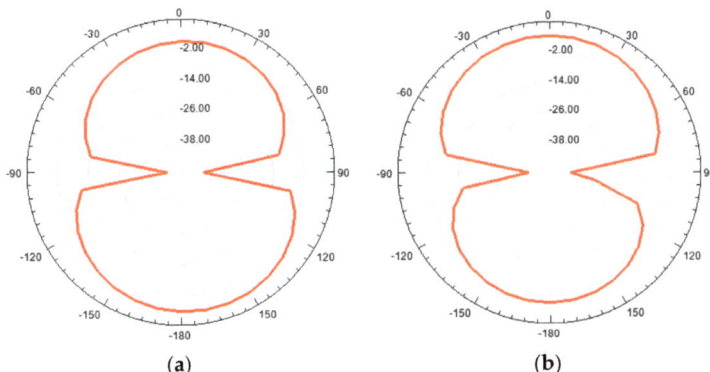

Figure 69. Simulation E-plane radiation pattern (gain scale) of the patch with DGS at the first resonance frequency. (**a**) Matryoshka geometry. (**b**) SRR geometry.

The main radiation characteristics, obtained by simulation, are summarized in Table 16.

Table 16. Summary of the simulated radiation pattern results at the first resonance frequency.

Parameter		Matryoshka	SRR
Direction of maximum radiation (θ) (Degree)		$\approx$180	$\approx$0
Maximum gain (dBi)		4.9	4.6
Half-power beamwidth (Degree)	H-Plane	140	137
	E-Plane	87	86
FBR (dB)		−3.2	2.8

A fair comparison of the miniaturization capability of the two DGS geometries under analysis (open Matryoshka and circular SRR) must consider unit-cells with the same dimension. The first resonance frequency of the microstrip patch with a DGS ground plane, as a function of the maximal dimension (the side of the square open Matryoshka geometry and diameter of the circular SRR), is shown in Figure 70.

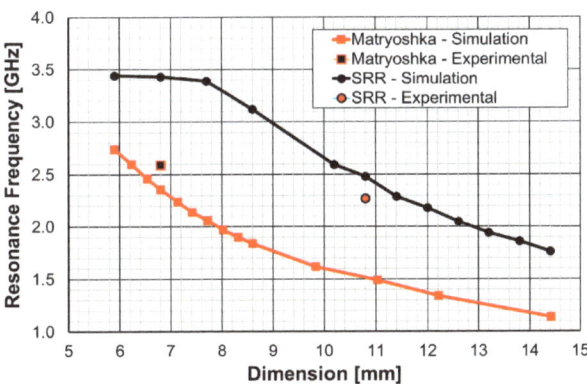

Figure 70. Simulation results for the first resonance frequency of the DGS unit-cell.

As can be observed, the open Matryoshka geometry provided much lower simulation results for the frequency of the first resonance. Moreover, the experimental results obtained for the two fabricated prototypes had a reasonable agreement with the simulations (a difference less than 9%). The average difference, for the resonance frequencies associated with each DGS geometry, was almost 1 GHz (0.94 GHz), which corresponded to 35.4%. This proves that the proposed open Matryoshka geometry had a much stronger miniaturization capability than the conventional circular SRR. As verified in Sections 3 and 4, the same conclusion was obtained for FSS [32] and filter [20] applications.

In a very recent work [28], a detailed analysis of the effects of a DGS with a Matryoshka unit-cell in the ground plane of a microstrip patch was carried out. A complete sensitivity analysis of the influence of the geometric parameters of the unit-cells in the antenna performance was conducted. The main conclusions obtained in [25] were confirmed and were supported by an extensive and systematic analysis, with simulations and experimental validation. The emphasis was on the comparison of the miniaturization capabilities of the open and closed complementary Matryoshka geometries. As an example, some results obtained for an optimized configuration are reproduced below.

The two configurations of the microstrip patch with a DGS ground plane, that is, with open and closed Matryoshka cells, are shown in Figure 71.

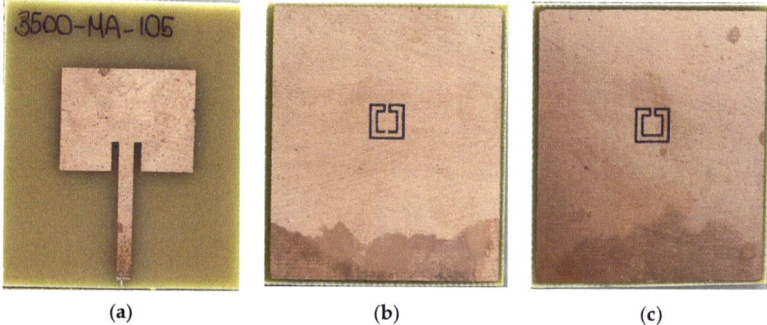

Figure 71. Microstrip patch antenna prototypes with DGS complementary Matryoshka cells. (**a**) Patch side. (**b**) Ground-plane side with an open Matryoshka geometry cell. (**c**) Ground-plane side with a closed Matryoshka geometry cell.

The |S21| of a 50 Ohm microstrip line with a DGS with open and closed Matryoshka square geometries (Figure 72) (L_1 = 7.5 mm, L_2 = 4.5 mm, w = 0.5 mm, and s = g = 0.5 mm), as a function of frequency, is shown in Figure 73. The measurement setups used are shown in Figure 74.

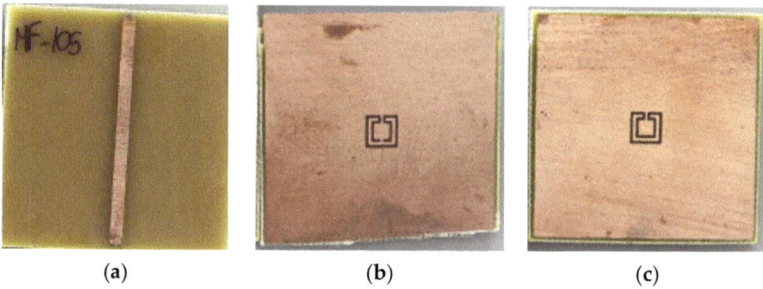

Figure 72. Microstrip line prototypes with DGS complementary Matryoshka cells. (**a**) Microstrip line side. (**b**) Ground-plane side with an open Matryoshka geometry cell. (**c**) Ground-plane side with a closed Matryoshka geometry cell.

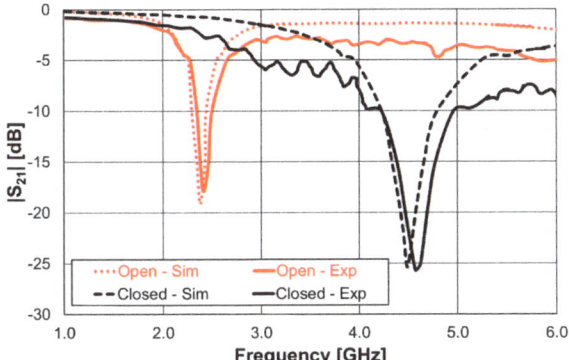

Figure 73. |S21| results for the microstrip line with a DGS with open and closed square Matryoshka cells.

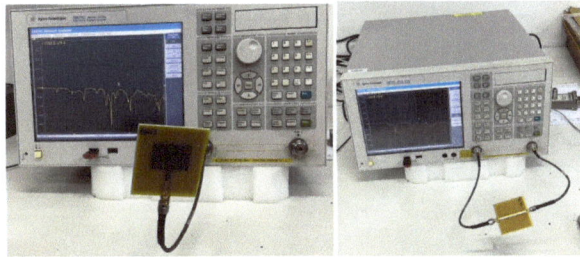

Figure 74. Setups used for the experimental characterization of the microstrip patch and microstrip line with a DGS with open and closed square Matryoshka cells.

There was a good agreement between simulation and experimental |S21| results. Although the open and closed Matryoshka unit-cells had the same dimensions, the first resonance of the open structure (2.40 GHz) was 47.4% below the first resonance of the closed structure (4.57 GHz). However, the closed structure had a much wider −10 dB bandwidth (560 MHz compared with 150 MHz). As shown in Figure 75, this reduction led also to a reduction of the first resonance frequency of the patch with the open Matryoshka DGS.

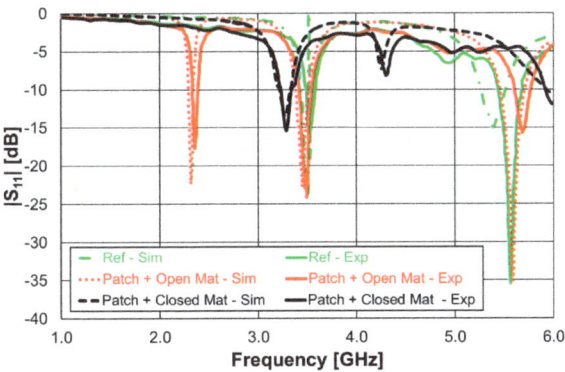

Figure 75. Amplitude of the input reflection coefficient of the patch without and with the DGS ground plane.

The input reflection coefficient of the common patch (without DGS) is also shown for reference in Figure 75. In this case, the microstrip patch antenna with open Matryoshka DGS presented the first resonance frequency at 2.35 GHz, which was 28.5% below the first resonance of the microstrip patch antenna with closed Matryoshka DGS (3.29 GHz) and 33.2% below the first resonance of the microstrip patch antenna with a solid ground plane (3.52 GHz).

The surface current distribution, shown in Figure 76, provided physical insight into the antenna's radiation mechanisms. The current distribution on the common patch (without DGS) is also shown, for reference.

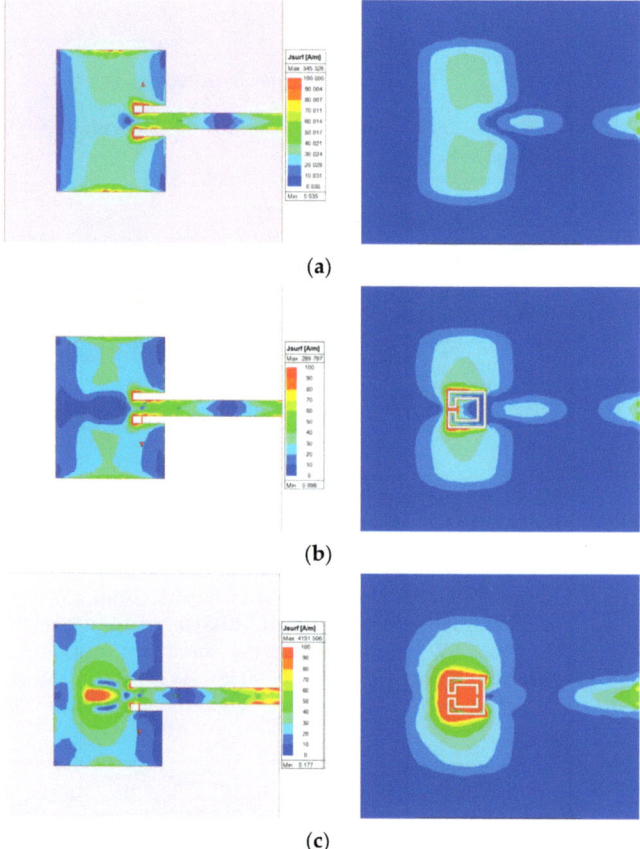

Figure 76. Current distribution on the microstrip patch and ground plane at the first resonance frequency. (**a**) Common patch. (**b**) Patch with closed Matryoshka DGS. (**c**) Patch with open Matryoshka DGS.

As expected, the presence of the DGS enormously changed the current distribution, not only on the ground plane but also on the patch. This change was more effective for the open Matryoshka geometry. The almost constant current distribution along the common patch width was strongly perturbed by the DGS.

The 3D radiation patterns of the patch with a DGS, with closed and open Matryoshka geometries, are shown in Figure 77. Again, the radiation pattern of the common patch (without DGS) is also shown, for reference.

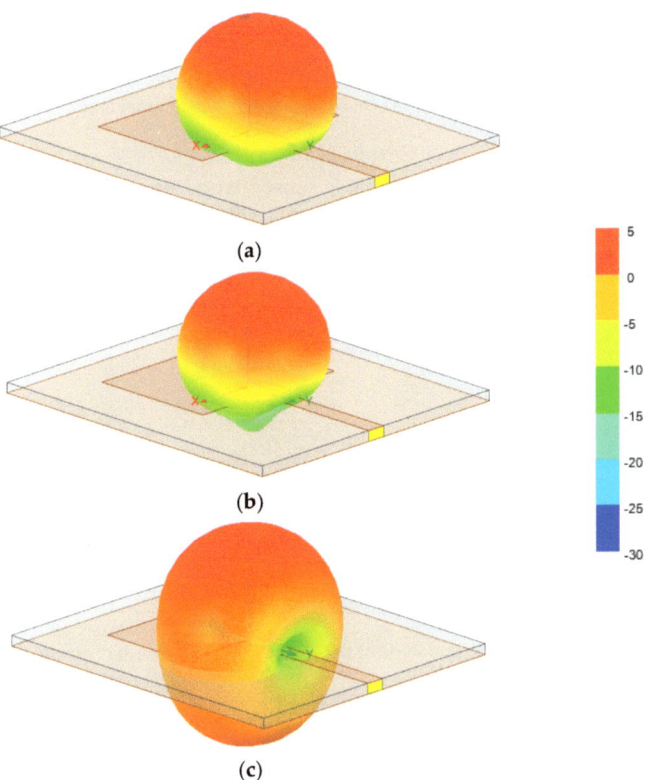

Figure 77. Three-dimensional radiation pattern (gain scale) of the microstrip patch at the first resonance frequency. (**a**) Common patch. (**b**) Patch with closed Matryoshka DGS. (**c**) Patch with open Matryoshka DGS.

The maximum gains for the common patch, the patch with closed Matryoshka DGS, and the patch with open Matryoshka DGS were 4.24 dBi, 3.55 dBi, and 2.40 dBi, respectively. Naturally, the increase in the back radiation caused by the DGS implied a decrease in the maximum gain, especially for the open Matryoshka DGS.

A summary of the main results obtained for the patch antenna without DGS and with DGS with open and closed Matryoshka geometries is presented in Table 17.

Table 17. Summary of the main patch antenna characteristics.

Parameter	Without DGS	With Open Matryoshka DGS	With Closed Matryoshka DGS
First resonance frequency (GHz)	3.52	2.35	3.29
−10 dB bandwidth (MHz)	92	53	108
−10 dB bandwidth (%)	2.6	2.3	3.3
Gain (dBi)	4.24	2.40	3.55

It can be concluded that both Matryoshka DGS geometries provided miniaturization, but the open structure was much more effective. However, both Matryoshka DGS geometries caused a decrease in the gain, being more pronounced for the open structure. The open structure also provided a narrower impedance bandwidth.

6. Examples of Application as a Sensor

If a material under test (MUT) is incorporated in a filter and the filter changes its frequency response according to the characteristics of the MUT, this filter can be used as a sensor [79]. Based on this idea, three practical sensors were proposed.

6.1. Alcohol Concentration Sensor

A new and simple sensor, based on a microstrip filter with an open Matryoshka configuration, was proposed in [34]. The proposed sensor was designed, fabricated, and successfully applied to detect the alcohol content of a liquid. Photos of the prototype and of the measurement setup are shown in Figure 78. A small acrylic container with internal dimensions $43.7 \times 43.7 \times 30.0$ mm^3 (57.29 mL capacity) and 3.0 mm and 1.0 mm thick side and bottom walls, respectively, was placed over the filter, centered on the Matryoshka geometry.

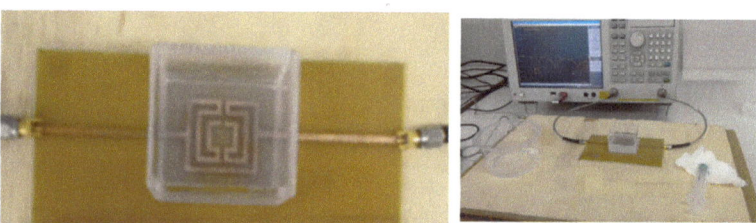

Figure 78. Prototype of the alcohol concentration sensor and of the measurement setup.

The experimental results obtained for the first resonance frequency, as a function of the alcohol concentration, for three different volumes of liquid, are shown in Figure 79.

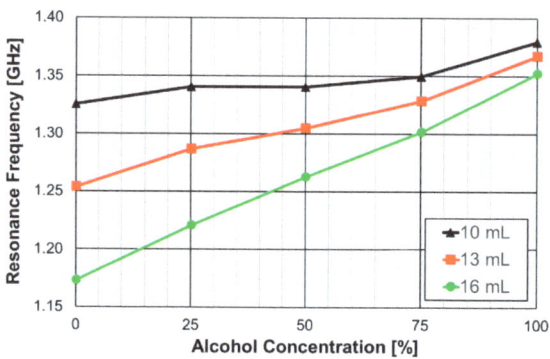

Figure 79. Resonance frequency for different alcohol concentrations and volumes.

For a 16 mL volume of the MUT (about 8.4 mm of liquid in the container), the response was almost linear. It was clear that the larger the MUT volume, the better the sensitivity (slope of the curve). Based on the design procedure described in Section 4.1, other prototypes can be fabricated to develop sensors to be used on other frequency bands. Moreover, other liquids can be characterized, based on calibration curves previously validated.

6.2. Sucrose Level and Water Content Sensors

In [80], a sensor based on a microstrip filter with a closed Matryoshka geometry DGS was used to obtain the sucrose level of an aqueous solution. Based on the analysis and design procedure described in Section 4.3, the prototype, shown in Figure 80, was developed. It is based on configuration 1 described in Table 14.

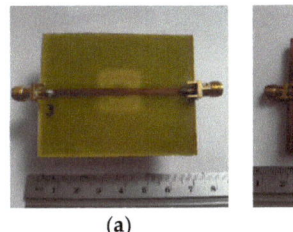

(a)   (b)

Figure 80. Prototype of the closed Matryoshka geometry DGS sensor. (**a**) Bottom view. (**b**) Top view with container.

A small acrylic container with internal dimensions of $30 \times 30 \times 15$ mm^3 (13.5 mL capacity) was placed over the filter, centered on the Matryoshka geometry. The filter is used upside down, that is, with the DGS on the top side.

The calibration curve obtained for the determination of the sucrose level in an aqueous solution is shown in Figure 81.

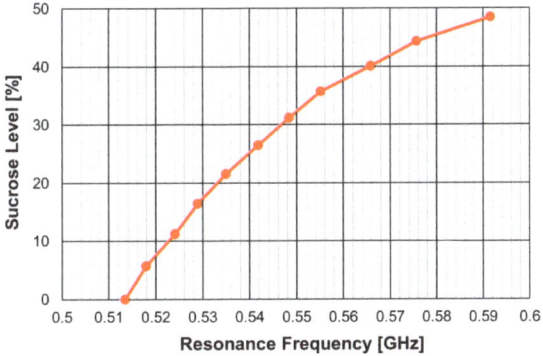

Figure 81. Calibration curve for a sucrose level content sensor.

A similar sensor is proposed in [41] to determine the distilled water content in a solution of isopropyl alcohol and distilled water. It is based on configuration 2 described in Table 14. The corresponding calibration curve is reproduced in Figure 82.

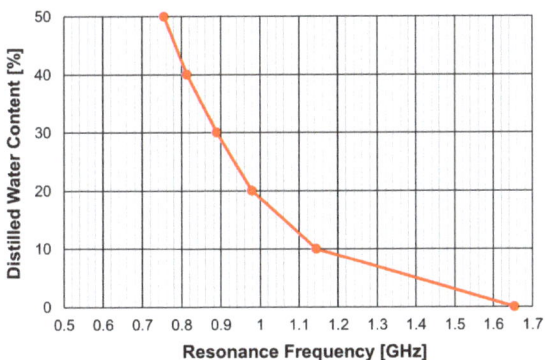

Figure 82. Calibration curve for distilled water content sensor.

6.3. Soil Moisture Sensor

In [31,44], a new soil moisture sensor based on a filter with a Matryoshka geometry DGS is described. The filter configuration is represented in Figure 83.

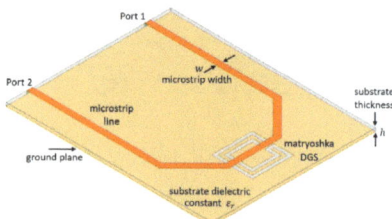

Figure 83. Structure of the filter configuration used as a soil moisture sensor.

A closed Matryoshka geometry DGS with L_1 = 20.0 mm, L_2 = 14.0 mm, w = 2.0 mm, and g = 1.0 mm was used. Photos of the prototype and of the measurement setup are shown in Figure 84.

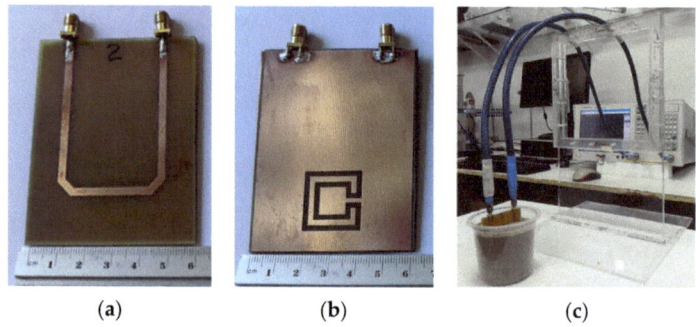

Figure 84. Photos of the experimental validation process. (**a**) Microstrip line side view. (**b**) DGS side view. (**c**) Measurement setup.

Two types of soil were measured: a sandy soil usually used in civil construction, with about 98% sand, and a garden soil rich in organic substances. The corresponding experimental results are shown in Figure 85. So far, the resonance frequency has been used, but in this case, the frequency points where |S21| reaches the −6 dB level was used because it is more stable [31,44].

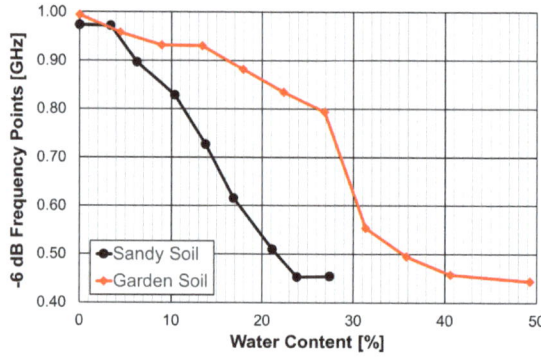

Figure 85. Experimental resonance frequency results for sandy and garden soils.

As the water content increased, the sandy soil absorbed less water and the sample saturated more quickly, starting from approximately 24%. On the other hand, garden soil absorbs more water, and the saturation point occurred at about 40%.

7. Conclusions and Perspectives of Future Developments

This paper reviewed the research activities developed on the topic of printed planar resonator structures nested inside each other, called Matryoshka geometries. These research activities started about 10 years ago, and most of the work has been performed at the Group of Telecommunications and Applied Electromagnetism (GTEMA) from the Instituto Federal da Paraíba in João Pessoa, Brazil. Although it is still a recent topic, much varied work has been done to describe the different types of structures, discover and explain their properties, obtain physical insight on the working principles, and explore potential applications.

The main characteristics of Matryoshka geometries stem from their meandered nature and the fact that the area occupied is defined by the external ring where the other interconnected rings are nested inside. These characteristics render them strong miniaturization capability and selective multi-resonances that are highly attractive to be used in many microwave devices. The initial application as FSS was quickly extended to other applications, such as filters, antennas, and different types of sensors.

The Matryoshka geometry was described initially, and the closed and open configuration were introduced. The physical parameters that describe the geometry were defined. The main common characteristics were analyzed, and the working principles were studied to provide physical insight on their behavior.

The initial application of Matryoshka geometries, as FSS unit-cells, was fully examined. Closed and open configurations were analyzed in detail, and comparison with simple rings was provided. The sensitivity of the transmission coefficient to each of the physical parameters was evaluated. To overcome the polarization dependence of simple FSS configurations, polarization-independent Matryoshka geometry configurations were introduced and analyzed. The combination of an FSS with a crossed dipole was proposed to enhance the multi-frequency response. Complimentary configurations, where the Matryoshka geometries were implemented with slots, were also analyzed. FSS unit-cells that use PIN diodes to provide reconfigurability were proposed and studied. It was effectively verified that in all the configurations studied, the Matryoshka-geometry-based configurations exhibited the expected multi-resonance behavior, and a remarkable miniaturization was provided.

Microstrip filters that use both square and circular rings with the open Matryoshka geometry were thoroughly examined. Configurations with different dimensions and number of rings were studied. The use of open Matryoshka geometries to implement DGS (in the ground plane) in microstrip filters was also proposed and analyzed. A combination of an open Matryoshka DGS configuration with a dielectric disk resonator was proposed to further enhance the miniaturization capability.

The use of DGS configurations with Matryoshka geometry in the ground plane of microstrip patch antennas was proposed and examined in detail. It was concluded that there was a remarkable miniaturization capability, but a decrease in gain and impedance bandwidth was inevitably observed.

Some applications of the devices based on configurations with Matryoshka geometries have already been envisaged and reported. Microstrip filters with open Matryoshka geometry were used to determine the alcohol concentration present in a liquid solution. The quality of the estimation increased as the volume of the sample increased. A microstrip filter with closed Matryoshka geometry DGS was used in a sensor conceived to obtain the sucrose content level and the distilled water content level of aqueous solutions. The corresponding calibration curves were proposed. Another microstrip filter with close Matryoshka geometry was used to obtain the percentage of moisture in soil samples. These applications of Matryoshka-based geometry configurations in sensors are very promising because they are simple, low cost, and potentially accurate.

Although many configurations based on Matryoshka geometries have been proposed and some potential applications of them as FSSs, filters, antennas, and sensors have been explored, it is clear that much more remains to be investigated. This is natural, since it is just a ten-year-old activity, and, therefore, many possibilities can be envisaged. Possible topics of future work are better compression of higher order resonances; the effect of the strip (stopband), or slot (passband) width; and, in the case of open rings, the positioning of the gap. Some work is already being carried out, including passband filters in multilayer structures, for which the first results should be published soon. Similarly, double-sided FSSs are being implemented. The miniaturization and multi-resonance capabilities of the Matryoshka geometries can be even more effective in 3D configurations, taking advantage of the now affordable 3D printing technology. Moreover, the use of new materials, as well as metamaterials, can offer specific characteristics that need to be explored. With adequate supporting materials, as well as fabrication and measurement facilities, the designs and applications, now limited to the microwave region, can be extended to the millimeter wave and Terahertz frequency bands. Optimized Matryoshka-geometry-based DGS sensors for many specific applications are also something that can be developed in the short term.

We would also like to highlight that the applications developed so far are conceptual and limited to manufacturing processes available on a laboratory scale. Much more can be done from the concepts presented. Moreover, the exploration of the mechanical and electrical properties of new flexible mesh composite materials can enhance the reconfigurability of these structures [81].

To conclude, we highlight that the investigation of the applications of Matryoshka geometries is an open field, whose results obtained so far encourage other groups to get involved in this research effort.

Author Contributions: Conceptualization, A.G.N., J.C.e.S. and J.N.d.C.; methodology, A.G.N., J.C.e.S. and J.N.d.C.; software, A.G.N., J.C.e.S. and J.N.d.C.; validation, A.G.N., J.C.e.S. and J.N.d.C.; formal analysis, A.G.N., J.C.e.S., J.N.d.C. and C.P.; investigation, A.G.N., J.C.e.S. and J.N.d.C.; resources, A.G.N., J.C.e.S. and J.N.d.C.; data curation, A.G.N., J.C.e.S., and J.N.d.C.; writing—original draft preparation, A.G.N., J.C.e.S., J.N.d.C. and C.P.; writing—review and editing, A.G.N., J.C.e.S., J.N.d.C. and C.P.; visualization, A.G.N., J.C.e.S., J.N.d.C. and C.P.; supervision, A.G.N., J.C.e.S. and J.N.d.C.; project administration, A.G.N., J.C.e.S. and J.N.d.C.; funding acquisition, A.G.N., J.C.e.S., J.N.d.C. and C.P. All authors have read and agreed to the published version of the manuscript.

Funding: This work was funded by Instituto Superior Técnico (IST), Instituto de Telecomunicações (IT) and Fundação para a Ciência e a Tecnologia (FCT) under grant UIDB/50008/2020. This work was supported in part by the Brazilian National Council for Scientific and Technological Development (CNPq)-Brazil, under project 309412/2021-8; in part by the Federal Institute of Paraíba (IFPB), under projects 21/2022 and 22/2022; the Electrical Engineering Graduation Program, PPGEE-IFPB; and in part by the Project FAPESQ/CAPES Nº 18/2020.

Acknowledgments: The authors acknowledge the dedicated work done by the MSc students listed in the references.

Conflicts of Interest: The authors declare no conflicts of interest.

References

1. Moon, S.; Khanna, S.K.; Chappell, W.J. Multilayer Silicon RF System-in-Package Technique Using Magnetically Aligned Anisotropic Conductive Adhesive. In Proceedings of the IEEE MTT-S International Microwave Symposium, Boston, MA, USA, 7–12 June 2009. [CrossRef]
2. Hu, F.; Hao, Q.; Lukowiak, M.; Sun, Q.; Wilhelm, K.; Radziszowski, S.; Wu, Y. Trustworthy Data Collection from Implantable Medical Devices Via High-Speed Security Implementation Based on IEEE 1363. *IEEE Trans. Inf. Technol. Biomed.* **2010**, *14*, 1397–1404. [CrossRef]
3. Ahmad, R.; Kuppusamy, P.; Potter, L.C. Nested Uniform Sampling for Multiresolution 3-D Tomography. In Proceedings of the IEEE International Conference on Acoustics, Speech and Signal Processing, Dallas, TX, USA, 14–19 March 2010. [CrossRef]
4. Ghali, C.; Hamady, F.; Elhajj, I.H.; Kayssi, A. Matryoshka: Tunneled Packets Breaking the Rules. In Proceedings of the International Conference on High Performance Computer & Simulation, Istanbul, Turkey, 4–8 July 2011. [CrossRef]

5. Raj, N.N.; Brijitta, J.; Ramachandran, D.; Rabel, A.M.; Jayanthi, V. Synthesis and Characterization of Matryoshka Doll-Like Silica Nanoparticles. In Proceedings of the International Conference on Advanced Nanomaterials & Emerging Engineering Technologies, Chennai, India, 24–26 July 2013. [CrossRef]
6. Ghosh, S.; Hiser, J.D.; Davidson, J.W. Matryoshka: Strengthening Software Protection via Nested Virtual Machines. In Proceedings of the IEEE/ACM International Workshop on Software Protection, Florence, Italy, 19 May 2015. [CrossRef]
7. Leroux, S.; Bohez, S.; De Coninck, E.; Verbelen, T.; Vankeirsbilck, B.; Simoens, P.; Dhoedt, B. Multi-Fidelity Matryoshka Neural Networks for Constrained IoT Devices. In Proceedings of the International Joint Conference on Neural Networks, Vancouver, BC, Canada, 24–29 July 2016. [CrossRef]
8. Kyaw, Z.; Qi, S.; Gao, K.; Zhang, H.; Zhang, L.; Xiao, J.; Wang, X.; Chua, T.-S. Matryoshka Peek: Toward Learning Fine-Grained, Robust, Discriminative Features for Product Search. *IEEE Trans. Multimed.* **2017**, *19*, 1272–1284. [CrossRef]
9. Richter, S.R.; Roth, S. Matryoshka Networks: Predicting 3D Geometry via Nested Shape Layers. In Proceedings of the IEEE/CVF Conference on Computer Vision and Pattern Recognition, Salt Lake City, UT, USA, 18–23 June 2018. [CrossRef]
10. Shinmura, S. Cancer Gene Analysis of Microarray Data. In Proceedings of the IEEE/ACIS International Conference on Big Data, Cloud Computing Data Science & Engineering, Yonago, Japan, 12–13 July 2108. [CrossRef]
11. Malik, L.; Patro, R. Rich Chromatin Structure Prediction from Hi-C Data. *IEEE/ACM Trans. Comput. Biol. Bioinform.* **2019**, *16*, 1448–1458. [CrossRef]
12. Sun, M.; Pu, L.; Zhang, J.; Xu, J. Matryoshka: Joint Resource Scheduling for Cost-Efficient MEC in NGFI-Based C-RAN. In Proceedings of the IEEE International Conference on Communications, Shanghai, China, 20–24 May 2019. [CrossRef]
13. Pozza, M.; Nicholson, P.K.; Lugones, D.F.; Rao, A.; Flinck, H.; Tarkoma, S. On Reconfiguring 5G Network Slices. *IEEE J. Sel. Areas Commun.* **2020**, *38*, 1542–1554. [CrossRef]
14. Zhu, Y.; Zhao, Z.; Du, L. Matryoshka Phononic Crystals for Anchor-Loss Reduction of Lamb Wave Resonators. In Proceedings of the IEEE International Conference on Nanotechnology, Montreal, QC, Canada, 29–31 July 2020. [CrossRef]
15. Huang, H.; Zhong, J.; Taku, K. Matryoshka Attack: Research on an Attack Method of Recommender System Based on Adversarial Learning and Optimization Solution. In Proceedings of the International Conference on Wavelet Analysis and Pattern Recognition, Adelaide, Australia, 2 December 2020. [CrossRef]
16. Drimmel, R.; Sozzetti, A.; Schröder, K.-P.; Bastian, U.; Pinamonti, M.; Jack, D.; Huerta, M.A.H. A Celestial Matryoshka: Dynamical and Spectroscopic Analysis of the Albireo System. *Mon. Not. R. Astron. Soc.* **2019**, *502*, 328–350. [CrossRef]
17. Ferreira, H.P.A. Matryoshka: A Geometry Proposal for Multiband Frequency Selective Surfaces. Master's Thesis, Instituto Federal da Paraíba, João Pessoa, Brazil, March 2014. Available online: https://repositorio.ifpb.edu.br/handle/177683/274 (accessed on 20 June 2023). (In Portuguese)
18. Neto, A.G.; D'Assunção Junior, A.G.; Silva, J.C.; Silva, A.N.; Ferreira, H.P.A.; Lima, I.S.S. A Proposed Geometry for Multi-Resonant Frequency Selective Surfaces. In Proceedings of the European Microwave Conference, Rome, Italy, 6–9 October 2014. [CrossRef]
19. Cruz, J.N. Characterization of a FSS with Open Matryoshka Geometry. Master's Thesis, Instituto Federal da Paraíba, João Pessoa, Brazil, July 2015. Available online: https://repositorio.ifpb.edu.br/handle/177683/240 (accessed on 20 June 2023). (In Portuguese)
20. Mariano, J.G.O. Implementation of Planar Filters Based on a Matryoshka Geometry. Master's Thesis, Instituto Federal da Paraíba, João Pessoa, Brazil, March 2017. Available online: https://repositorio.ifpb.edu.br/handle/177683/298 (accessed on 20 June 2023). (In Portuguese)
21. Sousa, T.R. Development of Frequency Selective Surfaces Based on a Polarization Independent Matryoshka Geometry. Master's Thesis, Instituto Federal da Paraíba, João Pessoa, Brazil, January 2019. Available online: https://repositorio.ifpb.edu.br/handle/177683/878 (accessed on 20 June 2023). (In Portuguese)
22. Neto, J.A.S. Use of Metamaterial Resonators to Improve the Performance of Planar Filters. Master's Thesis, Federal University of Rio Grande do Norte, Natal, Brazil, December 2019. Available online: https://repositorio.ufrn.br/handle/123456789/30834 (accessed on 20 June 2023). (In Portuguese)
23. Coutinho, I.B.G. Development of Frequency Selective Surfaces with Combination of Crossed Dipole and Matryoshka Geometries. Master's Thesis, Instituto Federal da Paraíba, João Pessoa, Brazil, February 2020. Available online: https://repositorio.ifpb.edu.br/handle/177683/1319 (accessed on 20 June 2023). (In Portuguese)
24. Alencar, M.O. Development of a Bandpass FSS Based on a Matryoshka Geometry. Master's Thesis, Instituto Federal da Paraíba, João Pessoa, Brazil, April 2020. Available online: https://repositorio.ifpb.edu.br/handle/177683/998 (accessed on 20 June 2023). (In Portuguese)
25. Santos, M.G.A.R. Research on the Use of Resonators with Matryoshka Geometry in Planar Antennas. Master's Thesis, Instituto Federal da Paraíba, João Pessoa, Brazil, June 2020. Available online: https://repositorio.ifpb.edu.br/handle/177683/1776 (accessed on 20 June 2023). (In Portuguese)
26. Neto, A.F. Planar Filters Based on a Matryoshka Geometry with Rectangular and Circular Rings. Master's Thesis, Instituto Federal da Paraíba, João Pessoa, Brazil, July 2020. Available online: https://repositorio.ifpb.edu.br/handle/177683/999 (accessed on 20 June 2023). (In Portuguese)
27. Duarte, L.M.S. Reconfigurable Frequency Selective Surfaces Using PIN Diodes and RF Inductors Combining Crossed Dipole and Matryoshka Geometries. Master's Thesis, Instituto Federal da Paraíba, João Pessoa, Brazil, July 2021. Available online: https://repositorio.ifpb.edu.br/handle/177683/1819 (accessed on 20 June 2023). (In Portuguese)

28. Souto, A.H.P. Analysis of the Radiation Characteristics of Microstrip Patch Antennas in the 3.5 GHz Frequency Band Using Resonators with Matryoshka Geometry. Master's Thesis, Instituto Federal da Paraíba, João Pessoa, Brazil, March 2023. Available online: https://repositorio.ifpb.edu.br/handle/177683/3135 (accessed on 20 June 2023). (In Portuguese)
29. Abreu, F.A.T. Matryoshka Geometry Used in a DGS Sensor. Master's Thesis, Instituto Federal da Paraíba, João Pessoa, Brazil, May 2023. Available online: https://repositorio.ifpb.edu.br/handle/177683/3267 (accessed on 20 June 2023). (In Portuguese)
30. Santos, D.A. DGS Filters Based on the Matryoshka Geometry. Master's Thesis, Instituto Federal da Paraíba, João Pessoa, Brazil, May 2023. Available online: https://repositorio.ifpb.edu.br/handle/177683/3266 (accessed on 20 June 2023). (In Portuguese)
31. Albuquerque, B.L.C. Development of a Soil Humidity Sensor Using a DGS Based on the Matryoshka Geometry. Master's Thesis, Instituto Federal da Paraíba, João Pessoa, Brazil, July 2023. Available online: https://repositorio.ifpb.edu.br/handle/177683/3259 (accessed on 20 June 2023). (In Portuguese)
32. Neto, A.G.; D'Assunção Junior, A.G.; Silva, J.C.; Cruz, J.N.; Silva, J.B.O.; Ramos, N.J.P.L. Multiband Frequency Selective Surface with Open Matryoshka Elements. In Proceedings of the European Conference on Antennas and Propagation, Lisbon, Portugal, 12–17 April 2015; Available online: https://ieeexplore.ieee.org/stamp/stamp.jsp?tp=&arnumber=7228440 (accessed on 20 June 2023).
33. Neto, A.G.; Silva, J.C.; Carvalho, J.N.; Cruz, J.N.; Ferreira, H.P.A. Analysis of the Resonant Behavior of FSS Using Matryoshka Geometry. In Proceedings of the IEEE/SBMO MTT-S International Microwave and Optoelectronics Conference, Porto de Galinhas, PE, Brazil, 3–6 November 2015. [CrossRef]
34. Neto, A.G.; Costa Junior, A.G.; Moreira, C.S.; Sousa, T.R.; Coutinho, I.B.G. A New Planar Sensor Based on the Matryoshka Microstrip Resonator. In Proceedings of the IEEE/SBMO MTT-S International Microwave and Optoelectronics Conference, Águas de Lindoia, SP, Brazil, 27–30 August 2017. [CrossRef]
35. Ferreira, H.P.A.; Serres, A.J.R.; Assis, F.M.; Carvalho, J.N. A Multi-Resonant Circuit Based on Dual-Band Matryoshka Resonator for Chipless RFID Tag. In Proceedings of the European Conference on Antennas and Propagation, London, UK, 9–13 April 2018. [CrossRef]
36. Neto, A.G.; Sousa, T.R.; Silva, J.C.; Mamedes, D.F. A Polarization Independent Frequency Selective Surface Based on the Matryoshka Geometry. In Proceedings of the IEEE/MTT-S International Microwave Symposium, Philadelphia, PA, USA, 10–15 June 2018. [CrossRef]
37. Neto, A.G.; Neto, A.F.; Andrade, M.C.; Silva, J.C.; Carvalho, J.N. Stop-Band Compact Filter with Reduced Transition Region for Application in the 2.4 GHz Frequency Band. In Proceedings of the Brazilian Symposium of Telecommunications and Signal Processing, Campina Grande, PB, Brazil, 16–19 September 2018. Available online: https://biblioteca.sbrt.org.br/articles/1708 (accessed on 20 June 2023). (In Portuguese)
38. Neto, A.G.; Silva, J.C.; Coutinho, I.B.G.; Alencar, M.O.; Albuquerque, I.F.; Santos, B.L.G. Polarization Independent Triple-Band Frequency Selective Surface Based on Matryoshka Geometry. In Proceedings of the IEEE/SBMO MTT-S International Microwave and Optoelectronics Conference, Aveiro, Portugal, 10–14 November 2019. [CrossRef]
39. Neto, A.G.; Silva, J.C.; Serres, A.J.R.; Alencar, M.O.; Coutinho, I.B.G.; Evangelista, T.S. Dual-Band Pass-Band Frequency Selective Surface Based on the Matryoshka Geometry with Angular Stability and Polarization Independence. In Proceedings of the European Conference on Antennas and Propagation, Copenhagen, Denmark, 15–20 March 2020. [CrossRef]
40. Neto, A.G.; Silva, J.C.; Coutinho, I.B.G.; Alencar, D.M. Triple Stop-Band Frequency Selective Surface with Application to 2.4 GHz Band. J. Commun. Inf. Syst. 2020, 35, 77–85. [CrossRef]
41. Mamedes, D.F.; Neto, A.G.; Bornemann, J.; Silva, J.C.; Abreu, F.A.T. A Sensor Using a Matryoshka Geometry Defected Ground Structure. In Proceedings of the European Conference on Antennas and Propagation, Madrid, Spain, 27 March–1 April 2022. [CrossRef]
42. Bhope, V.; Harish, A.R. Polarization Insensitive Miniaturized Multiband FSS using Matryoshka Elements. In Proceedings of the IEEE International Symposium on Antennas and Propagation and USNC-URSI Radio Science Meeting, Denver, CO, USA, 10–15 July 2022. [CrossRef]
43. Neto, A.G.; Carvalho, J.N.; Silva, J.C.; Lima, C.G.M.; Raimundo, R.A.; Macedo, D.A. Miniaturization of DGS Filter Based on Matryoshka Geometry Using Calcium Cobaltite Ceramic. In Proceedings of the IEEE/SBMO MTT-S International Microwave and Optoelectronics Conference, Castelldefels, Barcelona, Spain, 5–9 November 2023.
44. Albuquerque, B.L.C.; Nóbrega, G.K.; Ferreira, M.S.; D'Andrea, A.; Carvalho, J.N.; Neto, A.G. A New Soil Moisture Sensor Based on Matryoshka DGS. In Proceedings of the IEEE/SBMO MTT-S International Microwave and Optoelectronics Conference, Castelldefels, Barcelona, Spain, 5–9 November 2023.
45. Silva, J.C.; Neto, A.G.; Abreu, F.A.T.; Santos, D.A.; Peixeiro, C.; Lopes, M.E.S. Comparison of Two Sensors with Matryoshka Defected Ground Structures. In Proceedings of the IEEE/SBMO MTT-S International Microwave and Optoelectronics Conference, Castelldefels, Barcelona, Spain, 5–9 November 2023.
46. Neto, A.G.; Costa, J.A.; Cavalcante, G.A.; Henrique, R.L.; Santos, L.K.L. Numerical Results for a Pass-Band Filter Based on a Matryoshka Geometry Using Three Metal Layers. In Proceedings of the Symposium on Research, Innovation and Postgraduation at Federal Institute of Paraiba, João Pessoa, PB, Brazil, 22–24 November 2023. (In Portuguese)
47. Baena, J.D.; Bonache, J.; Martín, F.; Sillero, R.M.; Falcone, F.; Lopetegi, T.; Laso, M.A.G.; García-García, J.; Gil, I.; Portillo, M.F.; et al. Equivalent-Circuit Models for Split-Ring Resonators and Complementary Split-Ring Resonators Coupled to Planar Transmission Lines. IEEE Trans. Microw. Theory Tech. 2005, 53, 1451–1461. [CrossRef]

48. Kim, C.-S.; Park, J.-S.; Ahn, D.; Lim, J.-B. A Novel 1-D Periodic Defected Ground Structure for Planar Circuits. *IEEE Microw. Guid. Wave Lett.* **2000**, *10*, 131–133. [CrossRef]
49. Hong, J.-S.; Lancaster, M.J. *Microstrip Filters for RF/Microwave Applications*; John Wiley & Sons: New York, NY, USA, 2001.
50. Available online: https://www.ansys.com/products/electronics/ansys-hfss (accessed on 1 October 2023).
51. Munk, B.A. *Frequency Selective Surfaces: Theory and Design*; John Wiley & Sons: New York, NY, USA, 2000.
52. Vardaxoglou, J.C. *Frequency Selective Surfaces: Analysis and Design*; Research Studies Press: Taunton, UK, 1997.
53. Panwar, R.; Lee, J.R. Progress in frequency selective surface-based smart electromagnetic structures: A critical review. *Aerosp. Sci. Technol.* **2017**, *66*, 216–234. [CrossRef]
54. Bohra, H.; Prajapati, G. Microstrip filters: A review of different filter designs used in ultrawide band technology. *Makara J. Technol.* **2020**, *24*, 79–86. [CrossRef]
55. Fu, W.; Li, Z.-M.; Cheng, J.-W.; Qiu, X. A Review of Microwave Filter Designs Based on CMRC. In Proceedings of the IEEE MTT-S International Wireless Symposium (IWS), Shanghai, China, 20–23 September 2020. [CrossRef]
56. Edwards, T.C.; Steer, M.B. *Foundations of Interconnected and Microstrip Design*, 3rd ed.; John Wiley& Sons: Chichester, UK, 2000.
57. Neto, A.G.; Carvalho, J.N.; Mariano, J.G.O.; Sousa, T.R. Analysis of the application of matryoshka geometry to microstrip filters. In Proceedings of the Brazilian Congress of Electromagnetism and Brazilian Symposium on Microwaves and Optoelectronics (MOMAG), Porto Alegre, RS, Brazil, 25–29 July 2016. (In Portuguese)
58. Pandit, S.; Ray, P.; Mohan, A. Compact MIMO antenna enabled by DGS for WLAN applications. In Proceedings of the IEEE International Symposium on Antennas and Propagation & USNC/URSI National Radio Science Meeting, Boston, MA, USA, 8–13 July 2018. [CrossRef]
59. Souza, F.A.A.; Campos, A.L.P.S.; Neto, A.G.; Serres, A.J.R.; Albuquerque, C.C.R. Higher Order Mode Attenuation in Microstrip Patch Antenna with DGS H Filter Specification from 5 to 10 GHz Range. *J. Microw. Optoelectron. Electromagn. Appl.* **2020**, *19*, 214–227. [CrossRef]
60. Zhou, J.J.; Rao, Y.; Yang, D.; Qian, H.J.; Luo, X. Compact wideband BPF with wide stopband using substrate integrated defected ground structure. *IEEE Microw. Wirel. Compon. Lett.* **2021**, *31*, 353–356. [CrossRef]
61. Cao, S.; Han, Y.; Chen, H.; Li, J. An Ultra-Wide Stop-Band LPF Using Asymmetric Pi-Shaped Koch Fractal DGS. *IEEE Access* **2017**, *5*, 27126–27131. [CrossRef]
62. Farooq, M.; Abdullah, A.; Zakir, M.A.; Cheema, H.M. Miniaturization of a 3-way power divider using defected ground structures. In Proceedings of the IEEE Asia-Pacific Microwave Conference (APMC), Singapore, 10–13 December 2019. [CrossRef]
63. Oh, S.; Koo, J.-J.; Hwang, M.S.; Park, C.; Jeong, Y.-C.; Lim, J.-S.; Choi, K.-S.; Ahn, D. An unequal Wilkinson power divider with variable dividing ratio. In Proceedings of the IEEE/MTT-S International Microwave Symposium, Honolulu, HI, USA, 3–8 June 2007. [CrossRef]
64. Mansour, E.; Allam, A.; Abdel-Rahman, A.B. A novel approach to non-invasive blood glucose sensing based on a defected ground structure. In Proceedings of the European Conference on Antennas and Propagation (EuCAP), Dusseldorf, Germany, 22–26 March 2021. [CrossRef]
65. Oliveira, J.G.D.; Duarte Junior, J.G.; Pinto, E.N.M.G.; Neto, V.P.S.; D'Assunção, A.G. A new planar microwave sensor for building materials complex permittivity characterization. *Sensors* **2020**, *20*, 6328. [CrossRef] [PubMed]
66. Dautov, K.; Hashmi, M.; Nauryzbayev, G.; Nasimuddin, N. Recent advancements in defected ground structure-based near-field wireless power transfer systems. *IEEE Access* **2020**, *8*, 81298–81309. [CrossRef]
67. Hekal, S.; Abdel-Rahman, A.B.; Jia, H.; Allam, A.; Barakat, A.; Pokharel, R.K. A novel technique for compact size wireless power transfer applications using defected ground structures. *IEEE Trans. Microw. Theory Tech.* **2017**, *65*, 591–599. [CrossRef]
68. Neto, A.G.; Silva, J.C.; Coutinho, I.B.G.; Camilo Filho, S.S.; Santos, D.A.; Albuquerque, B.L.C. A Defected Ground Structure Based on Matryoshka Geometry. *J. Microw. Optoelectron. Electromagn. Appl.* **2022**, *21*, 284–293. [CrossRef]
69. Peixeiro, C. Microstrip Antenna Papers in the IEEE Transactions on Antennas and Propagation. *IEEE Antennas Propag. Mag.* **2012**, *54*, 264–268. [CrossRef]
70. Arya, A.K.; Kartikeyan, M.V.; Patnaik, A. Defected Ground Structure in the Perspective of Microstrip Antennas: A Review. *Frequenz* **2010**, *64*, 79–84. [CrossRef]
71. Khandelwal, M.K.; Kanaujia, B.K.; Kumar, S. Defected Ground Structures: Fundamentals, Analysis, and Applications in Modern Wireless Trends. *Int. J. Antennas Propag.* **2017**, *2017*, 2018527. [CrossRef]
72. Sung, Y.J.; Kim, S.; Kim, Y.-S. Harmonics Reduction with Defected Ground Structure for a Microstrip Patch Antenna. *IEEE Antennas Wirel. Propag. Lett.* **2003**, *2*, 111–113. [CrossRef]
73. Mishra, B.; Singh, V.; Singh, R.K.; Singh, N.; Singh, R. A Compact UWB Patch Antenna with Defected Ground for Ku/K Band Applications. *Microw. Opt. Technol. Lett.* **2017**, *60*, 1–6. [CrossRef]
74. Guha, D.; Biswas, M.; Antar, Y.M.M. Microstrip Patch Antenna with Defected Ground Structure for Cross Polarization Supression. *IEEE Antennas Wirel. Propag. Lett.* **2005**, *4*, 455–458. [CrossRef]
75. Dwivedy, B. Consideration of Engineered Ground Planes for Planar Antennas: Their Effects and Applications, a Review. *IEEE Access* **2022**, *10*, 84317–84329. [CrossRef]
76. Pendry, J.B.; Holden, A.J.; Robbins, D.J.; Stewart, W.J. Magnetism from Conductors and Enhanced Nonlinear Phenomena. *IEEE Trans. Microw. Theory Tech.* **1999**, *47*, 2075–2084. [CrossRef]
77. Balanis, C.A. *Antenna Theory: Analysis and Design*, 4th ed.; John Wiley & Sons: New York, NY, USA, 2015; pp. 788–814.

78. Available online: https://www.ansys.com/ (accessed on 1 October 2023).
79. Fraden, J. *Handbook of Modern Sensors: Physics, Designs, and Applications*, 4th ed.; Springer: New York, NY, USA, 2010.
80. Patricio, M.C.Z.B.; Santos, P.S.; Silva, J.C.; Carvalho, J.N.; Neto, A.G. Determination of Sucrose Dilution Level in Aqueous Solution Using a DGS Sensor Based on Matryoshka Geometry. In Proceedings of the Brazilian Symposium of Microwaves and Optoelectronics (SBMO), Natal, RN, Brazil, 13–16 November 2022. (In Portuguese)
81. Zeng, C.; Liu, L.; Xin, X.; Zhao, W.; Lin, C.; Liu, W.; Leng, J. 4D printed bio-inspired mesh composite materials with high stretchability and reconfigurability. *Compos. Sci. Technol.* **2024**, *249*, 110503. [CrossRef]

Disclaimer/Publisher's Note: The statements, opinions and data contained in all publications are solely those of the individual author(s) and contributor(s) and not of MDPI and/or the editor(s). MDPI and/or the editor(s) disclaim responsibility for any injury to people or property resulting from any ideas, methods, instructions or products referred to in the content.

Article

Design of 2.45 GHz High-Efficiency Rectifying Circuit for Wireless RF Energy Collection System

Yanhu Huang [1], Jiajun Liang [1,2,*], Zhao Wu [1] and Qian Chen [1]

1. Guangxi University Key Lab of Complex System Optimization and Big Data Processing, Guangxi Applied Mathematics Center, Yulin Normal University, Yulin 537000, China; yanhuhuang@ylu.edu.cn (Y.H.); kianty@ylu.edu.cn (Z.W.); 19899218213@163.com (Q.C.)
2. School of Electronic and Information Engineering, South China University of Technology, Guangzhou 510641, China
* Correspondence: jiajunliang@ylu.edu.cn

Abstract: A 2.45 GHz high-efficiency rectifying circuit for a wireless radiofrequency (RF) energy collection system is proposed. The RF energy collection system is composed of a transmitting antenna, a receiving antenna, a rectifying circuit and a load. The designed receiving antenna is a kind of dual-polarised cross-dipole antenna; its bandwidth is 2.3–2.5 GHz and gain is 7.97 dBi. The proposed rectifying circuit adopts the technology of an output matching network, which can suppress the high-harmonic components. When the input power at 2.45 GHz is 13 dBm and the load is 2 kΩ, the highest conversion efficiency of RF-DC is 74.8%, and the corresponding maximum DC output voltage is 4.92 V. The experiment results are in good agreement with the simulation results, which shows a good application prospect.

Keywords: energy collection; rectifying circuit; receiving antenna; high efficiency

Citation: Huang, Y.; Liang, J.; Wu, Z.; Chen, Q. Design of 2.45 GHz High-Efficiency Rectifying Circuit for Wireless RF Energy Collection System. *Micromachines* **2024**, *15*, 340. https://doi.org/10.3390/mi15030340

Academic Editors: Piero Malcovati and Haejun Chung

Received: 5 December 2023
Revised: 27 February 2024
Accepted: 27 February 2024
Published: 29 February 2024

Copyright: © 2024 by the authors. Licensee MDPI, Basel, Switzerland. This article is an open access article distributed under the terms and conditions of the Creative Commons Attribution (CC BY) license (https://creativecommons.org/licenses/by/4.0/).

1. Introduction

With the rapid development of wireless communication technology, Wireless-Fidelity (WiFi), Bluetooth, the Internet of Things (IoT) and other communication technologies have entered into most areas of people's lives, and these mature technologies have brought a lot of convenience to people's lives. Currently, 5G mobile communication is being rapidly developed and promoted, and the era of the Internet of Things is coming; and the future 6G communication is also in full swing in academic research. This means that in the 5G and 6G era, the electromagnetic environment of people's living space will become more and more complex, and electromagnetic waves will be more densely distributed in all corners. Electromagnetic waves are both carriers of information transmissions and energy at the same time. How to effectively collect and utilize the electromagnetic waves in the environment are effective considerations for alleviating the problem of energy shortage, and it is also the development trend of the new institutional communication technology for the future. Microwave energy transmission and collection technology has been a research hotspot in recent years; short-distance wireless energy transmission technology has been commercially applied in the fields of new energy vehicles, cell phones, laptops, etc., while medium- and long-distance wireless energy transmission technology has not been popularized due to the existence of a number of unresolved technical problems. Wireless energy transmission technology is mainly composed of four components: a microwave transmitter, microwave transmitter antenna, microwave receiver antenna and microwave rectifier circuit, which involve key technologies including high-efficiency DC-RF conversion technology, a high-power transmitter antenna, a high-efficiency receiver antenna and a high-efficiency receiver rectifier circuit.

With the rapid development of information technology, consumer electronic products are widely used in daily life. In order to maintain the continuous and uninterrupted operation of these devices, a continuous energy supply must be provided. The conventional battery power supply method has the problems of a short service life and the high cost of battery replacement. Based on this, wireless RF energy collection technology is considered to be a better alternative method for the energy supply [1].

A wireless RF energy collection system is the form of collecting RF energy at different frequencies in the natural environment through a receiving antenna first, and then converting into DC energy through a rectifying circuit. As a new energy collection technology, it has great application value in the fields of low-power wireless sensor networks [2], biomedical sensors [3,4], Internet of Things mobile terminal devices [5–7], intelligent wearable devices [8–12], and so on. The traditional wireless RF energy collection system mainly includes an RF signal source, transmitting antenna, receiving antenna and rectifying circuit. The basic composition of the rectifying circuit mainly includes an input filter, a rectifying diode, a pass-through filter and a DC load. The ratio of the DC power output of the rectifying circuit to the received RF signal power is called the rectification efficiency or the conversion efficiency, which is the most important index in evaluating the performance of a wireless RF energy collection system.

In the rectifying circuit of the RF energy collection system, due to the nonlinear characteristics of the rectifying diode, high-harmonic components are generated, which will flow back to the antenna for radiation again, resulting in energy loss. In the traditional rectifying circuit, the low-pass filter is used to suppress the high-harmonic components generated by the nonlinear rectifying diode to improve the conversion efficiency of the rectifying circuit. A broadband matching network is used in some wide input power rectifying circuits to increase the conversion efficiency [13–15].

According to existing research results, with an input power greater than 15 dBm, a rectification efficiency of more than 50% can be obtained only in a relatively narrow bandwidth range. Once the optimal maximum input power value is exceeded, the rectification efficiency will decrease sharply [16,17]. In the actual electromagnetic environment, the wireless microwave signal is dynamic change. Therefore, when the power of the RF signal is received by the receiving antenna changes, the rectification circuit should also be required to maintain a stable conversion efficiency. In [18], a miniaturized RF energy collector for an Internet of Things (IoT) application was proposed. Considering the insensitivity to the ambient RF energy and the low-profile design, the proposed RF energy collector uses a two-layer printed circuit board (PCB) substrate consisting of an orthogonally deployed antenna with an LC balancer, impedance matching and a Dickson charge pump circuit. The double-polarized antenna adopts the bow-typed dipole antenna. The conversion efficiency of the rectifying circuit is up to 71.4% from any incident ambient RF signal. In [19], in order to solve the problem of a low and non-uniform electromagnetic energy density in the environment, a triband monopole rectifying antenna was designed. A single diode rectifier was selected to operate efficiently over a wide range of input power levels from −10 dBm to 5 dBm and provide a minimum conversion efficiency of 60%. In [20], a broadband circularly polarized rectifying antenna was proposed, and a wideband rectifier with branch coupler was designed, with a bandwidth of 1.7 to 2.6 GHz in a wide operating power and a wide output load range. The RF–DC conversion efficiency at the ambient wireless power level was effectively improved. In this rectifier, the peak efficiency was 69% at 12 dBm (at 1.85GHz) and 64% at 10 dBm (at 2.45 GHz).

There have been many related reports in the field of broadband high-efficiency rectifier circuit technology; however, the difficulty between the operating bandwidth and conversion efficiency has still not been completely solved. In [21], a single-layer coplanar waveguide wideband rectifier circuit was described, which uses a voltage multiplier tube connected in series with a wideband matching network, and in order to realize miniaturization, the wideband matching consists of a series dual-inductor lumped element that finally achieves a conversion efficiency of more than 45% in the frequency range of 0.1–2.5 GHz; although the method obtains a wide operating bandwidth, the in-band conversion efficiency is not high. In [22], a novel three-stage impedance matching technique was used to realize a compact broadband high-efficiency rectifier circuit. The technique first utilizes a linearly tapered transmission line to achieve impedance modulation for different input power levels, and then further uses a second-order circular impedance matching branch to achieve the final broadband impedance matching. Ultimately, the technique achieves a conversion efficiency higher than 50% at an input power of 10 dBm and a frequency range of 0.97–2.55 GHz. A compact broadband high-efficiency energy harvesting rectifier has been proposed in the literature [23]. In this structure, a high conversion efficiency over a wide frequency range was achieved by designing a compact broadband low-loss matching network and a direct current filter with a wide stopband. For the broadband low-loss matching network, low loss and compact dimensions were achieved by strategically selecting and designing three transmission line segments. The final rectifier bandwidth obtained was 41.5% (2.0–3.05 GHz), and the RF–DC power conversion efficiency was higher than 70% at an input power of 10 dBm; when the input power decreases to 0 dBm, the measured efficiency stays above 45% in the range of 1.9–3.05 GHz, and at an input power of 14 dBm, the maximum measured efficiency is 75.8%. In [24], a circularly polarized rectifier antenna that has the advantage of high efficiency over a wide range of input power and frequency was presented. The antenna consists of an efficient rectifier and a broadband circularly polarized antenna. In this rectifier, the matching performance of the circuit over a wide input power range and frequency range was greatly improved by introducing a new broadband impedance compression technique. Simulation and measurement results showed that the conversion efficiency reached more than 60% (up to 76%) in the input power range of 5–17 dBm and the frequency band of 1.7–2.9 GHz (mobile, Wi-Fi, and ISM bands), which is a significant improvement in the performance of this rectifier circuit in the two metrics of bandwidth and efficiency. In summary, although there have been scholars trying to overcome the bottleneck between the operating bandwidth and conversion efficiency of rectifier circuits, the existing techniques and methods have not been able to completely solve the problem.

In order to improve the receiving efficiency and obtain a relatively high conversion efficiency in the wide input power, a wireless RF energy collection system working at 2.45 GHz was studied and designed in this study. The receiving antenna is a kind of a dual-polarized cross-dipole antenna. In order to facilitate the integration with the receiving antenna and obtain a higher conversion efficiency, the proposed rectifying circuit uses the form of diode voltage doubling and lumped elements to build the matching filter network, so as to suppress the harmonic energy and improve the rectification efficiency. Finally, the whole system was processed and tested. The test results show that the rectification circuit has a good harmonic suppression effect. At the input power of 13 dBm, the maximum conversion efficiency of the rectifier antenna reaches 74.8%, and the conversion efficiency is larger than 40% in the input range of 0–18 dBm.

2. Rectifying Circuit Simulation and Design

Figure 1 is a schematic of the proposed wireless RF energy collection system; it mainly includes a receiving antenna and rectifying circuit. The basic composition of the rectifying circuit mainly includes an input matching network, rectifying diode (rectifier), output matching network and load. The two series microstrip branches, T1 and T2, and one parallel shorter branch, T3, constitute the H-typed input matching network. As seen, T1 is a 50 Ω microstrip line connected to the receiving antenna, and T2 is connected to the ground. The input impedance can be optimized by changing the sizes of the branches. A block capacitance C_1 is used to isolate the DC signal between the rectifying diode and the input port. Meanwhile, only the high-frequency signals are allowed to pass through the rectifying diode. Rectifying diode HSMS-286C is used in this circuit. HSMS-286C is composed of two HSMS-2860 diodes inside the same package, which are electrically connected in parallel. This kind of rectifying diode can achieve twice the rectification effect compared to a single diode.

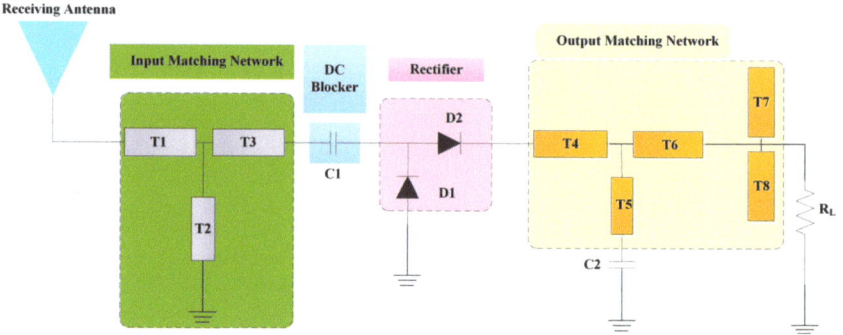

Figure 1. A schematic of the proposed wireless RF energy collection system.

An output matching network is used at the end of the circuit, which consists of an output matching circuit and a bypass capacitance C_2. The output matching network consists of two series microstrip lines, T4 and T6, a parallel short-circuited microstrip line, T5, which is connected in series with capacitor C_2, and two parallelly connected open-circuited microstrip lines, T7 and T8, at the end. By reasonably adjusting the sizes of these microstrip lines, a better matching can be obtained at a frequency of 2.45 GHz. The optimal value of C_1 is 33 pF, C_2 is 22 pF, and the load R is 2000 Ω.

Figure 2 shows the layout of the proposed rectifying circuit. Simulation software ADS2021 was used to simulate the circuit model. The printed circuit board selected for the rectifier circuit was Rogers 4350b, with a dielectric constant of 3.48, a loss angle of 0.017 and a thickness of 1.524 mm. The optimized parameters of the proposed rectifying circuit are shown in Table 1.

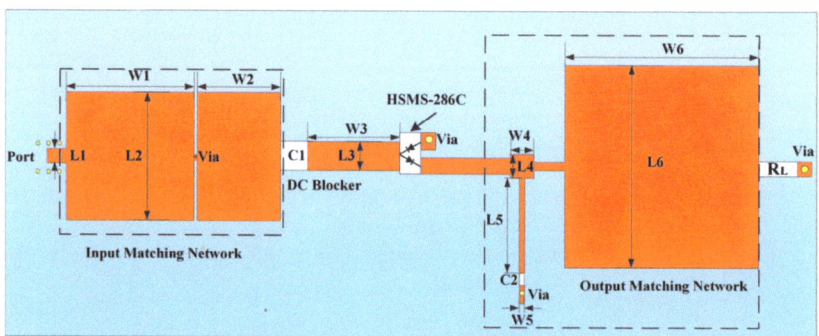

Figure 2. The layout of the proposed rectifying circuit.

Table 1. Optimized parameters of the proposed rectifying circuit.

Parameter	Unit	Parameter	Unit
L1	3 mm	W1	20 mm
L2	20 mm	W2	13 mm
L3	4.5 mm	W3	16 mm
L4	3.7 mm	W4	3.7 mm
L5	14.8 mm	W5	0.8 mm
L6	30.1 mm	W6	30.7 mm

As seen in Figure 3, the simulated conversion efficiency is changed significantly with the load. When the input power P_{in} is 13 dBm and the load R_L is 1000 Ω, the RF-to-DC conversion efficiency is 30%; under the same input power of 13 dBm with a load R of

2000 Ω, the RF-to-DC conversion efficiency is 74.8%. Therefore, it can be obtained that the output DC power of the rectifying circuit mainly depends on the output voltage and the output load, and the output load of the rectifying circuit is one of the key parameters affecting the overall conversion efficiency. The load can be adjusted to increase and improve the conversion efficiency within the predefined input power range.

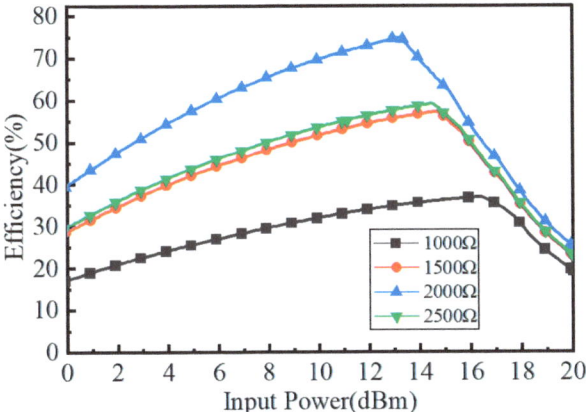

Figure 3. Rectifier efficiency versus input RF power P_{in}.

As can be seen from Figure 4, the simulated output voltage is changed with the input power. For a different load R, the output DC voltage is different. As can be seen, the output DC voltage is increasing from 0.5 V to 6.3 V when the input power increases from 0 dB to 20 dB. When the input power P_{in} is 13 dBm and the load R is 2000 Ω, the maximum output voltage is 4.92 V.

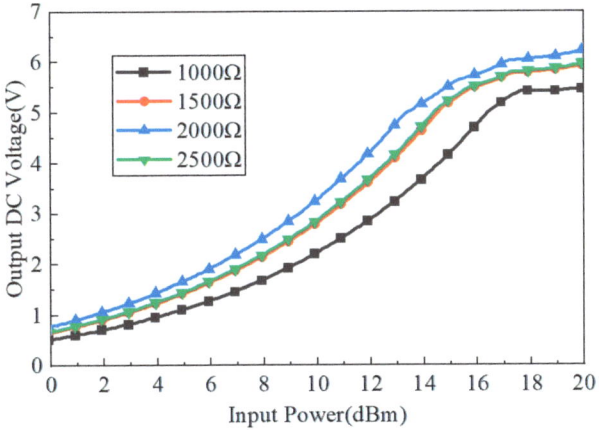

Figure 4. Output DC voltage versus input RF power P_{in}.

In the previous rectification antenna system, the conversion efficiency is closely related to the load and the input power, and the conversion efficiency is not the same for different loads. As seen from Figure 5, for a low input power, such as −5 dBm, the conversion efficiency increases nearly linearly as the input power increases. For an input power larger than 0 dBm, the conversion efficiency increase first with the input power, and then the conversion efficiency decreases; each input power corresponds to a load value with a maximum conversion efficiency. The proposed rectifying circuit can maintain a high rectification efficiency of more than 60% over the load resistance range from 1500 Ω to 2000 Ω. Figure 6 shows the output voltage under different load resistances and input powers. As seen from Figure 6, for a low input power, such as −5 dBm, the output voltage of the circuit changes very slightly, even though the load increases. When the input power increases, the output voltage of the circuit increases with it. For an input power larger than 0 dBm, the output voltage increases first with the load, and then the output voltage decreases; each input power corresponds to a load value with a maximum output voltage. As can be seen, the maximum output voltage occurs at 2000 Ω when the input power is 15 dBm.

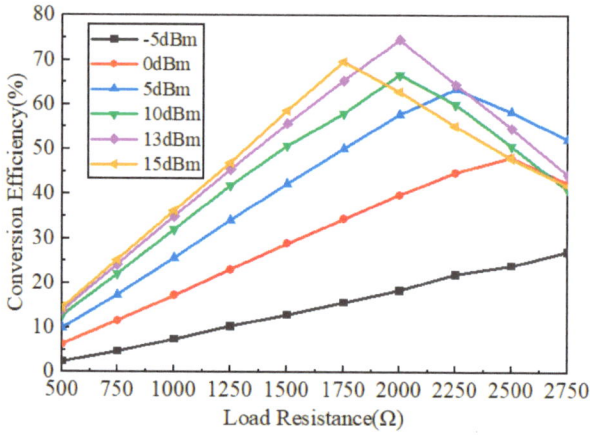

Figure 5. Conversion efficiency versus load resistance.

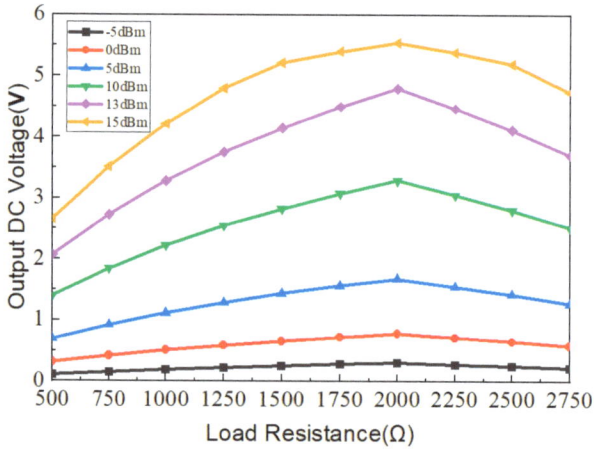

Figure 6. Output DC voltage versus load resistance.

As can be seen from Figure 7, by adding the output matching network, the input impedance can be adjusted. With the rectifying circuit with an output matching network, the measured results show an S_{11} less than −10 dB when the input power crosses from −9.6 dBm to −21.5 dBm. With the rectifying circuit without an output matching network, the measured S_{11} is less than −10 dB, using only the input power crossing from −14.8 dBm to −22.4 dBm. It is verified that the output matching network could expand the action operating range of the input power effectively.

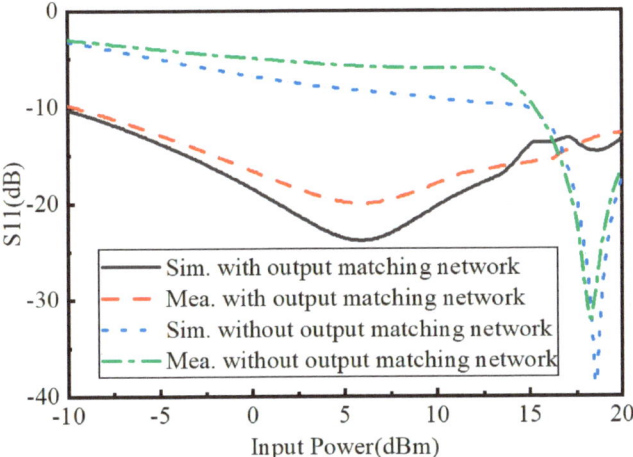

Figure 7. Measured and simulated S_{11} of the rectifiers with/without an output matching network versus input power.

3. Antenna Simulation and Design

To verify the proposed rectifying circuit, a broadband dual-polarized cross-dipole antenna with a vertical axial radiation pattern was designed. Figure 8 shows a 3D view of the proposed antenna. There are four identical square metal rings used as the radiators, where two square metal rings in the diagonal direction act as a dipole antenna. The double-linear polarization properties are obtained by orthogonally placing these two dipole antennas. The dipole antenna is fed by coaxial cable; the inner and external conductors of the coaxial line are connected to the two square metal rings (soldered via air bridge line), respectively. The coaxial feeder method guides the electromagnetic energy in the horizontal axial direction, and the maximum radiation pattern is pointed toward the vertical axial direction. Furthermore, for the dipole antenna located above the metal reflection ground, the vertical axial radiation is strengthened. The parameters of the proposed antenna are as follows: $H = 32$ mm, $D = 80$ mm, $a = 24$ mm, $b = 2$ mm, $s = 1.4$ mm, $c_1 = 2$ mm and $c_2 = 2$ mm.

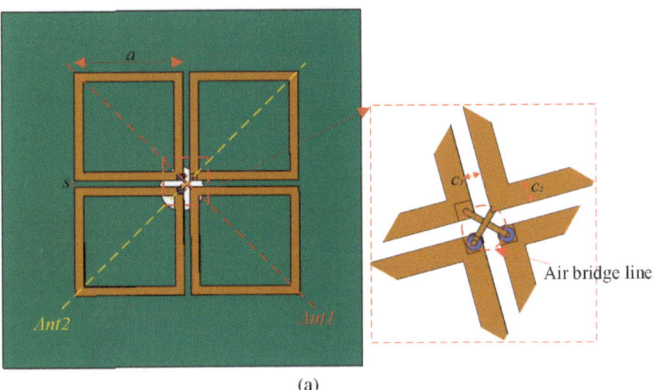

(a)

Figure 8. *Cont.*

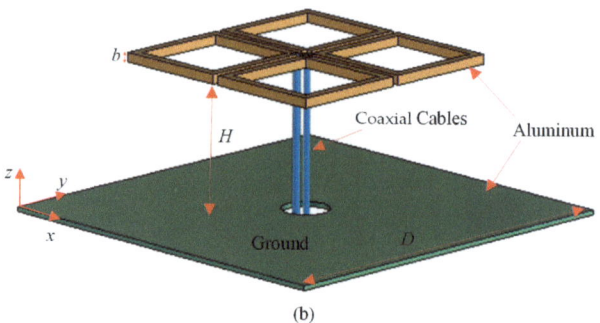

(b)

Figure 8. A 3D view of the proposed dual-polarized antenna. (**a**) Top view; (**b**) Side view.

Figure 9 shows the simulated and measured S-parameters of the proposed antenna; the relative impedance bandwidth ($S_{11} \leq -10$ dB) covers from 2.3 to 2.9 GHz. Figure 10 shows the 3D simulated radiation patterns of the proposed antenna. As can be seen from Figure 10, the two polarized antenna patterns are mutually orthogonal; the proposed antenna has a stable directional pattern, which is suitable for point-to-point wireless power transmission. Figure 11 shows the simulated and measured 2D radiation patterns at 2.45 GHz, and the simulation and test results are relatively similar. Figure 12 shows the simulated and measured 2D radiation patterns of the proposed antenna; both of the polarized antennas have a gain of about 7.97 dBi.

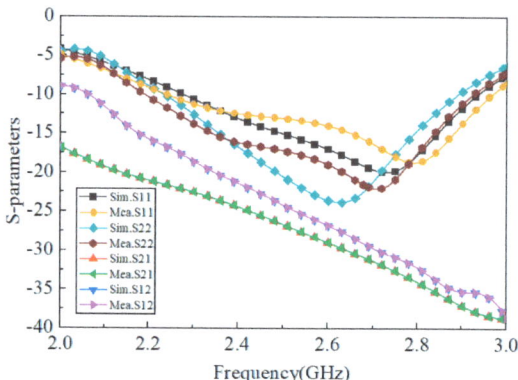

Figure 9. Simulated and measured S-parameters of the proposed antenna.

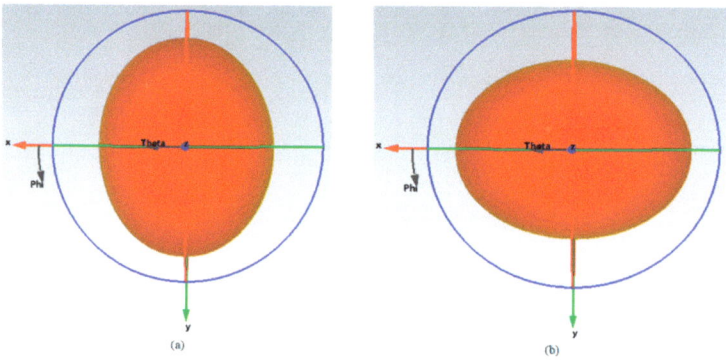

Figure 10. Simulated 3D radiation patterns of the proposed antenna. (**a**) Port 1 excited; (**b**) port 2 excited.

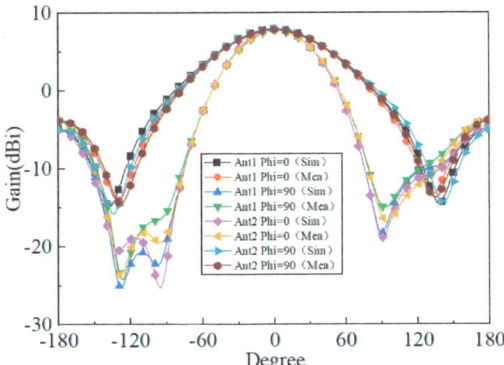

Figure 11. Simulated and measured 2D radiation patterns of the proposed antenna.

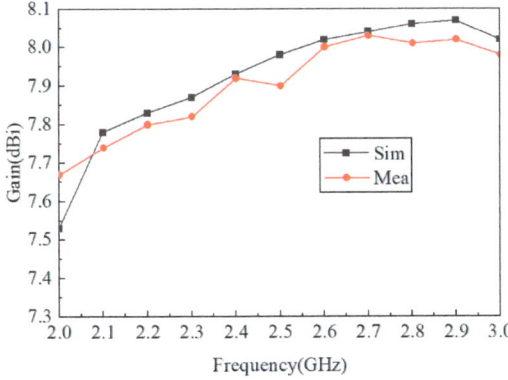

Figure 12. Simulated and measured gain of the proposed antenna.

4. Experiment Results

Figure 13 shows the fabricated rectifying antenna system and the antenna test setup in an anechoic chamber. The antenna radiation patterns were measured using the far-field test method The rectifying circuit was fabricated on a PCB Rogers 4350B with a dielectric constant of 3.48 and loss tangent of 0.017. The whole size of the rectifying circuit is 110 mm × 34 mm × 1.524 mm; the circuit port is connected with an SMA connector.

The dual-polarized cross-dipole antenna was mainly made of aluminium. Two rigid coaxial cables were used to feed the antenna, each antenna port was wired to connect a rectifier circuit. Four plastic clips were used to fix the adjacent square metal rings to avoid shaking. The dual-polarized cross-dipole antenna was connected by two rectifying circuits through its two antenna ports; the two branches of the rectifying circuit are named RA1 and RA2, respectively.

The wireless RF energy collection system experiment was conducted indoors; a measurement schematic and measurement setup are shown in Figure 14. The transmitting link consisted of a signal generator and transmitting horn antenna; the receiving link consisted of a dual-polarized cross-dipole antenna (receiving antenna), rectifying circuit and load.

Rectifying efficiency is one of the most important parameters for a rectifying circuit, which means the ability to convert the RF energy into DC power. The rectifying efficiency η can be calculated using the following equations:

$$\eta = \frac{P_{out}}{P_{in}} = \frac{V_{out}^2 / R_L}{P_{in}} \times 100\% \tag{1}$$

$$P_{in} = P_t \left(\frac{\lambda}{4\pi R}\right)^2 \mu G_r G_t \tag{2}$$

where P_{out} is the DC output power, P_{in} is the input RF power, V_{out} is the output voltage, R_L is the load of the rectifying circuit, μ is the polarization loss factor, G_t and G_r are the gains of the transmitting and receiving antennas, and λ is the free space wavelength of the operating frequency. In the actual test environment, the GIGOL DSG836A was used as the microwave source, the transmitting antenna used the standard gain horn, and the wireless RF energy collection system was placed in the far field of the horn antenna. The DC voltage output by the rectifying circuit in the wireless collecting system was measured on both ends of the resistive load (2000 Ω).

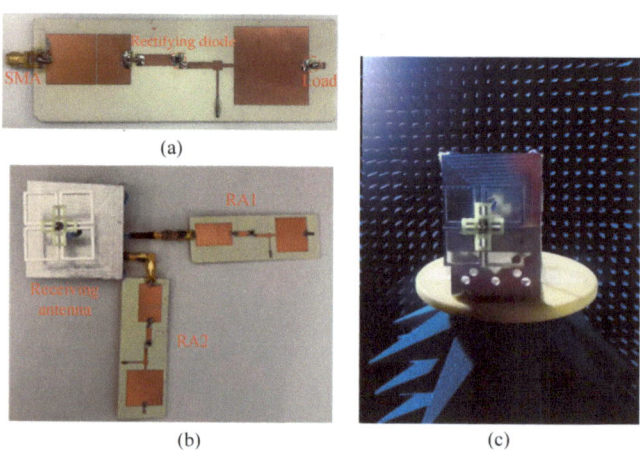

Figure 13. (a) Fabricated rectifying circuit. (b) Fabricated rectifying circuit connected with a dual-polarized antenna. (c) Antenna measurement setup.

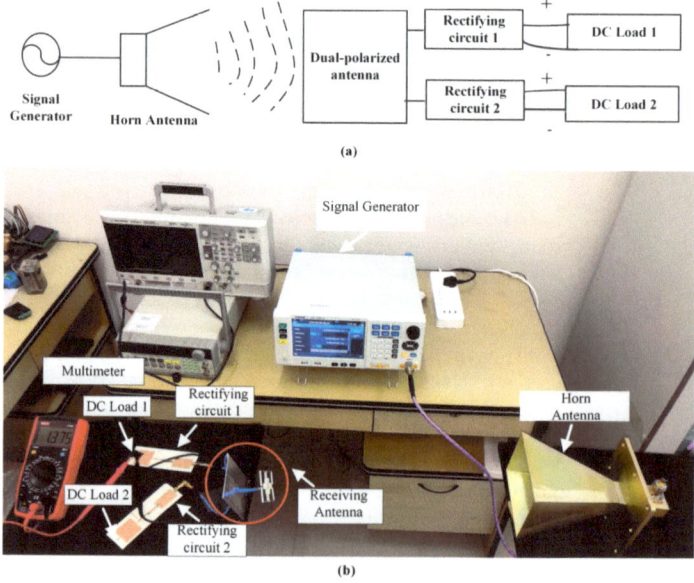

Figure 14. (a) Test schematic. (b) Measurement setup.

Figure 15 shows the simulated and measured efficiency of the rectifying circuit. In the simulation, the maximum efficiency reached 74.8% with the output matching network, when the input power was 13 dBm; the efficiency reached 74.9% without the output matching network, when the input power was 14.5 dBm. In the test, the efficiency of rectifier circuit 1 reached 68.03% with the output matching network, when the power was 13.6 dBm; the efficiency of rectifier circuit 1 reached 67.4% without the output matching network, when the power was 14.4 dBm. The efficiency of rectifier circuit 2 reached 57.2% with the output matching network, when the power was 14.5 dBm; the efficiency of rectifier circuit 2 reached 56% without the output matching network, when the power was 16 dBm.

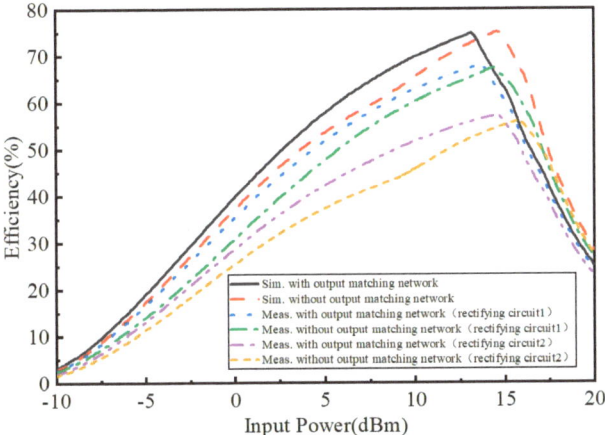

Figure 15. Simulated and measured efficiency of the rectifying circuit.

In Table 2, the effect of the rectifier circuit designed in this study is compared with those of rectifier circuits published in the literature. Compared with reference [9], this work has the advantages of a higher rectifying efficiency and lower input power. Compared with reference [10], this work has the advantages of a lower input power and smaller size. Compared with reference [15], this work has the advantage of a lower input power. Compared with reference [19], this work has the advantage of a higher rectifying efficiency. In brief, it can be seen that the designed 2.45 GHz rectifier circuit shows a high conversion efficiency.

Table 2. Rectifier performance comparison.

Ref.	Freq. (GHz)	S_{11} (dB)	Input Power (dBm)	Eff. (%)	Size of Rectenna (mm^2)
[9]	2.45	−10	20	53.56	/
[10]	2.45	−24	20	/	160 × 130
[15]	2.45	−21	22	75.7	60 × 60
[19]	1.8/2.1/2.45	−24	0	63.1	80.4 × 128.9
This work	2.45	−10.5	13	74.8	120 × 40

5. Conclusions

A 2.45 GHz high-efficiency rectifying circuit for a wireless RF energy collection system was proposed. The RF energy collection system is composed of a transmitting antenna, a receiving antenna, a rectifying circuit and a load. The designed receiving antenna is a kind of dual-polarized cross-dipole antenna; its bandwidth is 2.3–2.5 GHz and gain is 7.97 dBi. The proposed rectifying circuit adopts the technology of an output matching network, which can suppress the high-harmonic components. When the input power of 2.45 GHz is 13 dBm, and the load is 2000 Ω; the highest conversion efficiency of RF–DC is 74.8%, and the corresponding maximum DC output voltage is 4.92 V. The experiment results are in good agreement with the simulation results, which shows a good application prospect.

Author Contributions: Conceptualization, J.L.; methodology, Y.H.; validation, J.L. and Q.C.; writing—original draft preparation, Y.H. and J.L.; writing—review and editing, Z.W.; supervision, J.L. All authors have read and agreed to the published version of the manuscript.

Funding: This work was partially supported by the basic ability improvement project of young and middle-aged teachers in Guangxi Universities (2022KY0579); in part by the Guangxi Natural Science Foundation (2021GXNSFBA220003, 2021GXNSFAA075031) and in part by the Yulin Normal University College Students innovation and entrepreneurship Training Program project (No. 202310606020, No. S202210606109).

Data Availability Statement: Data are contained within the article.

Conflicts of Interest: The authors declare no conflicts of interest.

References

1. Reddy, M.V.; Hemanth, K.S.; Mohan, C.V. Microwave Power Transmission—A Next Generation Power Transmission System. *IOSR J. Electr. Electron. Eng.* **2013**, *4*, 24–28. [CrossRef]
2. Sogorb, T.; Llario, J.V.; Pelegri, J.; Lajara, R.; Alberola, J. Studying the Feasibility of Energy Harvesting from Broadcast RF Station for WSN. In Proceedings of the IEEE Instrumentation and Measurement Technology Conference, Victoria, BC, Canada, 12–15 May 2008; pp. 1360–1363.
3. Kiourti, A.; Nikita, K.S. A review of implantable patch antennas for biomedical telemetry: Challenges and solutions. *IEEE Antennas Wirel. Propag. Mag.* **2012**, *54*, 210–228. [CrossRef]
4. Huang, F.J.; Lee, C.M.; Chang, C.L.; Chen, L.K.; Yo, T.C.; Luo, C.H. Rectenna Application of Miniaturized Implantable Antenna Design for Triple-Band Biotelemetry Communication. *IEEE Trans. Antennas Propag.* **2011**, *59*, 2646–2653. [CrossRef]
5. Shafique, K.; Khawaja, B.A.; Khurram, M.D.; Sibtain, S.M.; Siddiqui, Y.; Mustaqim, M.; Chattha, H.T.; Yang, X. Energy harvesting using a low-cost rectenna for internet of things (IoT) applications. *IEEE Access* **2018**, *6*, 30932–30941. [CrossRef]
6. Jabbar, H.; Song, Y.S.; Jeong, T.T. RF energy harvesting system and circuits for charging of mobile devices. *IEEE Trans. Consum. Electron.* **2010**, *56*, 247–253. [CrossRef]
7. Zeadally, S.; Shaikh, F.K.; Talpur, A.; Sheng, Q.Z. Design architectures for energy harvesting in the Internet of Things. *Renew. Sustain. Energy Rev.* **2020**, *128*, 109901. [CrossRef]
8. Hyewon, L.; Jung-Sim, R. Wearable electromagnetic energy-harvesting textiles based on human walking. *Text. Res. J.* **2018**, *89*, 2532–2541. [CrossRef]
9. Shawalil, S.; Rani, K.N.A.; Rahim, H.A. 2.45 GHZ wearable rectenna array design for microwave energy harvesting. *Indones. J. Electr. Eng. Comput. Sci.* **2019**, *14*, 677–687. [CrossRef]
10. Chi, C.W. Design and modeling of a wearable textile rectenna array implemented on Cordura fabric for batteryless applications. *J. Electromagn. Waves Appl.* **2020**, *34*, 1782–1796. [CrossRef]
11. Borges, L.M.; Chávez-Santiago, R.; Barroca, N.; Velez, F.J.; Balasingham, I. Radio-frequency energy harvesting for wearable sensors. *Healthc. Technol. Lett.* **2015**, *2*, 22–27. [CrossRef]
12. Yang, G.; Maryam, R.; Seokheun, C. A wearable, disposable paper-based self-charging power system integrating sweat-driven microbial energy harvesting and energy storage devices. *Nano Energy* **2022**, *104*, 107923.
13. Shin, J.; Seo, M.; Choi, J.; So, J.; Cheon, C. A Compact and Wideband Circularly Polarized Rectenna with High Efficiency at X-Band. *Prog. Electromagn. Res.* **2014**, *145*, 163–173. [CrossRef]
14. Kang, Z.; Lin, X.; Tang, C.; Mei, P.; Liu, W.; Fan, Y. 2.45-GHz wideband harmonic rejection rectenna for wireless power transfer. *Int. J. Microw. Wirel. Technol.* **2017**, *9*, 977–983. [CrossRef]
15. Chou, J.H.; Lin, D.B.; Hsiao, T.W.; Chou, H.T. A Compact Shorted Patch Rectenna Design with Harmonic Rejection Properties for the Applications of Wireless Power Transmission. *Microw. Opt. Technol. Lett.* **2016**, *58*, 2250–2257. [CrossRef]
16. Singh, S.O.; Kumar, P.S.; Vikas, V.T. Highly Efficient Dual Diode Rectenna with an Array for RF Energy Harvesting. *Wirel. Pers. Commun.* **2023**, *131*, 2875–2896.
17. Trikolikar, A.; Lahudkar, S. Design & simulation of dual-band rectifier for ambient RF energy harvesting. *Int. J. Adv. Technol. Eng. Explor. (IJATEE)* **2021**, *8*, 1383.
18. Park, J.S.; Choi, Y.S.; Lee, W.S. Design of Miniaturized Incident Angle-Insensitive 2.45 GHz RF-Based Energy Harvesting System for IoT Applications. *IEEE Trans. Antennas Propag.* **2022**, *70*, 3781–3788. [CrossRef]
19. Surender, D.; Halimi, A.; Khan, T.; Talukdar, F.A.; Antar, Y. A triple band rectenna for RF energy harvesting in smart city applications. *Int. J. Electron.* **2023**, *110*, 789–803. [CrossRef]
20. Fakharian, M.M. A high gain wideband circular polarized rectenna with wide ranges of input power and output load. *Int. J. Electron.* **2021**, *109*, 83–99. [CrossRef]

21. Mansour, M.M.; Kanaya, H. High-Efficient Broadband CPW RF Rectifier for Wireless Energy Harvesting. *IEEE Microw. Wirel. Compon. Lett.* **2019**, *29*, 288–290. [CrossRef]
22. Joseph, S.D.; Huang, Y.; Hsu, S.S. Transmission Lines-Based Impedance Matching Technique for Broadband Rectifier. *IEEE Access* **2021**, *9*, 4665–4672. [CrossRef]
23. Wu, P.; Huang, S.Y.; Zhou, W.; Yu, W.; Liu, Z.; Chen, X.; Liu, C. Compact High-Efficiency Broadband Rectifier with Multi-Stage-Transmission-Line Matching. *IEEE Trans. Circuits Syst. II: Express Briefs* **2019**, *66*, 1316–1320. [CrossRef]
24. Du, Z.-X.; Bo, S.F.; Cao, Y.F.; Ou, J.-H.; Zhang, X.Y. Broadband Circularly Polarized Rectenna with Wide Dynamic-Power-Range for Efficient Wireless Power Transfer. *IEEE Access* **2020**, *8*, 80561–80571. [CrossRef]

Disclaimer/Publisher's Note: The statements, opinions and data contained in all publications are solely those of the individual author(s) and contributor(s) and not of MDPI and/or the editor(s). MDPI and/or the editor(s) disclaim responsibility for any injury to people or property resulting from any ideas, methods, instructions or products referred to in the content.

Article

A Novel Compact Broadband Quasi-Twisted Branch Line Coupler Based on a Double-Layered Microstrip Line

Fayyadh H. Ahmed *, Rola Saad and Salam K. Khamas

Communications Research Group, Department of Electronic and Electrical Engineering, University of Sheffield, Sheffield S1 3JD, UK; r.saad@sheffield.ac.uk (R.S.); s.khamas@sheffield.ac.uk (S.K.K.)
* Correspondence: fhahmed1@sheffield.ac.uk

Abstract: A novel quasi-twisted miniaturized wideband branch line coupler (BLC) is proposed. The design is based on bisecting the conventional microstrip line BLC transversely and folding bisected sections on double-layered substrates with a common ground plane in between. The input and output terminals, each with a length of $\lambda_g/4$, and the pair of quarter-wavelength horizontal parallel arms are converted into a Z-shaped meandered microstrip line in the designed structure. Conversely, the pair of quarter-wavelength vertical arms are halved into two lines and transformed into a periodically loaded slow-wave structure. The bisected parts of the BLC are placed on the opposite side of the doubled-layer substrate and connected through four vias passing through the common ground plane. This technique enabled a compact BLC size of 6.4×18 mm^2, which corresponds to a surface area miniaturization by ~50% as compared to the classical BLC size of 10×23 mm^2 at 6 GHz. Moreover, the attained relative bandwidth is 73.9% (4.6–10 GHz) for S11, S33, S21, and the phase difference between outputs ($\angle$S21 − $\angle$S41). However, if a coupling parameter (S41) of up to −7.5 dB is considered, then the relative bandwidth reduces to 53.9% (4.6–10 GHz) for port 1 as the input. Similarly, for port 3 as the input, the obtained bandwidth is 75.8% (4.5–10 GHz) for S33, S11, S43, and the phase difference between outputs ($\angle$S43 − $\angle$S23). Likewise, this bandwidth reduces to 56% (4.5–8 GHz) when a coupling parameter (S23) of up to −7.5 dB is considered. In contrast, the relative bandwidth for the ordinary BLC is 41% at the same resonant frequency. The circuit is constructed on a double-layered low-cost FR4 substrate with a relative permittivity of 4.3 and a loss tangent of 0.025. An isolation of −13 dB was realized in both S_{13} and S_{31} demonstrating an excellent performance. The transmission coefficients between input/output ports S_{21}, S_{41}, S_{23}, and S_{43} are between −3.1 dB to −3.5 dB at a frequency of 6 GHz. Finally, the proposed BLC provides phase differences between output ports of 90.5° and 94.8° at a frequency of 6 GHz when the input ports 1 and 3 are excited, respectively. The presented design offers the potential of being utilized as a unit cell for building a Butler matrix (BM) for sub-6 GHz 5G beamforming networks.

Keywords: branch line coupler (BLC); microstrip double-layered TL (MDL-TL); phased array antenna; slow wave structure

Citation: Ahmed, F.H.; Saad, R.; Khamas, S.K. A Novel Compact Broadband Quasi-Twisted Branch Line Coupler Based on a Double-Layered Microstrip Line. *Micromachines* **2024**, *15*, 142. https://doi.org/10.3390/mi15010142

Academic Editor: Haejun Chung

Received: 14 December 2023
Revised: 12 January 2024
Accepted: 15 January 2024
Published: 17 January 2024

Copyright: © 2024 by the authors. Licensee MDPI, Basel, Switzerland. This article is an open access article distributed under the terms and conditions of the Creative Commons Attribution (CC BY) license (https://creativecommons.org/licenses/by/4.0/).

1. Introduction

5G technology offers high channel capacity, a high data rate, and channel aggregation with low latency over MIMO fading environments. On the other hand, 5G components need to be compact to be incorporated into modern portable devices, such as smartphones and tablets, which tend to be slim and lightweight, while also requiring high processing capabilities [1].

Beam scanning antennas play a key role in 5G communication systems to attain the desired outputs. The use of beamforming feeding networks (BFNs) is an essential technique to obtain high directivity in a particular direction and improve the connection quality as well as coverage of 5G systems [2]. The function of BFNs is to adjust the phase and amplitude of feeding signals for phased array antenna systems [3]. The Butler matrix (BM) is one of the most common BFNs of 5G systems owing to distinctive features such as simplicity, lower cost, and an easy fabrication process. In addition, the BM does not need external biasing in its operation. Further, the BM operates as a reciprocal feeding system for transmitting and receiving signals in phased array antenna systems. The building block of the BM is the BLC, which can be utilized as a modulator, mixer, and phase shifter as well as its basic function as part of a feeding network for phased array antenna systems [2]. Therefore, this work focuses on the design of a compact BLC with an enhanced bandwidth. As illustrated in Figure 1, a BLC consists of four transmission lines grouped into two pairs, with each pair consisting of parallel horizontal and vertical lines. The characteristic impedance of the horizontal lines is $Z_o/\sqrt{2}$, while for the vertical counterparts, it is Z_o, where Z_o represents the characteristic impedance of the microstrip transmission line (MSTL). Each of the four transmission lines (TL) has a length of $\lambda_g/4$, where λ_g is the guided wavelength. Thereby, the dimensions of the BLC are primarily based on the operating frequency, and thus in low frequencies, the BLC extends over a large area of the host device board and hence leads to an increased size [4]. Another inherent issue with a conventional BLC is the limited bandwidth characteristics of no more than 50% that restrict their applications and require a large multi-sections circuit to gain wideband characteristics, which in turn increase the circuit's area [5].

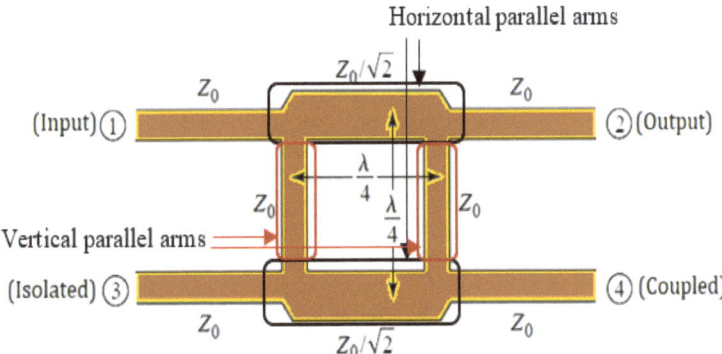

Figure 1. Geometry of a branch-line coupler.

On the other hand, in high-speed systems, microstrip lines find applications in transmitting signal pulses rather than analog microwave signals. These systems encompass various domains, including very high-speed computer logic (operating at GHz clock rates), high bit-rate digital communications, high-speed samplers for oscilloscopes or time-domain reflectometers, and radars [6]. One crucial characteristic of microstrip lines in these scenarios is propagation delay, which depends on the effective dielectric constant. However, there are situations where it becomes necessary to extend this delay. One approach to achieve this extension in effective propagation delay is by constructing a slow-wave transmission line. This can be accomplished, for instance, by introducing capacitive loading at intervals along the microstrip. In such cases, the delay is influenced by these capacitances, leading to a reduction in vp (velocity of propagation) and effectively slowing down the pulse when the capacitance (C) is increased [6].

Many efforts have been reported that address the above limitations. For example, the concept of coupled line unit cells was introduced in [4] to create a dual composite right/left-handed (D-CR/LH) unit cell, which results in a design that has a miniaturized area of ~52% of that of a conventional BLC at 1.8 GHz with a relative bandwidth of 18%. T-shaped slots and open stubs were employed in the horizontal and vertical arms of the BLC, resulting in a 30% bandwidth improvement and a 12.3% size reduction [5].

In another study, a double-layer board with slow-wave microstrip transmission lines and blind vias was used to achieve a 43% size reduction compared to a conventional design at the same resonant frequency. However, the increased number of vias led to higher insertion losses and design complexity [7]. A Koch fractal-shape BLC of various iterations was suggested in [8], where the sample was designed to operate at 2.4 GHz and offered a size reduction of ~81% in combination with a relative bandwidth of 33%.

In [9], a compact artificial transmission line was proposed for compact microwave components. The transmission line combines resonant-type composite right/left-handed transmission lines (CRLH TLs) with fractal geometry. Two sets of planar CRLH cell structures were provided: one based on a cascaded complementary single split ring resonator (CCSSRR), and the other based on complementary split ring resonators (CSRRs). A dual-band bandpass filter (BPF) and a monoband branch line coupler were designed based on the suggested artificial line.

The effectiveness of integrating CRLH TL and fractal geometry for designing compact broadband microwave devices was confirmed in [10]. In this study, a proposal was made for a compact balun with improved bandwidth, utilizing a completely artificial fractal-shaped composite right/left-handed transmission line (CRLHTL). Chip components were employed for the left-handed contribution, and fractal microstrip lines were utilized for the right-handed part, focusing on miniaturization. This innovative technology provided an extra degree of flexibility in crafting compact devices and demonstrates superiority over alternative methods.

In [11], open-ended stubs and transmission line meandering with a stepped impedance approach was proposed with a size reduction of ~61% and 50% compared to a conventional BLC, respectively. However, the narrow bandwidth of ~130 MHz represents a key limitation. A flexible coupler using a Teslin paper substrate was reported [12]. It replaced the conventional quarter-wavelength transmission lines with a collective of shunt open-stubs, series transmission lines, and meandered lines, resulting in a compact design with a surface area of $0.04\ \lambda_g^2$ and a 68% fractional bandwidth. Using a dual microstrip transmission line, the BLC size was reduced by 32% with a fractional bandwidth of 60% [13]. However, this approach had poor return losses over the operating bandwidth. To improve matching, T-shaped transmission lines were used, reducing fractional bandwidth and size to 50% and 44%, respectively. A compact BLC class introduced a prototype using open-circuited stubs to replace traditional quarter-wavelength transmission lines [14], resulting in a ~55.6% size reduction and achieving 11% and 50% fractional bandwidths for narrowband and wideband modes of operation, respectively.

In [15], a new configuration, BLC, is presented. The design applies two types of trapezoid-shaped resonators on the arms of the BLC to configure a wideband branch-line coupler. The proposed design achieved a size reduction of 79% compared to conventional couplers. In addition, it offers a fractional bandwidth of 22.2%. Ref. [16] used artificial transmission lines (ATL) for miniaturization. They replaced conventional transmission lines with right-handed transmission lines (RHTL) and constructed the branch-line coupler sides using cascaded T-Net RHTLs instead of quarter-wavelength transmission lines. This design achieved a 50% size reduction compared to the conventional BLC and a 33.3% fractional bandwidth (2.0–2.8 GHz). A simple method was used to improve bandwidth in [17]. By adding a single transmission line element to a conventional coupler, they increased bandwidth by approximately 25%. However, the proposed structure is larger at 25.7×22.8 mm^2 compared to the conventional coupler's 21.5×20.7 mm^2.

Triangular and trapezoidal resonators were added to the coupler for miniaturization and harmonic suppression [18]. The design achieved an 84% size reduction and wide harmonic suppression. However, it has a complex structure with a low-frequency band around 200 MHz, representing a 26% fractional bandwidth (FBW). A bandpass filter operating in three frequency bands utilized a dual-layer structure with distinct dielectric constants, as described in [19]. The dual-layer design was employed to diminish the overall size and enhance the isolation between the passbands. Consequently, the suggested configuration offers benefits such as a nearly 50% reduction in physical size and the alleviation of design constraints by utilizing the two substrates within a unified structure.

The majority of the aforementioned prototypes were based on composite right/left-handed structures to create branch lines, which might result in unfavorable characteristics that are associated with miniaturization such as shallow return losses for input ports, poorly isolated ports, and narrow bandwidths in some cases [3,4,6,8]. In addition, the structures of right/left-handed transmission lines probably increase the structure's complexity, which results in a challenging practical realization despite the overall size reduction.

In this study, a quasi-twisted shape branch line coupler is proposed, which is the longitudinal bisection of a conventional BLC into two sections and twisting each over the other. The structure is designed based on the microstrip double-layered TL (MDL-TL). The input/output transmission lines and horizontal arms of the BLC are built based on a Z-shape meandered section with round blend edges, while the $\lambda_g/4$ vertical arms of the BLC are adopted for the slow wave structure. The MDL-TLs are placed on two layers and connected using four conductive vias. A common ground plane is placed between the layers of the MDL-TLs, which incorporate circular slots around the vias to avoid shorting them to the common ground plane. The described configuration reduced the size of the conventional BLC by 49.9% and improved the relative bandwidth to 75.8%. The novel design is modelled and simulated using a computer simulation technology (CST) microwave studio and then fabricated and tested on a low-cost FR-4 substrate material demonstrating promising S-parameter results.

This paper is organized as follows: Section 2 explains the theoretical analysis and design procedures of developing a wideband MDL-TL and compares the achieved performance with that of a conventional microstrip line; Section 3 presents the analysis and design of a branch line coupler based on MDL-TL; finally, Section 4 presents the simulated and measured results demonstrating the novel BLC performance.

2. Theory

2.1. Selection of the Classical Microstrip Line Dimensions

The microstrip transmission line (MSTL) represents the building block of any passive and active microwave device due to a number of advantages, such as the easy fabrication process as well as the availability of numerous miniaturization approaches [20].

The width (W) of a microstrip line, situated on a thin grounded dielectric substrate with a height (h) and a dielectric constant (ε_r), can be calculated for a specific characteristic impedance (Z_0) as described in [20].

The microstrip physical length, l, that is required to generate a phase shift (delay) of θ can be determined as [20]:

$$\theta = \beta l = \sqrt{\varepsilon_e}\, k_o l \tag{1}$$

where f and c are the frequency and speed of light, respectively, and $k_o = \dfrac{2\pi f}{c}$, ε_e is the effective dielectric constant, which is in the range of $1 < \varepsilon_e < \varepsilon_r$.

2.2. The Impact of Right-Angled Bend on the Performance of the Microstrip Line

Complex microwave circuits usually comprise bend microstrip transmission lines, and quite often the width of the line does not change across the bend. Figure 2 illustrates a right-angled bend MSTL with its equivalent circuit. The bend MSTL generates capacitance, C_{bend}, and inductance, L_{bend}, which result in the gathering of additional charge at the line corner in particular at the outer edge of the bend area, whereas inductance arises due to the disruption of current flow [6]. Closed formulas for the capacitance and inductance of the bend are determined in [6].

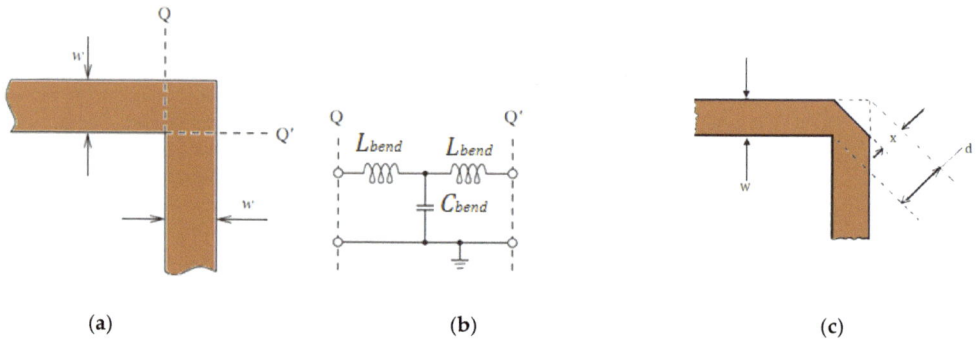

Figure 2. (a) Part of bend microstrip line, (b) Equivalent circuit, (c) Chamfered right-angled bend microstrip line.

Generally, performance and bandwidth enhancement of a microstrip line-based circuit is realized by compensating bend discontinuities. This is usually implemented by chamfering or rounding the corners, which leads to the minimizing of reactance. The percentage chamfer, M, is given by $(x/d) \times 100\%$, where x/d is chamfered to bend at a diagonal ratio at the bending corner as in Figure 2c and can be determined using [6]:

$$M = 52 + 65 e^{(-1.35 W/h)} \qquad (2)$$

which is valid for $W/h \geq 0.25$ and $\varepsilon_r \leq 25$.

In this work, the MDL-TL is used as a unit cell to build the BLC. This line is bent over a horizontal xy plane and vertical xz or yz planes when folded around double-layer substrates. The normal even-and odd-mode theory can be used to analyze the port parameters of a pair of quasi-twisted MDL-TLs.

The schematics of the meandered Z-shaped MSTL in even and odd modes are demonstrated in Figure 3a,b. The C_{mx} and C_{my} are the coupling capacitances in the x and y directions, respectively, factor 2 arises due to double-layered structure around the common ground plane, and C_{my} is much stronger than C_{mx}. Therefore, it can be neglected [21]. The difference in the characteristic impedance between the even and odd modes can be negligible since the coupling between top and bottom MSTL is very small, and they are given as in [21]:

$$Z_{0e} = Z_0 \sqrt{\frac{1+C}{1-C}}, \quad Z_{0o} = Z_0 \sqrt{\frac{1-C}{1+C}} \qquad (3)$$

where C is the coupling factor.

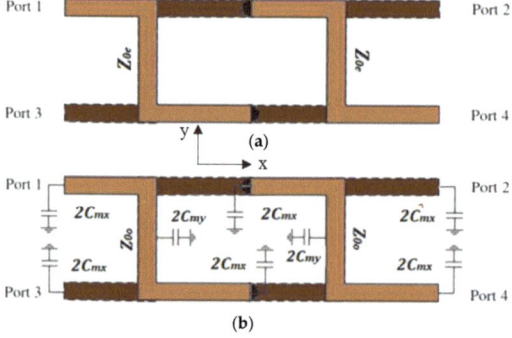

Figure 3. Double-layered MSTL equivalent model: (a) even mode; (b) odd mode.

2.3. Study of the Performance of the MDL-TL

This section aims to assess the effectiveness of the MDL-TL structure compared to a conventional microstrip transmission line by examining its S-parameters and comparing them with conventional ones. In order to create an MDL-TL structure from a conventional MSTL, four stages of modification are undertaken and can be observed in Figure 4. These stages include the one-layered conventional MSTL, the one-layered meandered MSTL without and with chamfering, and the double-layered with chamfering (MDL-TL). The effect of meandering of the structure and varying the line length over the *xy* plane and *x-z* plane was conducted on an FR-4 substrate with a dielectric permittivity of 4.4, a loss tangent of 0.025, and a substrate thickness of 0.8 mm at 6 GHz. This investigation includes a comparison of the key parameters of the four-line configurations: the reflection coefficient (S_{11}), transmission coefficient (S_{21}), and phase of S_{21} (or output phase).

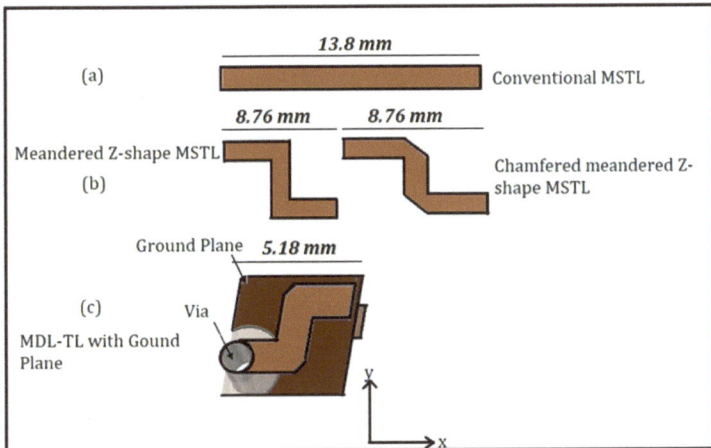

Figure 4. Various MSTL configurations: (**a**) single-layer straight MSTL; (**b**) single-layer meandered MSTL; (**c**) double-layer meandered MSTL.

In this study, a specific section of a $\lambda_g/2$ length transmission line at 6 GHz, which corresponds to a physical length of 13.8 mm, was chosen. The width of the line was set at 1.52 mm to maintain a characteristic impedance of 50 Ω. The transmission line was transformed into a meandered Z-shape and then into an MDL-TL configuration using vias with diameters equal to the width of the microstrip line. The meandered Z-shape had a length of 8.76 mm, while the MDL-TL configuration had a length of 5.18 mm, as illustrated in Figure 4.

Figure 5a demonstrates the reflection coefficient, S_{11}, of the four types of lines, where it can be noted how meandering the line without applying the chamfering technique deteriorates the impedance matching at the input ports by increasing S_{11} from −30 dB to −18 dB at a target frequency of 6 GHz. This is attributed to the emergence or occurrence of bending parameters like C_{bend} and L_{bend}, as explained earlier. However, it is also clear from Figure 5a how chamfering the corners provides discontinuity substitution and improves the matching again by reducing S_{11} to ~−38 dB. It can also be noticed that the S_{11} of the MDL-TL is as well matched as the conventional MSTL at 6 GHz. Moreover, the proposed MDL-TL structure offers as good matching as a conventional line as it has an almost equal and lower S_{11} than a conventional line, especially at a target frequency of around 6 GHz.

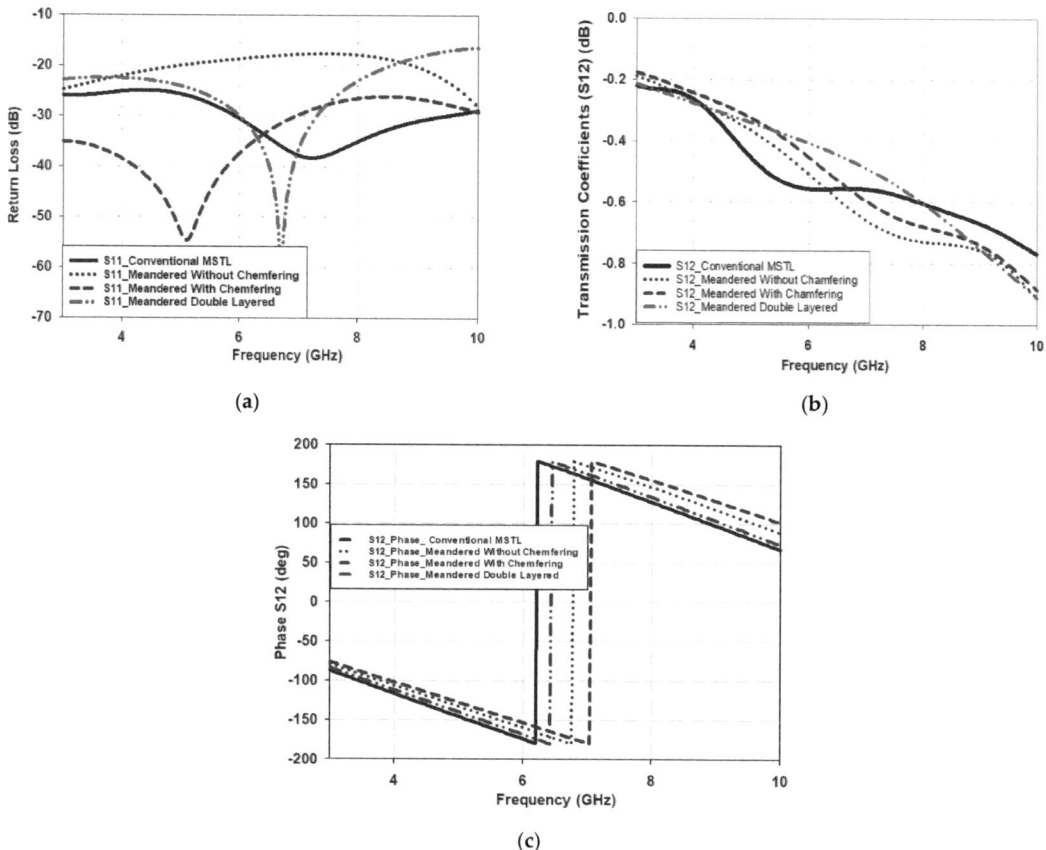

Figure 5. Performance of different MSTL configurations: (a) S11 (b) S12; (c) phase of S12 (output phase).

Figure 5b demonstrates the transmission coefficients of all line configurations, where it can be noted that adjusting the MSTL with the meandering, chamfering, etc. of a one-layered MSTL does not have much of an effect on the transmission coefficient since S_{21} is within −0.45 dB to −0.55 dB at 6 GHz. Furthermore, the S_{21} of the double-layered MSTL is higher by about 0.075 dB in the frequency range of 6 GHz to 7 GHz, which confirms a good performance compared to a conventional MSTL. On the other hand, the output phases of all configurations have approximately the same phase delay of 180° since the MSTL is modeled with an electrical length of $\lambda_g/2$ at 6 GHz, as presented in Figure 5c. It should be noted that the same results were obtained when port 2 is considered as the input port, and they have been omitted for brevity. The achieved results confirm that the proposed configuration can be utilized as a unit cell for compact-size structures with improved performance.

3. Design of Branch Line Coupler

3.1. Conventional Branch Line Coupler

Figure 6 shows a 10 mm × 23 mm conventional BLC designed at 6 GHz on an FR4 substrate with a thickness of 0.8 mm and dielectric constant (ε_r) of 4.4 with a loss tangent of 0.025, which are the same substrate specifications as the proposed BLC for an effective comparison. The 50 Ω microstrip line sections for ports 1, 2, 3, and 4 are designed with a width of 1.52 mm, and the $\lambda_g/4$ sections of the coupler with an impedance of $Z_0/\sqrt{2}$ (35.35 Ω) are of a microstrip line with a width of 2.62 mm, as shown in Figure 6.

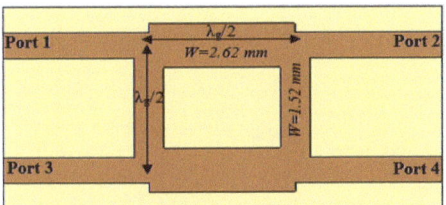

Figure 6. Structure of the conventional branch coupler operating at 6 GHz.

Figure 7 shows the simulated scattering parameters for the conventional branch line coupler operating at 6 GHz. The input reflection coefficient (S_{11}) and the isolation coefficient between input ports, S_{31}, are presented in Figure 7a, demonstrating a -30 dB excellent match for the conventional BLC at the frequency of interest, 6 GHz, alongside a relative impedance bandwidth of 42.2% (5.214–8 GHz), in addition to a perfect isolation of -50 dB. On the other hand, Figure 7b illustrates the transmission coefficient (S_{21}) and coupling coefficient (S_{41}), where it can be seen that the power is divided equally between the output ports (2 and 4) at 6.18 GHz with a value of -3.8 dB. However, the delivered power declines in both output ports (2, and 4) as the frequency increases, reaching -5.6 dB at the transmission port (port 2) and -6.38 dB at the coupled port (port 4), as shown in Figure 7b.

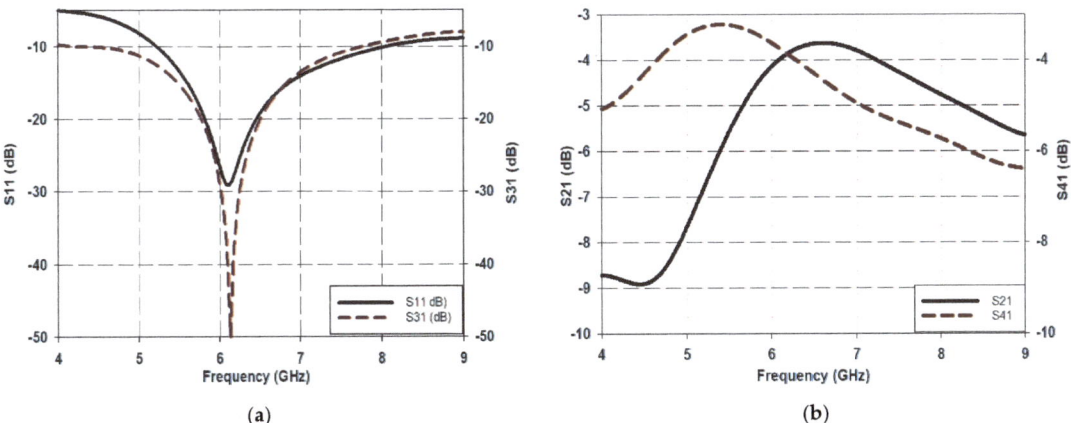

Figure 7. Scattering parameters of conventional BLC: (**a**) reflection coefficient (S_{11}) and isolation coefficient (S_{31}); (**b**) transmission coefficient (S_{21}) and coupling coefficient (S_{41}).

3.2. Proposed Quasi-Twisted Branch Line Coupler Structure

A branch line coupler incorporating an MDL-TL topology was designed on a double-layered 0.8-mm FR-4 substrate, as shown in Figure 8. The horizontal $\lambda_g/4$ dimensions and input/output ports' ends of the conventional branch line coupler of Figure 1 were transformed to an MDL-TL-based Z-shape, while the vertical $\lambda_g/4$ dimensions were converted to a periodically loaded open-stub configuration (slow wave structure) to achieve a compactness in the structure. Figure 8a presents a perspective of the proposed design. The preliminary dimensions of the suggested BLC were selected based on design equations of [20] for the conventional MSTL's width and length, respectively. These dimensions were then slightly optimized using CST. The bending of the meandered Z-shape of the

MDL-TL generates a certain reactive component that negatively affects the transmission line performance, such as reflection and transmission coefficients. Therefore, to compensate for this reactance, a 1.5 mm radius round-chamfering was implemented at the bending of the MDL-TL. A summary of the designed BLC specifications is presented in Table 1. The proposed branch line coupler configuration has a quasi-twisted structure as illustrated in Figure 8b.

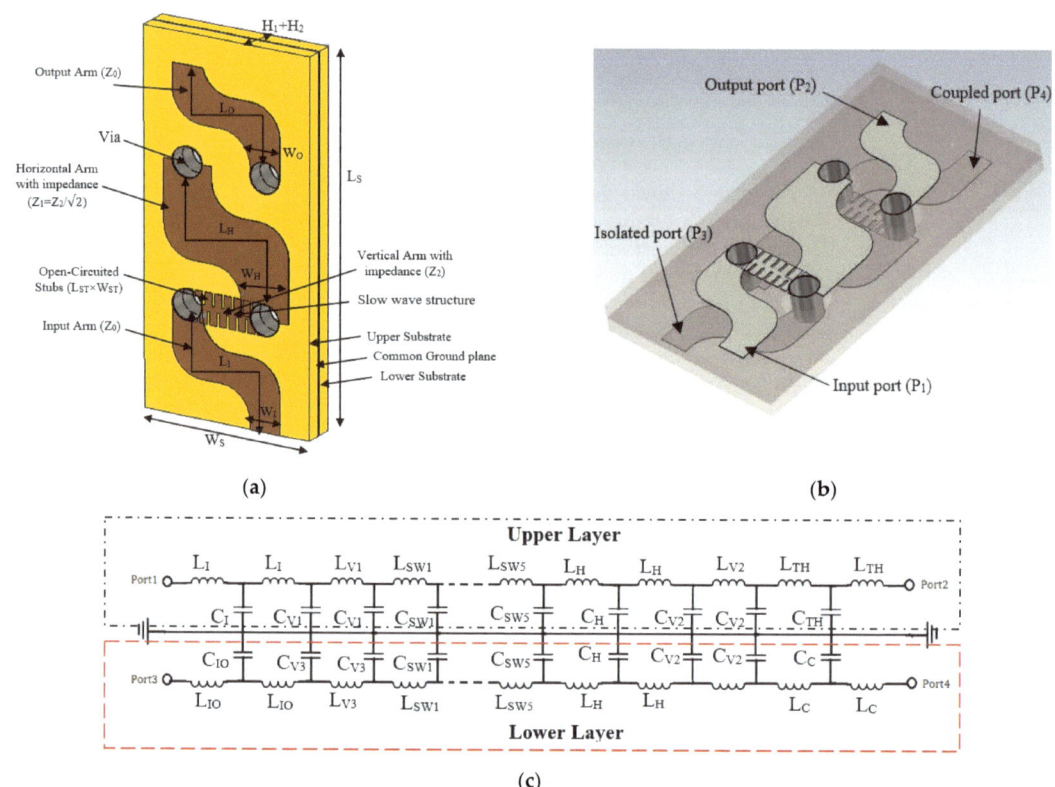

Figure 8. Proposed quasi-twisted branch line coupler: (**a**) perspective view; (**b**) frame mode view; (**c**) LC equivalent circuit.

Table 1. The design specifications of the suggested BLC.

Description	Notation	Dimension (λ_g at 6 GHz)	Description	Notation	Dimension (λ_g at 6 GHz)
Lengths of input arm (P_1) and Isolated arm (P_3)	L_I	0.33	Coupled arm (P_4) length	L_c	0.359
All ports' width	W_I	0.054	Via radius	V_R	0.027
Horizontal arm (port) length	L_H	0.311	Substrates' thickness	$H_1 \& H_2$	0.0286
Horizontal arm (port) Width	W_H	0.086	Substrates' Length	L_S	0.645
Output arm (P_2) length	L_o	0.305	Substrates' Width	W_S	0.304
Horizontal arm (port) Width	W_H	0.086	Open-circuited stub length	L_{ST}	0.022
Open-circuited stub width	W_{ST}	0.0134	Actual circuit area	$L_c \times W_c$	0.359×0.23

The approximate equivalent LC circuit of the proposed coupler is drawn as shown in Figure 8c. The circle was drawn for the proposed design based on mapping several equivalent circles to various components. These include the equivalent circle for the radial bend, representing the meandered transmission part [6,19], the equivalent circuit for the upper and lower slow-wave structures [6] as well as the equivalent circuit for four vias [22]. The complete equivalent circuit connects all the equivalent circuits together for the two layers of the proposed model.

In the Figure 8c circuit, L_I, C_I, L_{TH}, C_{TH}, L_{IO}, C_{IO}, and L_C, C_C represent the inductances and capacitances of the equivalent circuit of the input through isolated and coupled ends, respectively. Meanwhile, L_{SW1} to L_{SW5} and C_{SW1} to C_{SW5} represent the equivalent circuit inductances and capacitances of the five sections of the slow wave structures on both the upper and lower substrates, respectively. Finally, L_{V1}, C_{V1}, L_{V2}, C_{V2}, L_{V3}, C_{V3}, L_{V4}, and C_{V4} represent the equivalent circuit inductances and capacitances of the four vias of the proposed design.

The proposed design concept is inspired by twisted cable shapes, aiming to substantially reduce the blank space occupied by traditional BLC circuitry. This was accomplished by dividing the conventional BLC horizontally and interweaving the upper and lower segments in a twisted fashion, utilizing a Z-shaped meandering technique with traditional MSTL components. The design procedure of the suggested BLC in Figure 8 involves two key steps as follows.

Step 1: As shown in Figure 8, the characteristic impedance of the input, output, coupled, and isolated ports was selected to be $Z_0 = 50\ \Omega$. Also, using the equation in [20] the relevant lines' $\lambda_g/4$ MSTL (the four input/output ports), shown in Figure 8, are designed with a width of 1.53 mm. On the other hand, the width of the horizontal characteristic's impedance of $Z_1 = Z_0/\sqrt{2}$, is calculated as 2.63 mm and further optimized to 2.42 mm to realize the optimal s-parameter's performance. The vertical $\lambda_g/4$ length arms with an impedance of $Z_2 = Z_0$ are designed to occupy an optimal small area. The internal and external corners of all the Z-shaped sections are round-chamfered with a radius of 1.5 mm to improve the reflection and transmission coefficients, as described in Section 2. All inputs and outputs ending with a length of $\lambda_g/4$ are converted to a meandered Z-shape. This is undertaken to eliminate the blank space occupied by these ends and achieve the most compact BLC possible. The length, L_o, of the output port (P_2) is less than $\lambda_g/4$, which in turn is shorter than the L_c length of the coupled port (P_4), as shown in Figure 8b. These modifications in the lengths are necessary to tune the differences between the output ports' phases to ~90°. The proposed structure, demonstrated in Figure 8, provides a novel BLC configuration, which is significantly miniaturized by ~50% as compared to a conventional BLC operating in the same frequency of 6 GHz.

Step 2: The idea behind a slow wave structure involves the incorporation of shunt capacitors at regular intervals along the length of the transmission line, as illustrated in Figure 9a,b [23]. This technique results in decreasing both the characteristic impedance and phase velocity, as can also be modelled from Equations (7) and (8) below [23].

$$Z_{0\_Loaded} = \sqrt{\frac{L}{C + \frac{C_p}{d}}} \quad (4)$$

$$V_{p-Loaded} = \frac{1}{\sqrt{L\left(C + \frac{C_p}{d}\right)}} \quad (5)$$

where C_p represents the periodically added capacitor at a distance, d, along the transmission line. Z_{0_loaded} and Z_0 denote the characteristic impedance of the loaded and unloaded lines, respectively. On the other hand, reducing the phase velocity facilitates the achievement of an effectively longer electrical length by utilizing a physically shorter length.

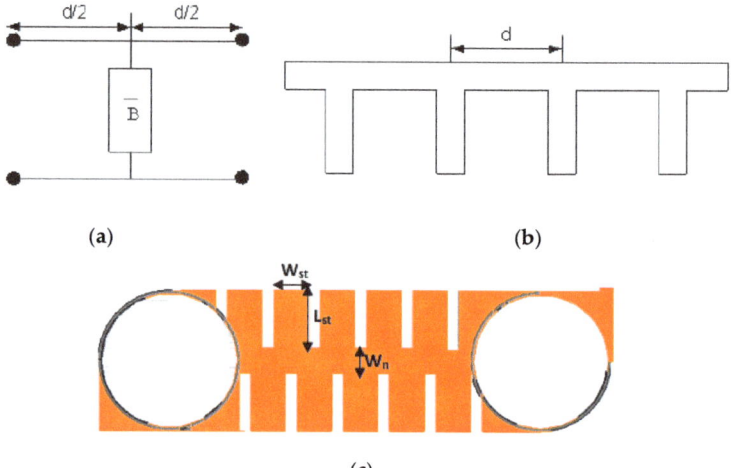

Figure 9. (a) A single periodic section circuit diagram of periodically loaded transmission line; (b) schematic diagram of periodically loaded line with open stub used as loaded capacitance; and (c) the proposed slow wave structure.

In the proposed design, a periodically loaded slow wave structure, as shown in Figure 9c, is adopted to design a compact $\lambda_g/4$ line with an enhanced bandwidth to accommodate the space limitation introduced due to the folding of the BLC halves and the use of a double-layered Z-shaped meandering technique to realize compactness. Additionally, to obtain a line with a specific characteristic impedance, the loading section (W_n) should possess a higher characteristic impedance, such that its characteristic impedance is decreased to the desired characteristic impedance, usually of 50 Ω impedance, after loading.

The relation between the physical length, l_p, and the electric length, l_e, of the periodically loaded stub TL is given as [23]:

$$l_e = l_p \left(\frac{\omega_0}{V_{p-Loaded}} \right) \quad (6)$$

The added capacitance C_p, in terms of the known parameters of the loaded and unloaded TL, is given as [24]:

$$C_p = \frac{\phi}{N\omega_0} \left(\frac{Z_0^2 - Z_{o\_loaded}^2}{Z_0^2 Z_{o\_loaded}} \right) \quad (7)$$

where ω_0 is the angular frequency, and N is an integer that refers to the stub section's number.

On the other hand, open stubs with a length of multiples of a quarter wavelength were added in parallel to one pair of branch-line coupler sides, operating as a parallel stub transformer as reported in [25]. Therefore, the characteristic impedance of the added stubs attenuates the maxima in return loss characteristics (S_{11}) when the frequency deviates, which, in turn, broadens the bandwidth.

In the proposed configuration, shown in Figure 8, the slow wave structure is accomplished by inserting alternate slots around the conventionally straight MSTL with respective length and width of 2.62 mm and 1.52 mm, which creates rectangular stubs around both sides of the line as shown in Figure 9c. To create rectangular stubs around the MSTL, slots were placed in an alternate configuration, resulting in stub dimensions of $W_{st} \times L_{st} = 0.234$ mm² on both sides of the line as depicted in Figure 9c. The addition of these slots transforms the 50 Ω MSTL to a narrow MSTL with a width of $W_n = 0.271$ mm that provides a characteristic impedance of 105.9 Ω, which compensates for the shunt impedances of the periodic stubs and results in an overall characteristic impedance that is close to 50 Ω.

4. Fabrication and Measurements

The proposed novel miniaturized BLC, shown in Figure 10, was fabricated on an FR4 substrate with a dielectric constant of 4.4 and a thickness of 0.8 mm, where a thin ground plane of 70 μm thickness was inserted between the two FR4 substrates, forming a sandwich-like structure. Both substrates were truncated at their corners using cross-sectional areas of 2 × 5 mm² each to expose the ground plane and enable the SMA connector to be easily connected, as illustrated in Figure 10b.

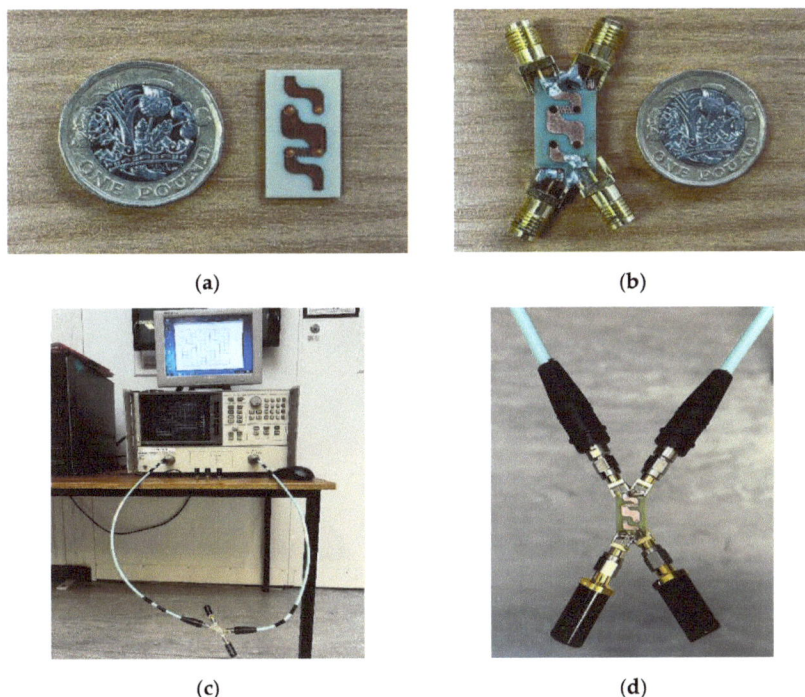

Figure 10. (a) BLC prototype, (b) BLC prototype with SMA connectors, (c) measurement setup system, and (d) termination of un-fed ports.

The fabricated novel BLC was measured using an HP 8720B vector network analyzer (VNA) from Test Equipment Center, Inc., Gainesville, FL, USA, as shown in Figure 10c. The reflection coefficient and isolation coefficient between input ports 1 and 3 as well as the transmission and coupling coefficients were measured by connecting the relevant ports to the VNA, while the remaining ports were terminated by a 50 Ω load to prevent additional mismatching and increase the measurements' reliability, as illustrated in Figure 10d.

As per the design specifications of the proposed BLC, the required phase difference between the output signals is 90°. This phase difference can be verified by measuring the phases of the transmission coefficients, S_{21} and S_{43}, and coupling coefficients, S_{41} and S_{23}, at the output ports 2 and 4. Once these measurements are carried out, the phase difference can be determined as follows:

$$\varphi = \begin{cases} \angle S_{21} - \angle S_{41} & \text{for input from port 1} \\ \angle S_{43} - \angle S_{23} & \text{for input from port 3} \end{cases} \quad (8)$$

The performance of the proposed BLC is evaluated by observing the four-port S-parameters' magnitudes and phase differences as shown in Figure 11. The four principal scattering parameters considered in the analysis are as follows: the reflection coefficient (RC) S_{11}, transmission coefficient (TC) S_{21}, isolation coefficient (IC) S_{31}, and coupling coefficient (CC) S_{41}, when port 1 is excited as the input port. On the other hand, when port 3 is excited, the required scattering parameters, S_{13}, S_{23}, S_{33}, and S_{43}, are considered. Ports 1 and 3 were chosen as they are located on opposite sides of the coupler and are designated as input ports. In addition, from Figure 11a, it is evident that a good agreement was accomplished between the simulated and measured reflection coefficients. For example, the measured −10 dB S_{11} bandwidth extends from 4.6 GHz to 10 GHz, which corresponds to a relative bandwidth of 73.9% compared to a typical bandwidth of ~40% from an identical traditional branch line coupler. As a result, the proposed configuration offers a substantial bandwidth enhancement.

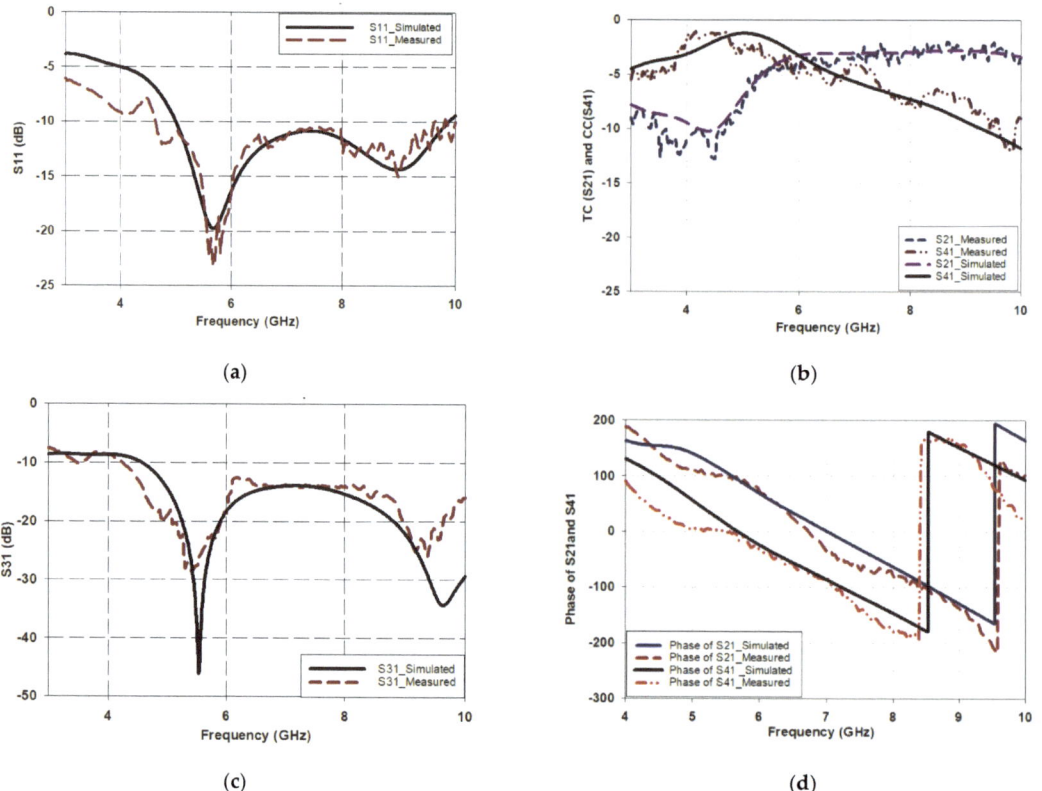

Figure 11. Four-port S-parameters for the proposed BLC when port 1 is excited: (**a**) S11, (**b**) S21 and S41 (**c**) S31, and (**d**) phase of S21 and S41.

The transmission and coupling coefficients, S_{21} and S_{41}, respectively, are presented in Figure 11b with good agreement between measurements and simulations. From these results, it can be observed that both the transmission and coupling coefficients are −3.9 dB at the desired frequency of 6 GHz, which is close to the ideal value of −3 dB. Notably, S_{21} remains higher than −3.5 dB for the entire operating band ranging from 5.8 GHz to 10 GHz. However, the coupling coefficient, S_{41}, gradually degrades with increasing frequency, possibly due to vias and substrate losses since the input port and coupled ports are located on opposite sides of different substrates. Despite this degradation, the power delivered to port 4 remains greater than -7.5 dB up to a frequency of 8 GHz.

Furthermore, the isolation coefficient between the input ports, S_{31}, as shown in Figure 11c demonstrates a magnitude of less than -13 dB throughout the operating band, signifying a good isolation between the input ports. This ensures that the proposed BLC meets the necessary design specifications for optimal performance.

Finally, Figure 11d presents the difference between the phases at the output ports 2 and 4 for input excitations from port 1. A 90° phase difference is required between the output signals. At a design frequency of 6 GHz, the phase difference between the output ports for inputs from port 1 is ($\angle S_{21} - \angle S_{41} = 90.50°$), which satisfies the required phase difference for typical BLC design specifications. In addition, the phase difference error (PDE) is 0.5° for the designed frequency of 6 GHz, which is marginal.

Figure 12 presents the proposed BLC performance when port 3 is excited. From Figure 12a, it is evident that the -10 dB S_{33} bandwidth extends from 4.5 GHz to 10 GHz, which corresponds to a relative bandwidth of 75.8% compared to ~40% for the traditional branch line coupler based on the same design specifications and operating frequency of 6 GHz. It should be noted that a marginal difference of 1.9% occurs between the reflection coefficients' bandwidths of S_{11} and S_{33}. The transmission coefficient, S_{23}, and the coupling coefficient, S_{43}, for port 3 excitations are illustrated in Figure 12b. At the target band, i.e., at 6 GHz, the transmission and the coupling coefficients of the proposed BLC design are -3.9 dB, which is close to the ideal value of -3 dB. Notably, the transmission coefficient, S_{43}, remain higher than -3.5 dB for the entire operating band ranging from 5.8 GHz to 10 GHz, and this behavior is consistent throughout this wide operating frequency range. However, the coupling coefficient, S_{23}, gradually degrades with increasing frequency, and this is due to the use of vias as well as losses inherited from the lossy FR4 substrates, as the input port and coupled ports are located on opposite sides of different substrates. Despite this degradation, the power delivered to port 4 remains greater than -7.5 dB up to a frequency of 8 GHz. In Figure 12c, the isolation coefficient between input ports, S_{13}, is depicted. It is evident that S_{13} remains less than -14 dB throughout the operating band. It can be concluded that the proposed miniaturized BLC has excellent performance in terms of scattering parameters (S-parameters) compared to a traditional BLC design. Additionally, based on the obtained results, it can also be confirmed that the ports are reciprocal and have the same S-parameters characteristics for all ports. This consolidates the principle that the proposed miniaturized BLC can be used as a unit cell for constructing a Butler matrix, which in turn has potential use in the development of phased array antenna systems.

Figure 12d illustrates the phase difference between output ports 2 and 4 for input excitations from port 3. At a frequency of 6 GHz, the phase difference between output ports for input from port 3 ($\angle S_{43} - \angle S_{23} = 94.8°$) meets the requirements of a BLC coupler for good performance. However, a slight discrepancy between the simulated and measured phases is observed for input port 3. This difference may be attributed to fabrication tolerance and the lump solder for SMA feeders. Nevertheless, the phase difference error (PDE) at a design frequency of 6 GHz is 4.8° for input port 3 excitation.

Table 2 provides a comparison between the performance of the proposed BLC and those of previously published BLC designs. Most of the designs presented in Table 2 were focused on either improving the bandwidth or reducing the size of the BLC. The proposed work, however, achieves both bandwidth enhancement and size reduction, which is crucial for the design of a 5G system. Furthermore, when compared to the literature, the proposed miniaturized wide band BLC offers other advantages such as design simplicity, ease of fabrication, smaller phase difference errors, and equal power distribution among output ports at a design frequency of 6 GHz.

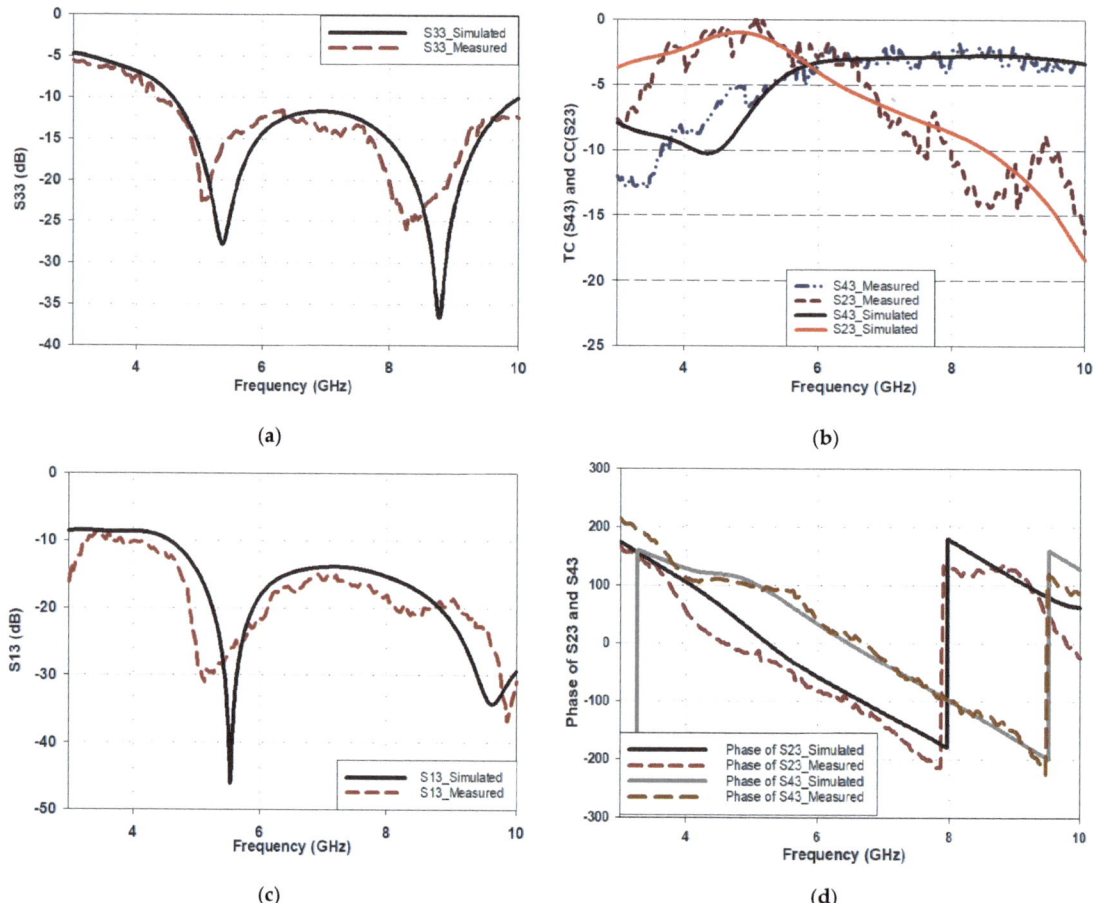

Figure 12. Performance of the proposed BLC when port 3 is excited: (**a**) S33, (**b**) S43 and S23 (**c**) S13, and (**d**) phase of S23 and S43.

Table 2. Comparison of proposed BLC performance with respect to published designs available in the literature.

Ref.	Operating Frequency (GHz)	Relative Bandwidth (%)	Size (λ_g^2)	S_{11} (dB)	S_{21} (dB)	S_{41} (dB)	S_{31} (dB)	PDE Error (Deg)
[4]	3.5	30.22	0.27 × 0.16	−27.47	−4.4	−3.1	−26.2	3.54
[5]	2.36	31.77	0.25 × 0.29	−25.5	−2.9	−3.2	−27.5	1
[7]	1.77	7.2	0.62 × 0.62	−35.9	−3.1	−3	−37.9	-
[8]	0.9	67.7	0.2 × 0.2	−19.89	−3.69	−3.67	−17.5	-
[9]	3	50.4	0.45 × 0.25	−21	−0.95	−10.5	−29	±10
[1]	3.5	40.2	0.55 × 0.65	−12	−3.19	−2.8	−15.2	−2
[11]	0.95	22.2	0.17 × 0.08	−30	2.3	−3.2	29	-
[12]	2.4	33.3	0.2 × 0.35	−20	12.5	−2.5	−20	±13
[13]	2	13 (S11 < −20)	0.3 × 0.3	−30	−3.5	−3.5	−27	±5
This Work	6	75.8	0.3 × 0.74	−14	−3.17	−3.17	−18.9	0.5

5. Conclusions

A novel quasi-twisted miniaturized wideband branch line coupler was introduced, which employs a Z-shaped meandered microstrip line with a slow wave structure on a double-layer substrate. The performance of the proposed design was evaluated, demonstrating agreement between the simulation and measurements and showing advantages over conventional microstrip branch line couplers in terms of size reduction and band improvement.

The novel quasi-twisted miniaturized wideband branch line coupler operates within the frequency band of 4.5 GHz to 10 GHz, offering a high relative bandwidth of up to 75.8% and reducing the size by 49.9% compared to conventional branch line couplers with a general bandwidth of 42.2%. A comparison between the proposed branch line coupler and the conventional design indicates that the former exhibits favorable scattering parameter specifications. Furthermore, the obtained results suggest that the ports of the branch line coupler demonstrate reciprocity, benefiting from consistent S-parameter characteristics across all ports. As a result, the proposed branch line coupler holds the potential to be integrated into a compact and wideband Butler matrix for future deployment in phased antenna array systems for potential use in 5G applications.

Author Contributions: Conceptualization, F.H.A.; Methodology, F.H.A.; Software (CST Studio Suite 2023), F.H.A.; Investigation, F.H.A.; Writing—review & editing, R.S. and S.K.K.; Supervision, R.S. and S.K.K. All authors have read and agreed to the published version of the manuscript.

Funding: This research received no external funding.

Data Availability Statement: Data are contained within the article.

Conflicts of Interest: The authors declare no conflict of interest.

References

1. Kiani, S.H.; Altaf, A.; Anjum, M.R.; Afridi, S.; Arain, Z.A.; Anwar, S.; Khan, S.; Alibakhshikenari, M.; Lalbakhsh, A.; Khan, M.A.; et al. MIMO antenna system for modern 5G handheld devices with healthcare and high rate delivery. *Sensors* **2021**, *21*, 7415. [CrossRef] [PubMed]
2. Vallappil, A.K.; Rahim, M.K.A.; Khawaja, B.A.; Aminu-Baba, M. Metamaterial based compact branch-line coupler with enhanced bandwidth for use in 5G applications. *Appl. Comput. Electromagn. Soc. J. (ACES)* **2020**, *35*, 700–708.
3. Zhang, J.; Ge, X.; Li, Q.; Guizani, M.; Zhang, Y. 5G millimeter-wave antenna array: Design and challenges. *IEEE Wirel. Commun.* **2016**, *24*, 106–112. [CrossRef]
4. Gomha, S.; El-Rabaie, E.-S.M.; Shalaby, A.-A.T.; Elkorany, A.S. Design of new compact branch-line coupler using coupled line dual composite right/left-handed unit cells. *J. Optoelectron. Adv. Mater.* **2015**, *9*, 836–841.
5. Abdulbari, A.A.; Rahim, S.K.A.; Abd Aziz, M.Z.A.; Tan, K.G.; Noordin, N.; Nor, M. New design of wideband microstrip branch line coupler using T-shape and open stub for 5G application. *Int. J. Electr. Comput. Eng.* **2021**, *11*, 1346–1355. [CrossRef]
6. Edwards, T.C.; Steer, M.B. *Foundations for Microstrip Circuit Design*; John Wiley & Sons: Hoboken, NJ, USA, 2016.
7. Alhalabi, H.; Issa, F.; Pistono, E.; Kaddour, D.; Podevin, F.; Baheti, A.; Abouchahine, S.; Ferrari, P. Miniaturized branch-line coupler based on slow-wave microstrip lines. *Int. J. Microw. Wirel. Technol.* **2018**, *10*, 1103–1106. [CrossRef]
8. Chen, W.L.; Wang, G.M. Design of novel miniaturized fractal-shaped branch-line couplers. *Microw. Opt. Technol. Lett.* **2008**, *50*, 1198–1201. [CrossRef]
9. Xu, H.-X.; Wang, G.-M.; Liang, J.-G. Novel composite right-/left-handed transmission lines using fractal geometry and compact microwave devices application. *Radio Sci.* **2011**, *46*, 1–11. [CrossRef]
10. Xu, H.-X.; Wang, G.-M.; Chen, X.; Li, T.-P. Broadband balun using fully artificial fractal-shaped composite right/left handed transmission line. *IEEE Microw. Wirel. Compon. Lett.* **2011**, *22*, 16–18. [CrossRef]
11. Gomha, S.; EL-Sayed, M.; Shalaby, A.A.T.; Ahmed, S. Miniaturization of Branch-line couplers using open stubs and stepped impedance unit cells with meandering transmission lines. *Circuits Syst. Int. J. (CSIJ)* **2014**, *1*, 13–26.
12. Kumar, K.V.P.; Alazemi, A.J. A flexible miniaturized wideband branch-line coupler using shunt open-stubs and meandering technique. *IEEE Access* **2021**, *9*, 158241–158246. [CrossRef]
13. Kumar, M.; Islam, S.N.; Sen, G.; Parui, S.K.; Das, S. Design of miniaturized 10 dB wideband branch line coupler using dual feed and T-shape transmission lines. *Radioengineering* **2018**, *27*, 207–213. [CrossRef]
14. Nie, W.; Xu, K.-D.; Zhou, M.; Xie, L.-B.; Yang, X.-L. Compact narrow/wide band branch-line couplers with improved upper-stopband. *AEU-Int. J. Electron. Commun.* **2019**, *98*, 45–50. [CrossRef]
15. Siahkamari, H.; Jahanbakhshi, M.; Al-Anbagi, H.N.; Abdulhameed, A.A.; Pokorny, M.; Linhart, R. Trapezoid-shaped resonators to design compact branch line coupler with harmonic suppression. *AEU-Int. J. Electron. Commun.* **2022**, *144*, 154032. [CrossRef]

16. Wang, X.-Z.; Chen, F.-C.; Chu, Q.-X. A Compact Broadband 4 × 4 Butler Matrix with 360° Continuous Progressive Phase Shift. *IEEE Trans. Microw. Theory Tech.* **2023**, *71*, 3906–3914. [CrossRef]
17. Mextorf, H.; Schernus, W. A Novel Branch-Line Coupling Topology. *IEEE Trans. Microw. Theory Tech.* **2023**, *71*, 3644–3649. [CrossRef]
18. Roshani, S.; Yahya, S.I.; Roshani, S.; Rostami, M. Design and fabrication of a compact branch-line coupler using resonators with wide harmonics suppression band. *Electronics* **2022**, *11*, 793. [CrossRef]
19. Majidifar, S.; Hayati, M. New approach to design a compact triband bandpass filter using a multilayer structure. *Turk. J. Electr. Eng. Comput. Sci.* **2017**, *25*, 4006–4012. [CrossRef]
20. Pozar, D.M. *Microwave Engineering*; John Wiley & Sons: Hoboken, NJ, USA, 2011.
21. Tian, H.; Gao, J.; Su, M.; Wu, Y.; Liu, Y. A novel 3D two-ways folded microstrip line and its application in super-miniaturized microwave wireless components. *J. Electromagn. Waves Appl.* **2015**, *29*, 364–374. [CrossRef]
22. LaMeres, B.J. *Characterization of a Printed Circuit Board via*; University of Colorado at Colorado Springs: Colorado Springs, CO, USA, 2000.
23. Rawat, K.; Ghannouchi, F. Design of reduced size power divider for lower RF band using periodically loaded slow wave structure. In Proceedings of the 2009 IEEE MTT-S International Microwave Symposium Digest, Boston, MA, USA, 7–12 June 2009; pp. 613–616.
24. Eccleston, K.W.; Ong, S.H. Compact planar microstripline branch-line and rat-race couplers. *IEEE Trans. Microw. Theory Tech.* **2003**, *51*, 2119–2125. [CrossRef]
25. Mayer, B.; Knochel, R. Branchline-couplers with improved design flexibility and broad bandwidth. In Proceedings of the IEEE International Digest on Microwave Symposium, Dallas, TX, USA, 8–10 May 1990; pp. 391–394.

Disclaimer/Publisher's Note: The statements, opinions and data contained in all publications are solely those of the individual author(s) and contributor(s) and not of MDPI and/or the editor(s). MDPI and/or the editor(s) disclaim responsibility for any injury to people or property resulting from any ideas, methods, instructions or products referred to in the content.

Article

Investigation of a Circularly Polarized Metasurface Antenna for Hybrid Wireless Applications

Bikash Ranjan Behera [1], Mohammed H. Alsharif [2,*] and Abu Jahid [3,*]

1. Department of Electronics and Communication Engineering, Vel Tech Rangarajan Dr. Sagunthala R&D Institute of Science and Technology (Deemed-To-Be-University), Chennai 600062, Tamil Nadu, India; drbikash@veltech.edu.in
2. Department of Electrical Engineering, College of Electronics and Information Engineering, Sejong University, Seoul 05006, Republic of Korea
3. School of Electrical Engineering and Computer Science, University of Ottawa, 25 Templeton St., Ottawa, ON K1N 6N5, Canada
* Correspondence: malsharif@sejong.ac.kr (M.H.A.); ajahi011@uottawa.ca (A.J.)

Citation: Behera, B.R.; Alsharif, M.H.; Jahid, A. Investigation of a Circularly Polarized Metasurface Antenna for Hybrid Wireless Applications. *Micromachines* **2023**, *14*, 2172. https://doi.org/10.3390/mi14122172

Academic Editor: Mark L. Adams

Received: 13 November 2023
Revised: 25 November 2023
Accepted: 27 November 2023
Published: 29 November 2023

Copyright: © 2023 by the authors. Licensee MDPI, Basel, Switzerland. This article is an open access article distributed under the terms and conditions of the Creative Commons Attribution (CC BY) license (https://creativecommons.org/licenses/by/4.0/).

Abstract: The increasing prevalence of the Internet of Things (IoT) as the primary networking infrastructure in a future society, driven by a strong focus on sustainability and data, is noteworthy. A significant concern associated with the widespread use of Internet of Things (IoT) devices is the insufficient availability of viable strategies for effectively sustaining their power supply and ensuring their uninterrupted functionality. The ability of RF energy-harvesting systems to externally replenish batteries serves as a primary driver for the development of these technologies. To effectively mitigate concerns related to wireless technology, it is imperative to adhere strictly to the mandated limitations on electromagnetic field emissions. A TA broadband polarization-reconfigurable Y-shaped monopole antenna that is improved with a SADEA-tuned smart metasurface is one technique that has been proposed in order to accomplish this goal. A Y-shaped printed monopole antenna is first taken into consideration. To comprehend the process of polarization reconfigurability transitioning from linear to circular polarization (CP), a BAR 50-02 V RF PIN Diode is employed to shorten one of the parasitic conducting strips to the ground plane. A SADEA-driven metasurface, which utilizes the artificial intelligence-driven surrogate model-assisted differential evolution for antenna synthesis, is devised and positioned beneath the radiator to optimize performance trade-offs while increasing the antenna's gain and bandwidth. The ultimate prototype achieves the following: an impedance bandwidth of 2.58 GHz (3.27–5.85 GHz, 48.45%); an axial bandwidth of 1.25 GHz (4.19–5.44 GHz, 25.96%); a peak gain exceeding 8.45 dBic; and when a highly efficient rectifier is integrated, the maximum RF-DC conversion efficiency of 73.82% and DC output of 5.44 V are obtained. Based on the results mentioned earlier, it is considered appropriate to supply power to intelligent sensors and reduce reliance on batteries via RF energy-harvesting mechanisms implemented in hybrid wireless applications.

Keywords: monopole antenna; polarization; artificial intelligence (AI); SADEA; intelligent metasurfaces; IoT applications; RF energy harvesting; smart sensors

1. Introduction

Recent years have seen a surge in the popularity of low-power embedded devices and sensors for consumer and industrial needs based on the Internet of Things (IoT), prompting researchers to focus on developing alternatives to the battery-based power supplies [1]. It unleashed the potential of ambient RF energy harvesting [2], which makes use of the electromagnetic spectrum, an interesting prospect from the perspective of the device, since it has the ability to reduce the cost with a limited need for routine maintenance. It concerns how RF energies are used in real-time scenarios. Since RF energy harvesting requires a connection to and interaction with ambient electromagnetic waves [2], RF front-ends are considered a necessary part of the process. Evaluating the performance of the components based on their circular polarization (CP) characteristics allows for improved signal matching to be accomplished [3,4]. First and foremost, it must be ensured that incoming signals from LTE, ISM, WLAN, Wi-MAX, and 5G, are correctly received regardless of the antenna's direction. Thus, the front-end you choose is quite crucial [5,6]. Hence, we settled on the printed monopole antenna (PMAs) due to its small size, ease of analysis, good radiation efficiency, satisfactory radiation pattern, and sensitivity in the time domain.

Antenna technology has undergone rapid development in recent years. When designing effective antennas, the phenomenon of metamaterials was proposed [7], by taking into account the viewpoints of contemporary applications. It refers to something that is beyond the unusual electromagnetic (EM) properties that can be modified to take on new forms, i.e., an artificially driven engineered metasurface. In general, no material on this earth is thought to exhibit negative permeability. Every natural substance on the earth has positive permittivity, permeability, refractive index, etc. However, there is evidence of unusual negative qualities that manipulate waves and enhance the performance of antennas [8]. The evolution of the communication system has been facilitated by the emergence of several application areas due to such materials, particularly in the case of antennas [8]. As evidenced in the literature of more recent times, high-gain antennas [9] are frequently regarded as critical parts of wireless communication networks. Its importance comes from the fact that they increase signal strength, which they accomplish by minimizing interference and free space path loss (FSPL). So, the antenna performance evaluations often focus heavily on CP antenna gain. The gain has been enhanced in this article using MS approaches [9]. However, improving impedance bandwidth and CP characteristics (CP bandwidth, CP gain, and antenna efficiency) has not been investigated w.r.t. performance trade-offs [10–31]. That is why prospective applications call for a single antenna element with certain features, most notably the ability to provide enhanced CP. Vias [10], non-metasurfaces [11–15], and electromagnetic metasurfaces [16–31] are a few methods that have been used previously for the implementation of enhancing the antenna's performance. So, acquiring a broadband CP is one of the requirements for a successful polarization system, targeting IoT. The works reported in [10–29] are needed to pursue a surrogate model-assisted differential evolution for antenna synthesis, that is, a SADEA-driven smart metasurface, for attaining better outcomes.

In this paper, we look at whether it is possible to design a reconfigurable, broadband, printed monopole antenna by using the surrogate model-assisted differential evolution for antenna synthesis, that is, a SADEA-driven metasurface. Here, a BAR 50-02 V RF PIN Diode is used to short partial ground to one of the parasitic conducting strips (i.e., PCS_L) and create a broadside directional pattern while also incorporating the metasurface of evenly and symmetrically spaced unit cells. Thus, the antenna's gain was increased over its operating frequency range, and its impedance and axial bandwidths were also enhanced, contributing to stable antenna performance [32]. In a concurrent way, all three aspects of CP analysis [33] were addressed; (a) electric field distribution, (b) surface current distribution, and (c) far-field radiation are modelled and interpreted in a comprehensive analogy. Ahead of it, (d) amplitude-phase responses are also examined over the desired operating frequency band. So, the fact that CP is provided in the form of distinctive gain-bandwidth product (GBR) connections aids in persuading the reader that they exist. Towards these ends, the authors followed the DAVI principle: Designing, Analyzing, Validating, and Implementing. A flowchart showing the boarding process involved during this present work is shown in Figure 1.

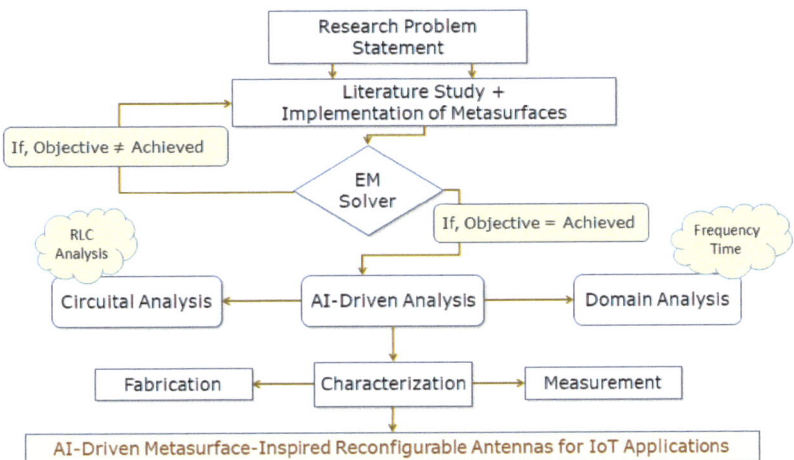

Figure 1. Flowchart explaining the workflow of the proposed work reported for the IoT application.

2. Antenna Design

The proposed antenna is fabricated on an FR-4 substrate by incorporating the twin parasitic conducting strips (PCSs) 1.16 mm from the upper edges of the partial ground plane, 21.6 mm in length and 1.84 mm in width, where the authors extended the ground plane following the performance trade-offs [33]. The communication in between the partial ground plane and one of the parasitic conducting strips (PCS_L) is crucial for CP features, achieved by shorting the partial ground and PCS_L using a BAR 50-02 V PIN Diode. A 50-Ω microstrip feedline is used for the input excitation. The proposed antenna is presented in Figure 2, with state-of-the-art results in Figures 3–12. During the OFF-state, it attains an impedance bandwidth (IBW) of 620 MHz with no CP bandwidth, whereas, in the ON-state, it attains an impedance bandwidth (IBW) of 2.11 GHz with a CP bandwidth (ARBW) of 460 MHz. The average antenna gain in both states of operation is around 3.2 dBi. With the implementation of a SADEA-driven metasurface reflector, it is possible to obtain broadband CP, CP gain of >8.2 dBic, antenna efficiency of >75%, and directional pattern.

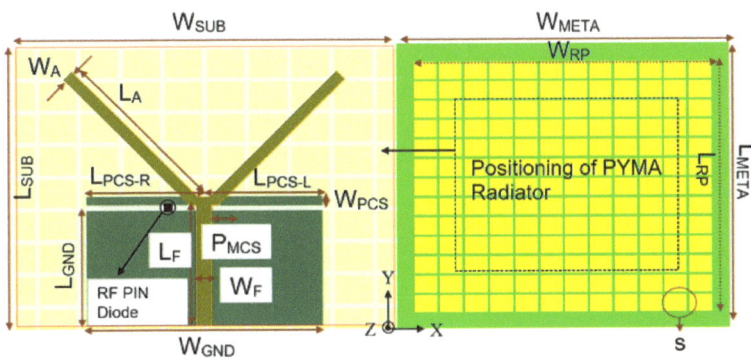

Top View (Transition from LP-to-CP)   Top View (Metasurface)

Dimensions: W_{SUB}=1.33λ_0, L_{SUB}=0.9λ_0, W_{GND}=0.83λ_0, L_{GND}=0.4λ_0, L_{PCS-R}=L_{PCS-L}=0.36λ_0, W_{PCS}=0.03λ_0, P_{MCS}=0.05λ_0, W_A=0.04λ_0, L_A=0.6λ_0, L_F=0.42λ_0, W_F=0.05λ_0, W_{RP}=1.78λ_0, L_{RP}= 1.48λ_0 and S=0.1λ_0 × 0.06λ_0 where, λ_0 is at 5 GHz.

(A)

Figure 2. Cont.

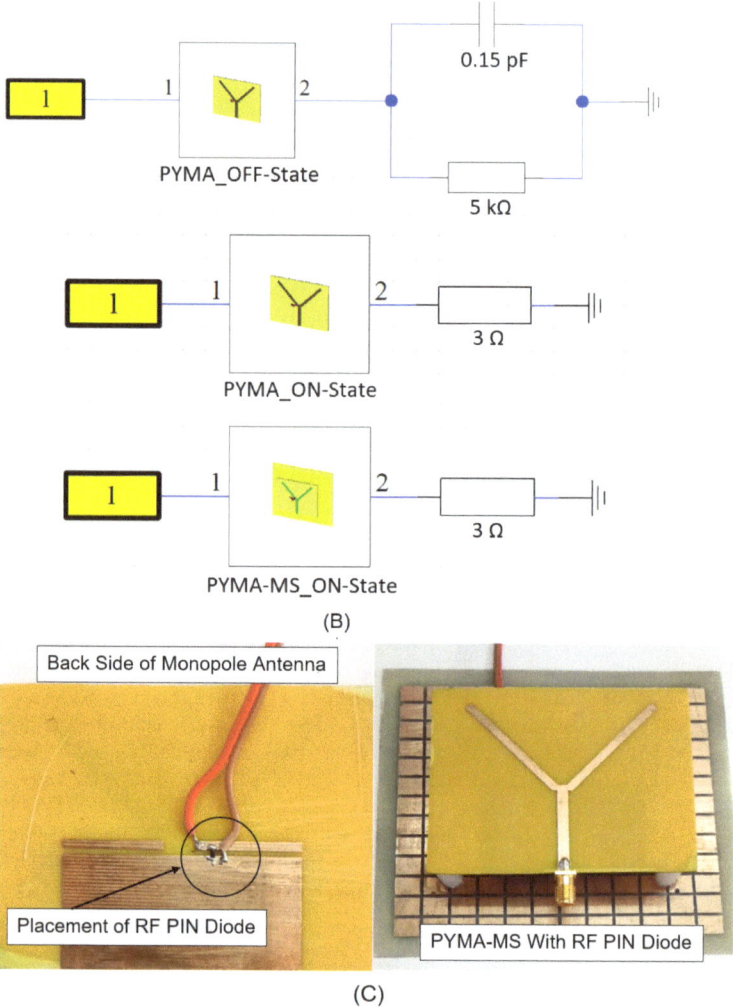

Figure 2. Design methodology of the proposed antenna: (**A**) schematic, (**B**) biasing, and (**C**) prototype.

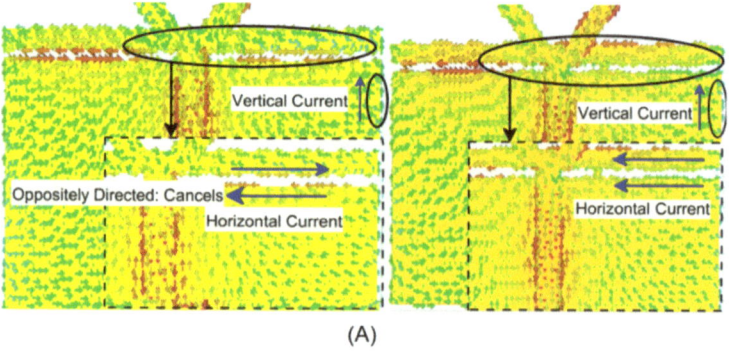

Figure 3. *Cont.*

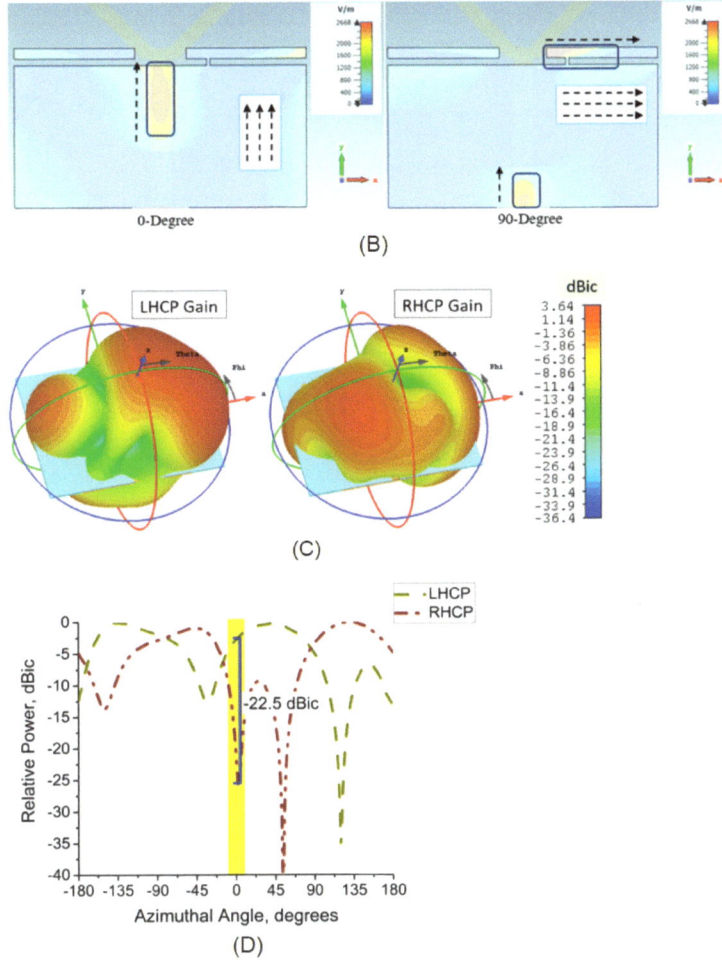

Figure 3. CP at 5 GHz: (**A**) 1st method; (**B**) 2nd method; and (**C**,**D**) 3rd method (3 fundamental aspects of CP analysis).

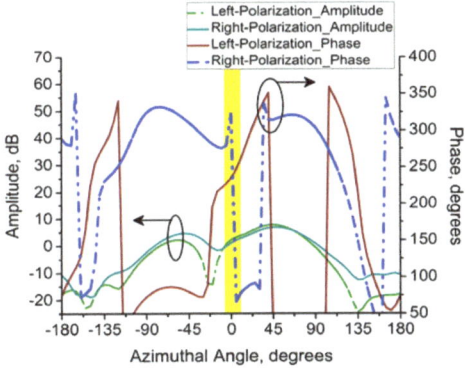

Figure 4. CP at 5 GHz (cont.): Amplitude and phase responses.

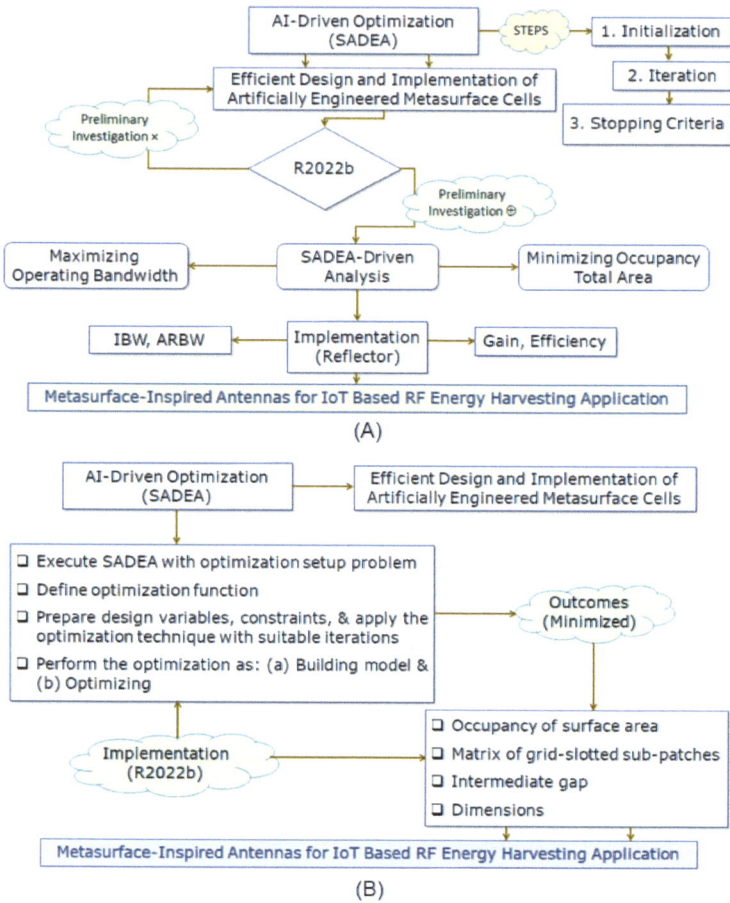

Figure 5. (**A**) SADEA: Process and (**B**) SADEA: Implementation using MATLAB 2022b platform.

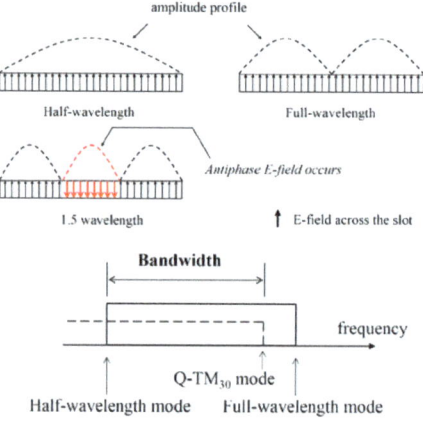

Figure 6. Generation of higher-order modes of artificially engineered metasurface (RMS).

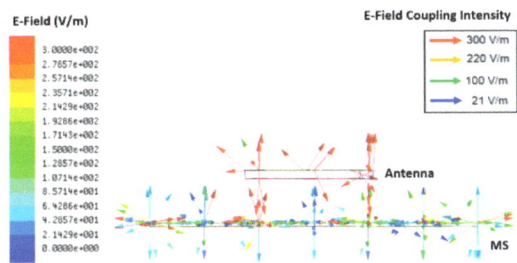

Figure 7. Coupling phenomena of artificially engineered metasurface (RMS).

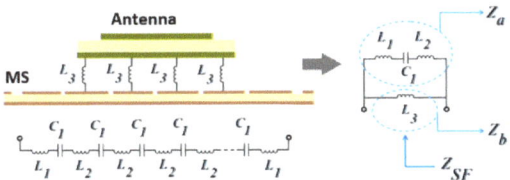

Figure 8. Circuit representation of artificially engineered metasurface (RMS).

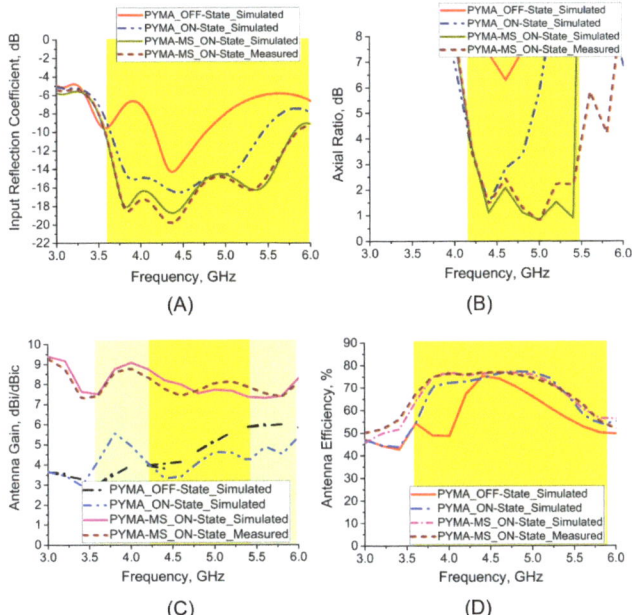

Figure 9. Antenna performances: (**A**) IBW, (**B**) ARBW, (**C**) antenna gain, and (**D**) antenna efficiency.

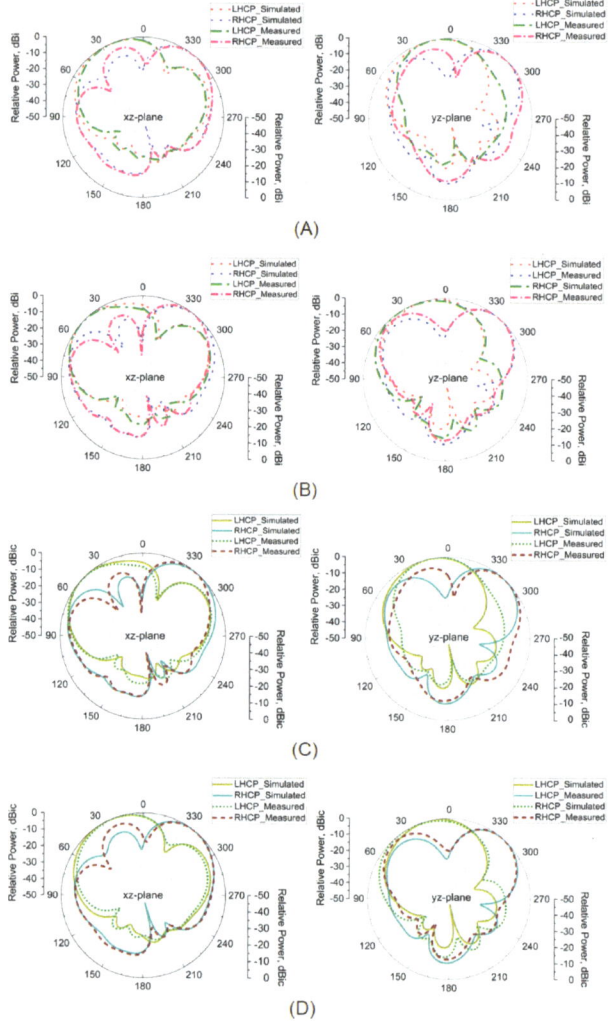

Figure 10. Radiation patterns: (**A**) 4.5 GHz (OFF-state), (**B**) 5 GHz (OFF-state), (**C**) 4.5 GHz (ON-state), and (**D**) 5 GHz (ON-state).

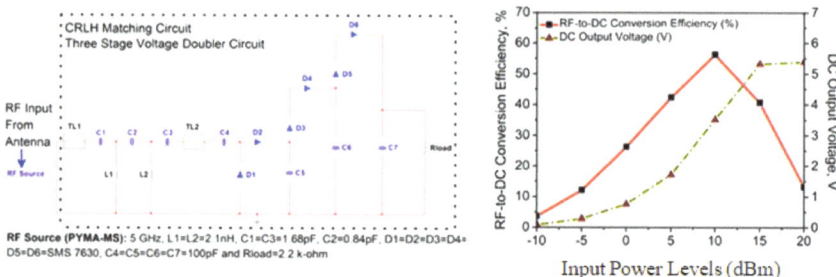

Figure 11. Rectifier circuit-I with its outcomes for the proposed AI-tuned CP metasurface antenna.

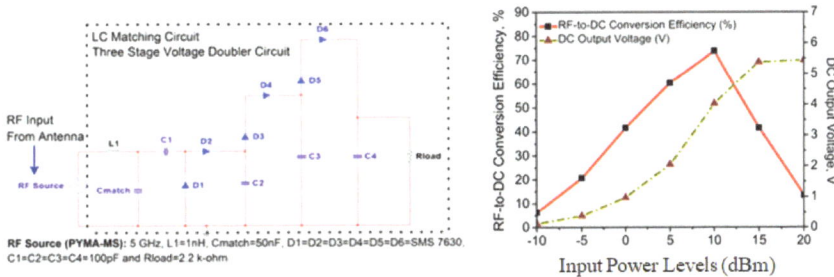

Figure 12. Rectifier circuit-II with its outcomes for the proposed AI-tuned CP metasurface antenna.

3. CP Mechanism: CEM Approach

Figures 3A–D and 4 show the different responses to CP analysis [33] persuaded at the 5 GHz band. Here, the surface current distribution is the CP analysis's initial and most fundamental aspect. Thus, in the OFF-state, parasitic conducting strips (PCSs) play no role in the non-active device without a connection to the partial ground plane. It correlates to the fact that the surface currents caused by the horizontal margins of the partial ground plane flow in the opposite direction. Due to mutual cancellation, only vertical surface currents on the printed monopole arm remain. This results in a wave with linear polarization (LP). When one of the parasitic conducting strips (PCS_L) is connected to the partial ground plane by the BAR50-02 V RF PIN Diode, the surface currents on the PCS_L and the partial ground plane are rearranged so that the resultant currents on the upper edge of the PCS_L and the lower edge of the partial ground plane do appear in the same direction. So, this accounts for horizontal surface currents while the ON-state is active. As a result of it, the viability of CP features depends on the presence of horizontal and vertical components, which can also be referred to as horizontal and vertical currents, making for the viability of CP.

Analysis of the CP can also be performed by looking out at the electric field distribution as the second primitive method. Therefore, the CP reconfigurable antenna realizes LHCP in the +z direction (i.e., outward in nature) when PCS_L is coupled to the partial ground plane with the BAR50-02 V RF PIN Diode. At 5 GHz, the rotation of electric field vectors goes from anti-clockwise to clockwise, which causes the phase to shift from 0° to 90°. Thus, the formation of LHCP (left-handed circular polarization) is verified through the presence of an orthogonal change in the electric field pattern.

The normalized radiation pattern constitutes the third and final primitive method for CP analysis. Here, a correlation is found w.r.t. the relative power at 5 GHz. The proposed antenna possesses exceptional LHCP properties. Techniques like the first, second, and third have been used to investigate the CP's explanation reported here. The amplitude and phase responses are presented in a much better manner so that the CP features can be understood.

4. AI-Tuned Design and Implementation of Metasurface (RMS)

Antenna Toolbox[TM] in MATLAB provides users access to the various functions and applications that may be used to design, analyze, and visualize antenna elements. These elements include the metasurface unit cells, which are an essential component of artificially engineered metasurfaces [32]. The question is whether it can be generated utilizing parametrically defined elements, arbitrary planar structures, or isolated three-dimensional forms. Because it uses an EM solver, it is just as competent as the methods of moments of computing all of the desired characteristics to carry out the final verification. The toolbox allows manually optimizing the antenna designs [34] so that they can be optimized in the most effective way possible. An approach for designing an antenna that is powered by artificial intelligence (AI) is known as SADEA, which stands for the surrogate model-assisted differential evolution for antenna synthesis. SADEA is an acronym for this method. It is based on the theories of machine learning [34] and evolutionary computation.

The global optimization is performed with the help of SADEA, and statistical learning techniques are used to build a surrogate model. Thus, it is of the utmost importance, in the context of the surrogate model-assisted optimization approach, that the process by which the surrogate modeling and optimization are made to cooperate successfully is the surrogate modeling and the optimization [35]. In addition to that, some ideas taken from an evolutionary search framework that takes into account surrogate models have been incorporated into SADEA [36,37]. Both the search engine that SADEA uses, called differential evolution (DE), and the machine learning technique that is used for surrogate modeling, known as a Gaussian process (GP), are described in [38,39]. So, in the case of metasurface-inspired antenna design, SADEA optimization appears to be utilized only in extremely unusual circumstances and is reported as very rare in the literature to design artificially engineered metasurfaces. The MATLAB R2022b platform initiates the process of designing an RMS layer to improve the antenna with consideration of the IoT application perspective, which can be seen in Figure 5.

By using the AI-driven SADEA method, the optimization goal is to achieve improvement in the outcomes: (a) maximizing operating bandwidth with center frequency at 5 GHz and (b) minimizing the area of occupation for the case of the designed metasurface layer (RMS). Thus, a critical observation is made during this analysis, as RMS plays a vital role in maintaining the antenna's performance, considering the trade-offs from an application perspective. Figure 9 shows the SADEA-optimized outcomes (simulation), along with its interpretation of the measured outcomes. So, in continuation with the motive of attaining improved outcomes, the execution is performed for the RMS reflector (with the optimized dimensions for the size of grid-slotted sub-patch cells, intermediate gap, and overall size of the RMS layer) at the height of $0.33\lambda_o$ below the Y-shaped monopole antenna to obtain good operating bandwidth, a high CP antenna gain, and enhanced directional features.

Here, the reflector layer is made in the form of RMS, which has a surface area of $1.78\lambda_o \times 1.48\lambda_o$, shown in Figure 2. It consists of grid-slotted sub-patches of the 12×12 cells, as each cell is $0.1\lambda_o \times 0.06\lambda_o$ and has an intermediate gap of $0.016\lambda_o$. Further, the sub-patches are well positioned on a rectangular-shaped PEC body having the dimensions of $2\lambda_o \times 1.65\lambda_o \times 0.02\lambda_o$ (λ_o = 5 GHz). They are merged to form a rectangular-shaped metasurface reflector, which is loaded with a polarization-reconfigurable monopole antenna.

Carrying out the analysis further, in Equation (1), other parameters, like the thickness of the substrate (h_{sub}), the relative permittivity of the substrate (ϵ_r), and λ_o, each have a role in determining the effective gap ($h_{air-gap}$) in between the radiator and AI-tuned RMS. The theoretical $h_{air-gap}$ is 21.85 mm, but the simulated $h_{air-gap}$ measures 20 mm. This study also shows the final mathematical formula, which is then used to look at how to place reflectors in relation to the radiator, taking into account how often it works and not its structure. Its building blocks are a gridlike pattern of tiny elements printed on a grounded slab, with or without shorting vias. Through this analytical approach, one can easily approximate the $h_{air-gap}$:

$$h_{air-gap} = 0.42\lambda_o - h_{sub}\sqrt{\epsilon_r} \qquad (1)$$

The average CP gain increases by a factor of 3.65 times, from 2.35 dBic to 8.58 dBic. More importantly, the antenna's IBW increases by a factor of 1.08 times, from 2.11 GHz to 2.28 GHz, and its ARBW increases by a factor of 2.7 times, from 460 MHz to 1.25 GHz. This is motivated by geometric understanding, as the development of such antenna performances aims to realize the performance trade-offs. So, the proposed RMS unit cells are shown to concentrate radiation fields for the occurrence of quasi-TM_{30} modes (i.e., the transformation of TM_{10} to TM_{30} modes) using the finite integral technique (FIT), resulting in a more uniform E-field distribution and ultimately, better impedance performance over a wide frequency range. Higher modes [40] are shown to be caused by RMS unit cells in the form of grid-slotted sub-patches in Figure 6. Here, the impedance matching is improved in the antenna design by minding the gaps between sub-patches where radiation can behave differently [41–45], which holds the capacity of manipulating these EM waves towards constructive interference.

The coupling phenomena and circuit model of the proposed antenna are depicted in Figures 7 and 8. The RMS layer is made up of a series of inductive and capacitive layers that are stacked one on top of the other. Layer L_1 serves as the RMS external radiating border and is connected to the inductive unit cells that are located on layer L_2, which capacitive layer C_1 then separates. Following it, the inductance L_3 layer establishes connections with the ground and the RMS layers, because the overall inductances, which include the dielectric substrate of the proposed antenna as well as the air substrate that is located in between the dielectric substrate of the antenna and the metallic RMS component, are theoretically represented by L. To simplify the analysis, a generalized equivalent circuit will be stated as parallel impedances ($Z_a || Z_b$). The equivalent surface impedance is calculated, and its frequency dependency w.r.t. RLC is shown in Equations (2) and (3):

$$\eta_{eq} = \frac{j\omega L_3(1 - \omega^2 L_1 C_1 - \omega^2 L_2 C_1)}{1 - \omega^2 L_1 C_1 - \omega^2 L_2 C_1 - \omega^2 L_3 C_1} \quad (2)$$

$$f_R = \frac{1}{2\pi\sqrt{(L_1 + L_2 + L_3)C_1}} \quad (3)$$

5. CP Mechanism: GBR Approach

To further evaluate the CP, the gain-bandwidth product relationship, Equation (4) states:

$$C_{criteria} = F(BW_{3-dB}, G_{3-dB}) \quad (4)$$

Then,

$$C_{criteria} = BW_{3-dB} \times G_{3-dB} \quad (5)$$

Further,

$$C_{criteria} = \frac{BW_{3-dB} \times G_{3-dB}}{100} \quad (6)$$

Equation (6) can be re-written as follows:

$$C_{criteria-1} = \frac{BW_{3-dB} \times G_{3-dB(avg)}}{100} \quad (7)$$

$$C_{criteria-2} = \frac{BW_{3-dB} \times G_{3-dB(max)}}{100} \quad (8)$$

$$C_{criteria-3} = \frac{BW_{3-dB} \times G_{3-dB(min)}}{100} \quad (9)$$

$$C_{criteria-4} = \frac{BW_{3-dB} \times G_{3-dB(peak)}}{100} \quad (10)$$

The limitations of traditional comparison methods that were discovered earlier are addressed in a major way by $C_{criteria-1}$-$C_{criteria-4}$. In traditional comparison methods, these CP antennas are practically compared to only one feature. Equation (6) is the generalized equation, often known as the gain-bandwidth product (GBR), for evaluating the CP antennas. So, this particular study was presented in the literature for the very first time as a research evaluation. Comparative research with work already published in [10–29] is included in Table 1 to bolster the study by providing further evidence in this context. By including major evaluative components such as the CP bandwidth and CP gain, this article offers a novel perspective on the investigation of CP that is both informative and eye-opening. Further, the examination of CP features that were covered before is implemented in every case, even though the antenna shape/frequency at which it works differ from one another.

Table 1. Comparison of the performances of the proposed antenna with the ones reported in [10–29].

Ref.	IBW	ARBW	CP Gain	Peak Gain	$C_{\text{criteria-1}}$	$C_{\text{criteria-4}}$
[10]	7.29%	5.5%	6.5 dBic	7.1 dBic	0.35	0.39
[11]	28.6%	14.2%	4.4 dBic	4.4 dBic	0.62	0.62
[12]	31.6%	20.8%	6.9 dBic	6.9 dBic	1.43	1.43
[13]	35.6%	16.3%	5.5 dBic	5.5 dBic	0.89	0.89
[14]	29.6%	20.7%	7.44 dBic	7.44 dBic	1.54	1.54
[15]	12.68%	12.68%	6.8 dBic	6.8 dBic	0.86	0.86
[16]	20.89%	11.4%	5.2 dBic	7.5 dBic	0.59	0.85
[17]	19.6%	15.8%	4.25 dBic	6.5 dBic	0.67	1.02
[18]	33.7%	16.5%	5.8 dBic	5.9 dBic	0.95	0.97
[19]	17.8%	3.6%	7.8 dBic	8.2 dBic	0.28	0.29
[20]	27.5%	7.8%	6.1 dBic	5.8 dBic	0.47	0.45
[21]	20%	20%	6.1 dBic	6.6 dBic	1.22	1.32
[22]	8.4%	4%	6.1 dBic	6.95 dBic	0.24	0.27
[23]	31.8%	20.4%	7.47 dBic	8.05 dBic	1.52	1.64
[24]	15.1%	12.4%	6.15 dBic	6.5 dBic	0.76	0.81
[25]	27%	2.5%	3.02 dBic	3.02 dBic	0.07	0.07
[26]	1.55%	1.05%	7.8 dBic	8.2 dBic	0.07	0.08
[27]	25.6%	25.6%	7.3 dBic	6.25 dBic	1.6	1.86
[28]	23%	14%	—	—	—	—
[29]	7%	8.3%	—	—	—	—
Work	48.45%	25.96%	8.45 dBic	9.15 dBic	2.19	2.37

6. Measurement

In Figure 2C, the fabricated prototype of the proposed polarization-reconfigurable monopole antenna inspired by RMS is shown. Figures 9A and B shows that the simulated antenna measured impedance bandwidths (IBWs) as follows: 3.59–5.78 GHz, 2.19 GHz, 46.71% and 3.57–5.84 GHz, 2.27 GHz, 48.45%. The simulated and measured axial bandwidths (ARBWs) are as follows: 4.22–5.28 GHz, 1.06 GHz, 23.31% and 4.19–5.44 GHz, 1.25 GHz, 25.96%, respectively. Similarly, the average simulated and measured CP antenna gain lies between 7.5 and 9.35 dBic, with the average antenna efficiency of >75% in their operating bands, shown in Figures 9C and D. Furthermore, the radiation patterns at 4.5 and 5 GHz are presented in Figures 10A–D.

5G [2] is widely regarded as an advanced wireless technology that offers significant economic potential for facilitating the translation of ideas into practical solutions, thereby overcoming obstacles and fostering the creation of diverse strategies. However, the path loss associated can be detrimental to the overall system performance. So, a plausible solution to mitigate this will be to use a highly directive, efficient antenna with enhanced performance capabilities. Thus, to operate in the 5G band, these antennas must also have wide operating bandwidth (>1 GHz) covering the entire allotted band along with other performance trade-offs [32], reported here as a widespread solution [46–51].

7. RF Energy Harvesting: Simulation Perspective from IoT Application

The design and implementation of a front-end that operates with enhanced performances, due to the incorporation of RMS, and embedded with the multi-stage rectifier circuit system (GVDs) are said to be important to satisfy these types of trade-offs [2].

The overall results that are displayed in Figures 11 and 12 are appropriate for powering these sensors that can be found in low-power devices (such as wearables, medical, and healthcare plug-based kits), which need a consistent DC output voltage of 2.4–5.5 V to function properly. In a nutshell, the application of this particular technology will progressively become an essential component in the process of developing effective systems. Despite the various challenges it faces, once these obstacles are overcome, it will usher in an age of clean, green, and sustainable energy suitable for 5G and 6G applications [52,53].

To briefly summarize and assess the RF energy-harvesting capabilities, it is combined with the GVD rectifier circuit. Within this circuit, the theoretical analogies are supplied for the RF-DC power conversion efficiency (η_o, %) and DC output voltage (V_{out}, V). The three-stage Greinacher Voltage Doubler, complete with a CRLH- and LC-matching rectifier circuit, has been designed and incorporated into the proposed antenna. It is tested for input power levels (P_{in}) between 0 and +20 dBm, where $\eta_o > 60\%$ and $V_{out} > 2.1$ V at 5 dBm, when it is simulated at the ADS platform with a load resistance (R_{load}) of 2.2 kΩ. Table 2 contains an in-depth performance study, compared with some of the reported works in [46–51]. Finally, the η_o is calculated by using Equation (11), and multi-stage GVD configuration is explained based on Equations (12) and (13).

$$\eta_0(\%) = \frac{P_{load}}{P_{incident}} = \frac{V^2_{out}}{P_{in} \times R_{load}} \quad (11)$$

Theoretical insights into the proposed rectifier model are also studied prior to the simulation in the ADS environment with the LSSP scenario. So, in this case, each stage with its own GVD configuration is regarded as a single battery with an open circuit output voltage ($V_{o.c.}$), internal resistance (R_{int}), and load resistance (R_{load}). The DC output voltage is given by:

$$V_{out} = \frac{V_{o.c.}}{R_{int} + R_{load}} \times R_{load} \quad (12)$$

For n number of stages in series and connected to R_{load}, then V_{out} is represented as:

$$V_{out} = \frac{nV_{o.c.}}{nR_{int} + R_{load}} \times R_{load} \quad (13)$$

So, it is observed in our analysis that the total number of stages in the system has a substantial impact on the output voltage. The utilization of a partial ground plane in the proposed antenna was what resulted in the maximization of the captured energy, which can energize the sensors used in IoT applications. This realization of a higher amount of DC output voltage can be attributed to utilizing a partial ground plane concept. In the continuation of our inquiry, we have found that the application of RMS improved the gain of the RF front-end. When evaluating the effectiveness of a rectenna model in an RF energy harvesting system, one of the most important metrics to look at is the amount of power that is received by the antenna. This is because the power received by the antenna is directly proportional to the model's performance. Given the particular parameters of operating frequency and the availability of RF signals, the only feasible option to maximize the results of RF energy-harvesting is to increase the 3 dB gain of the CP-printed monopole antenna.

Table 2. Comparison of RF energy-harvesting features with the reported ones in [46–51].

Ref.	Gain	P_{in}	η_o	V_{out}
[46]	5.85 dBic (CP)	5 dBm	50%	—
[47]	6.9 dBi (LP)	5 dBm	—	0.1 V
[48]	7.3 dBi (LP)	5 dBm	14%	1.1 V
[49]	5.01 dBic (CP)	5 dBm	43%	1.16 V
[50]	5.5 dBi (LP)	5 dBm	5%	0.2 V
[51]	2.6 dBi (LP)	5 dBm	55%	—
Present Work	>8.45 dBic (CP)	5 dBm	43% (I)	1.7 V (I)
	>8.45 dBic (CP)	5 dBm	61% (II)	2.1 V (II)

8. Implementation Perspective: A Quick Overview

- This simple antenna employs the monopole and parasitic conducting strip configurations. Because the polarization pattern is meant to look like a CP, there is no need for electromagnetic circuits for it to work well.
- The incorporation of an RF PIN Diode (BAR 50-02V from Infineon Technologies) in a unique manner avoids using additional lumped parts due to planar configuration and less complex antenna design to attain polarization reconfigurability (i.e., LP to CP).

- CP is examined using a novel gain-bandwidth product. Also, four methodologies are shown to highlight the significance of CP feature analysis at the operating bandwidth.
- With the addition of a SADEA-driven metasurface (RMS) reflector in a single layer, there is a significant improvement in impedance and axial ratio bandwidth responses (broadband characteristics). An enhanced CP antenna accompanies this improvement gain > 8.45 dBic, which provides a superior performance in comparison to [10–29], from the perspective of IoT-inspired RF energy harvesting, aiming towards energy-efficient communication and computing technologies.
- Here, the η_o and V_{out} are computed by ADS. The proposed SADEA-driven metasurface antenna is tested with the CRLH- and LC-based Grienacher voltage doubler circuits (GVD). The antenna system has CRLH- and LC-based Grienacher voltage doubler circuits (GVD) built into it so that the potential utility of the system can be tested out in relation to an application that uses RF energy harvesting, which presents a more effective method towards implementing rectifiers than those stated in [46–51]. Also, in this case, the maximum V_{out} that can be attained is 5.39 V (for I) and 5.44 V (for II), and the maximum η_o that can be attained is 56.28% (for I) and 73.82% (for II), respectively. A comparative study in this regard is already given in Table 2.

9. AI-Enabled IoT Applications towards Smart Living: A Future Research Direction

Self-sustainable smart sensors through the RF energy-harvesting mechanisms can be a path-breaking innovation for elderly people, as they require less maintenance and supervision. One of the major concerns in the utilization of these sensors by elderly people is the way they are powered; there is always an increasing demand for these devices to be automated, especially for those who have memory problems. So, there exists significant research need for automating the battery charging process in these wearables, and on the other hand, the continuous availability of alternate energy sources from the open space, rendering a viable option for harnessing RF energy. In recent years the concept of energy harvesting has emerged as a promising solution, which, if incorporated into these low-powered smart devices, has a promising scope [52–54].

10. Conclusions

This work examines a SADEA-driven metasurface-inspired printed monopole antenna. It offers a broadened IBW and ARBW with a measured CP gain of >8.45 dBic and antenna efficiency of >75% in its operating bands. Along with their physical insights, intuition about the CP is also presented with three different aspects based on computational electromagnetism. A complete design is given for the wide-scale solution to a problem that the printed monopole antennas have, mainly when their CP characteristics are analyzed, as the unique gain-bandwidth product relationship, which is rarely seen in the literature, targeting IoT-inspired applications. As a result, the work that is being reported here combines electromagnetics with artificial intelligence-driven SADEA to achieve better results from an application perspective regarding CP characteristics. This is carried out with the goal of satisfying the primary needs of the human race, which are aimed at powering the sensors (known as RF energy harvesting).

Author Contributions: Conceptualization, B.R.B. and M.H.A.; methodology, B.R.B. and M.H.A.; writing—original draft preparation, B.R.B. and M.H.A.; writing—review and editing, B.R.B., M.H.A. and A.J.; project administration, M.H.A. and A.J.; funding acquisition, M.H.A., and A.J. All authors have read and agreed to the published version of the manuscript.

Funding: This research has received no external funding.

Institutional Review Board Statement: Not applicable.

Informed Consent Statement: Not applicable.

Data Availability Statement: Data are contained within the article.

Conflicts of Interest: The authors declare no conflict of interest.

References

1. Divakaran, S.K.; Krishna, D.D.; Nasimuddin, N. RF energy harvesting systems: An overview and design issues. *Int. J. Microw.-Comput.-Aided Eng.* **2019**, *29*, e21633. [CrossRef]
2. Behera, B.R.; Meher, P.R.; Mishra, S.K. Microwave antennas-An intrinsic part of RF energy harvesting systems: A contingent study about design methodologies and state-of-the-art technologies in current scenario. *Int. J. Microw.-Comput.-Aided Eng.* **2020**, *30*, e22148. [CrossRef]
3. Toh, B.Y.; Cahill, R.; Fusco, V.F. Understanding and measuring circular polarization. *IEEE Trans. Educ.* **2003**, *46*, 313–318.
4. Sahu, N.K.; Mishra, S.K. Compact dual-band dual-polarized monopole antennas using via-free metasurfaces for off-body communications. *IEEE Antennas Wirel. Propag. Lett.* **2022**, *21*, 1358–1362. [CrossRef]
5. Mouapi, A. Radiofrequency energy harvesting systems for internet of things applications: A comprehensive overview of design issues. *Sensors* **2022**, *22*, 8088. [CrossRef] [PubMed]
6. Behera, B.R.; Meher, P.R.; Mishra, S.K. Metasurface superstrate inspired printed monopole antenna for RF energy harvesting application. *Prog. Electromagn. Res. C* **2021**, *110*, 119–133. [CrossRef]
7. Milias, C.; Muhammad, B.; Kristensen, J.T.B.; Mihovska, A.; Hermansen, D.D.S. Metamaterial-inspired antennas: A review of the state of the art and future design challenges. *IEEE Access* **2021**, *9*, 89846–89865. [CrossRef]
8. Iqbal, K.; Khan, Q.U. Review of metasurfaces through unit cell design and numerical extraction of parameters and their applications in antennas. *IEEE Access* **2022**, *10*, 112368–112391. [CrossRef]
9. Esmail, B.A.F.; Koziel, S.; Szczepanski, S. Overview of planar antenna loading metamaterials for gain performance enhancement: The two decades of progress. *IEEE Access* **2022**, *10*, 27381–27403. [CrossRef]
10. Elahi, M.; Altaf, A.; Yang, Y.; Lee, K.Y.; Hwang, K.C. Circularly polarized dielectric resonator antenna with two annular vias. *IEEE Access* **2021**, *21*, 41123–41128. [CrossRef]
11. Lin, W.; Wong, H. Polarization reconfigurable wheel-shaped antenna with conical-beam radiation pattern. *IEEE Trans. Antennas Propag.* **2015**, *63*, 491–499. [CrossRef]
12. Lin, W.; Wong, H. Wideband circular-polarization reconfigurable antenna with L-shaped feeding probes. *IEEE Antennas Wirel. Propag. Lett.* **2017**, *16*, 2114–2117. [CrossRef]
13. Chen, Z.; Wong, H. Liquid dielectric resonator antenna with circular polarization reconfigurability. *IEEE Trans. Antennas Propag.* **2018**, *66*, 444–449. [CrossRef]
14. Yang, W.W.; Dong, X.Y.; Sun, W.J.; Chen, J.X. Polarization reconfigurable broadband dielectric resonator antenna with a lattice structure. *IEEE Access* **2018**, *6*, 21212–21219. [CrossRef]
15. Wang, S.; Yang, D.; Geyi, W.; Zhao, C.; Ding, G. Polarization-reconfigurable antenna using combination of circular polarized modes. *IEEE Access* **2021**, *9*, 45622–45631. [CrossRef]
16. Zhu, H.L.; Cheung, S.W.; Liu, X.H.; Yuk, T.I. Design of polarization reconfigurable antenna using metasurface. *IEEE Trans. Antennas Propag.* **2014**, *62*, 2891–2898. [CrossRef]
17. Yang, W.; Che, W.; Jin, H.; Feng, W.; Xue, Q. A polarization-reconfigurable dipole antenna using polarization rotation AMC structure. *IEEE Trans. Antennas Propag.* **2015**, *63*, 5305–5315. [CrossRef]
18. Wu, Z.; Li, L.; Li, Y.; Chen, X. Metasurface superstrate antenna with wideband circular polarization for satellite communication application. *IEEE Antennas Wirel. Propag. Lett.* **2016**, *15*, 374–377. [CrossRef]
19. Wu, F.; Luk, K. Single-port reconfigurable magneto-electric dipole antenna with quad-polarization diversity. *IEEE Trans. Antennas Propag.* **2017**, *65*, 2289–2296. [CrossRef]
20. Lin, W.; Wong, H.; Ziolkowski, R.W. Circularly polarized antenna with reconfigurable broadside and conical beams facilitated by a mode switchable feed network. *IEEE Trans. Antennas Propag.* **2018**, *66*, 996–1001. [CrossRef]
21. Lin, W.; Chen, S.L.; Ziolkowski, R.W.; Guo, Y.J. Reconfigurable, wideband, low-profile, circularly polarized antenna & array enabled by an artificial magnetic conductor ground. *IEEE Trans. Antennas Propag.* **2018**, *66*, 1564–1569.
22. Chen, X.; Zhao, Y. Dual-band polarization and frequency reconfigurable antenna using double layer metasurface. *AEUE-Int. J. Electron. Commun.* **2018**, *95*, 82–87. [CrossRef]
23. Zheng, Q.; Guo, C.; Ding, J. Wideband and low RCS circularly polarized slot antenna based on polarization conversion of metasurface for satellite communication application. *Microw. Opt. Technol. Lett.* **2018**, *60*, 679–685. [CrossRef]
24. Tran, H.H.; Park, H.C. Wideband reconfigurable antenna with simple biasing circuit and tri-polarization diversity. *IEEE Antennas Wirel. Propag. Lett.* **2019**, *18*, 2001–2005. [CrossRef]
25. Li, W.; Wang, Y.M.; Hei, Y.; Li, B.; Shi, X. A compact low-profile reconfigurable metasurface antenna with polarization and pattern diversities. *IEEE Antennas Wirel. Propag. Lett.* **2020**, *7*, 1170–1174. [CrossRef]
26. Chen, A.; Ning, X. A pattern and polarization reconfigurable antenna with metasurface. *Int. J. Microw.-Comput.-Aided Eng.* **2021**, *31*, e22312. [CrossRef]
27. Tran, H.H.; Bui, C.D.; Nguyen-Trong, N.; Nguyen, T.K. A wideband non-uniform metasurface-based circularly polarized reconfigurable antenna. *IEEE Access* **2021**, *9*, 42325–42332. [CrossRef]
28. Li, W.; Xia, S.; He, B.; Chen, J.; Shi, H.; Zhang, A.; Li, Z.; Xu, Z. A reconfigurable polarization converter using active metasurface and its application in horn antenna. *IEEE Trans. Antennas Propag.* **2016**, *64*, 5281–5290. [CrossRef]

29. Van Aardt, R.; Joubert, J.; Odendaal, J.W. A dipole with reflector-backed active metasurface for the linear-to-circular polarization reconfigurability. *Materials* **2022**, *15*, 3026. [CrossRef]
30. Yao, L.; Lin, H.; Koshelev, K.; Zhang, F.; Yang, Y.; Wu, J.; Kivshar, Y.; Jia, B. Full-stokes polarization perfect absorption with diatomic metasurfaces. *Nano Lett.* **2021**, *21*, 1090–1095.
31. Wilson, N.C.; Shin, E.; Bangle, R.E.; Nikodemski, S.B.; Vella, J.H.; Mikkelsen, M.H. Ultrathin pyroelectric photodetector with integrated polarization-sensing metasurface. *Nano Lett.* **2023**, *23*, 8547–8552. [CrossRef]
32. Mohanty, A.; Behera, B.R.; Esselle, K.P.; Alsharif, M.H.; Jahid, A.; Mohsan, S.A.H. Investigation of a dual-layer metasurface-inspired fractal antenna with dual-polarized/-modes for 4G/5G applications. *Electronics* **2022**, *11*, 2371. [CrossRef]
33. Meher, P.R.; Behera, B.R.; Mishra, S.K. A compact circularly polarized cubic DRA with unit-step feed for Bluetooth/ISM/Wi-Fi/Wi-MAX applications. *AEUE-Int. J. Electron. Commun.* **2021**, *128*, 153521. [CrossRef]
34. Grout, V.; Akinsolu, M.O.; Liu, B.; Lazaridis, P.I.; Mistry, K.K.; Zaharis, Z.D. Software solutions for antenna design exploration: A comparison of packages, tools, techniques, and algorithms for various design challenges. *IEEE Antennas Propag. Mag.* **2019**, *61*, 48–59. [CrossRef]
35. Liu, B.; Aliakbarian, H.; Ma, Z.; Vandenbosch, G.A.E.; Gielen, G.; Excell, P. An efficient method for antenna design optimization based on evolutionary computation and machine learning techniques. *IEEE Trans. Antennas Propag.* **2014**, *62*, 7–18. [CrossRef]
36. Liu, B.; Koziel, S.; Ali, N. SADEA-II: A generalized method for efficient global optimization of antenna design. *J. Comput. Des. Eng.* **2017**, *4*, 86–97. [CrossRef]
37. Liu, B.; Akinsolu, M.O.; Ali, N.; Abd-Alhameed, R. Efficient global optimisation of microwave antennas based on a parallel surrogate model-assisted evolutionary algorithm. *IET Microwaves Antennas Propag.* **2019**, *13*, 149–155. [CrossRef]
38. Akinsolu, M.; Liu, B.; Grout, V.; Lazaridis, P.I.; Mognaschi, M.E.; Barba, P.D. A parallel surrogate model assisted evolutionary algorithm for electromagnetic design optimization. *IEEE Trans. Emerg. Top. Comput. Intell.* **2019**, *3*, 93–105. [CrossRef]
39. Liu, B.; Akinsolu, M.O.; Song, C.; Hua, Q.; Excell, P.; Xu, Q.; Huang, Y.; Imran, M.A. An efficient method for complex antenna design based on a self adaptive surrogate model-assisted optimization technique. *IEEE Trans. Antennas Propag.* **2021**, *69*, 2302–2315. [CrossRef]
40. Zhang, J.; Akinsolu, M.O.; Liu, B.; Vandenbosch, G.A.E. Automatic AI-driven design of the mutual coupling reducing topologies for the frequency reconfigurable antenna arrays. *IEEE Trans. Antennas Propag.* **2021**, *69*, 1831–1836. [CrossRef]
41. Zheng, Q.; Guo, C.; Ding, J.; Akinsolu, M.O.; Liu, B.; Vandenbosch, G.A.E. A wideband low-RCS metasurface-inspired circularly polarized slot array based on the AI-driven antenna design optimization algorithm. *IEEE Trans. Antennas Propag.* **2022**, *70*, 8584–8589. [CrossRef]
42. Mohanty, A.; Behera, B.R.; Nasimuddin, N. Hybrid metasurface loaded tri-port compact antenna with gain enhancement and pattern diversity. *Int. J. Microw. Comput. Eng.* **2021**, *31*, e22795. [CrossRef]
43. Jhansi, Y.; Poojitha, P.; Mahigeethika, D.; Behera, B.R.; Pal, D. Investigation of a FSS inspired printed monopole antenna with gain enhancement for 5G applications. In Proceedings of the 2023 IEEE International Conference on Recent Advances in Electrical, Electronics, Ubiquitous Communication, and Computational Intelligence (RAEEUCCI), Chennai, India, 19–21 April 2023; pp. 1–4.
44. Sneha, D.S.; Rajesh, M.; Sravani, V.; Behera, B.R.; Pal, D. Non-uniform metasurface-based microstrip patch antenna with bandwidth and gain enhancement for wireless applications. In Proceedings of the 2023 IEEE International Conference on Recent Advances in Electrical, Electronics, Ubiquitous Communication, and Computational Intelligence (RAEEUCCI), Chennai, India, 19–21 April 2023; pp. 1–4.
45. Xu, K. Silicon electro-optic micro-modulator fabricated in standard CMOS technology as components for all silicon monolithic integrated optoelectronic systems. *J. Micromech. Microeng.* **2021**, *21*, 054001. [CrossRef]
46. Jie, A.M.; Nasimuddin Karim, M.F.; Chandrasekaran, K.T. A wide-angle circularly polarized tapered-slit patch antenna with a compact rectifier for energy harvesting systems. *IEEE Antennas Propag. Mag.* **2019**, *61*, 94–111. [CrossRef]
47. Bakkali, A.; Pelegri-Sebastia, J.; Sogorb, T.; Llario, V.; Bou-Escriva, A. A dual-band antenna for RF energy harvesting systems in wireless sensor networks. *J. Sens.* **2016**, *5725836*. [CrossRef]
48. Singh, N.; Kanaujia, B.K.; Beg, M.T.; Khan, T.; Kumar, S. A dual polarized multiband rectenna for RF energy harvesting. *AEUE-Int. J. Electron. Commun.* **2018**, *93*, 123–131. [CrossRef]
49. Surender, D.; Halimi, M.A.; Khan, T.; Talukdar, F.A.; Antar Yahia, M.M. Circularly polarized DRA rectenna for 5G and Wi-Fi bands RF energy harvesting in smart city applications. *IETE Tech. Rev.* **2022**, *39*, 880–893. [CrossRef]
50. Dardeer, O.M.; Elsadek, H.A.; Abdallah, E.A.; Elhennawy, H.M. A dual band circularly polarized rectenna for RF energy harvesting applications. *ACES J.* **2019**, *34*, 1594–1600.
51. Chandrasekaran, K.T.; Agarwal, K.; Alphones, A.; Mitra, R.; Karim, M.F. Compact dual-band metamaterial-based high-efficiency rectenna: An application for ambient electromagnetic energy harvesting. *IEEE Antennas Propag. Mag.* **2020**, *62*, 18–29. [CrossRef]
52. Alsharif, M.H.; Kelechi, A.H.; Albreem, M.A.; Chaudhry, S.A.; Zia, M.S.; Kim, S. Sixth generation (6G) wireless networks: Vision, research activities, challenges and potential solutions. *Symmetry* **2020**, *12*, 676. [CrossRef]

53. Arun Kumar, A.; Nanthaamornphong, A.; Selvi, R.; Venkatesh, J.; Alsharif, M.H.; Uthansakul, P.; Uthansakul, M. Evaluation of 5G techniques affecting the deployment of smart hospital infrastructure: Understanding 5G, AI and IoT role in smart hospital. *Alex. Eng. J.* **2023**, *83*, 335–354. [CrossRef]
54. Behera, B.R.; Mishra, S.K. Investigation of a broadband circularly polarized printed monopole antenna for RF energy harvesting application. In Proceedings of the 2022 URSI Regional Conference on Radio Science (URSI-RCRS), Indore, India, 1–4 December 2022; pp. 1–4.

Disclaimer/Publisher's Note: The statements, opinions and data contained in all publications are solely those of the individual author(s) and contributor(s) and not of MDPI and/or the editor(s). MDPI and/or the editor(s) disclaim responsibility for any injury to people or property resulting from any ideas, methods, instructions or products referred to in the content.

Communication

A Novel High-Isolation Dual-Polarized Patch Antenna with Two In-Band Transmission Zeros

Fuwang Li [1], Yi-Feng Cheng [1,*], Gaofeng Wang [1,*] and Jiang Luo [1,2]

[1] The Shaoxing Integrated Circuit Institute, Hangzhou Dianzi University, Hangzhou 310018, China; 212040136@hdu.edu.cn (F.L.); luojiang@hdu.edu.cn (J.L.)
[2] State Key Laboratory of Millimeter Waves, Southeast University, Nanjing 210096, China
* Correspondence: chengyifeng2013@gmail.com (Y.-F.C.); gaofeng@hdu.edu.cn (G.W.)

Abstract: In this study, we present a novel dual-polarized patch antenna that exhibits high isolation and two in-band transmission zeros (TZs). The design consists of a suspended metal patch, two feeding probes connected to an internal neutralization line (I-NL), and a T-shaped decoupling network (T-DN). The I-NL is responsible for generating the first TZ, and its decoupling principles are explained through an equivalent circuit model. Rigorous design formulas are also derived to aid in the construction of the feeding structure. The T-DN realizes the second TZ, resulting in further improvement of the decoupling bandwidth. Simulation and experimental results show that the proposed antenna has a wide operating bandwidth (2.5–2.7 GHz), high port isolation (>30 dB), and excellent efficiency (>85%).

Keywords: high-isolation; transmission zero; dual-polarized; patch antenna; equivalent circuit model

Citation: Li, F.; Cheng, Y.-F.; Wang, G.; Luo, J. A Novel High-Isolation Dual-Polarized Patch Antenna with Two In-Band Transmission Zeros. *Micromachines* **2023**, *14*, 1784. https://doi.org/10.3390/mi14091784

Academic Editor: Haejun Chung

Received: 9 August 2023
Revised: 9 September 2023
Accepted: 14 September 2023
Published: 18 September 2023

Copyright: © 2023 by the authors. Licensee MDPI, Basel, Switzerland. This article is an open access article distributed under the terms and conditions of the Creative Commons Attribution (CC BY) license (https://creativecommons.org/licenses/by/4.0/).

1. Introduction

A dual-polarized antenna array is widely adopted in current wireless communication systems due to its remarkable potential to increase the channel capacity and combat the multipath fading effect. Port isolation is a critical factor in evaluating the performance of dual-polarized antennas, as it determines the degree of independence between the orthogonal polarizations [1]. To improve port isolation, various techniques [2–11] have been proposed in recent years. For example, a single-layer, dual-port, and dual-mode antenna with enhanced port isolation is proposed in [2]. High isolation is realized by reducing surface waves between antennas. In [3], the port isolation is improved for dual-polarized stepped-impedance slot antenna by using shorting pins. C-shaped structures and square rings are designed to enhance the isolation between stacked microstrip patch antenna arrays [4]. In [5,6], dielectric superstrates and Defected Ground Structure (DGS) are utilized to improve the E-plane and H-plane isolation. Cross-polarization levels are suppressed by using decoupling strips and nested structures in [7]. A dual-feed technique is proposed in [8] to achieve high isolation (over 30 dB) between two antenna ports. In addition, complementary magneto–electric coupling feeding methods are employed in [9] to achieve high isolation and low cross-polarization. By introducing an air bridge as an inductor to compensate for the capacitance load, high isolation between the two polarization ports is realized [10]. In [11], by adding shorting vias and additional ground, the mutual coupling and cross-polarization have been significantly suppressed. However, the abovementioned decoupling methods for dual-polarized antennas have some limitations, such as complex decoupling structure, narrow bandwidth, and low radiation efficiency.

This paper proposes a novel high-isolation dual-polarized patch antenna with two in-band transmission zeros. Slots and probes are commonly used to feed patch antennas. In this design, rectangular probes are used to feed the suspended patch, which results in extended operating bandwidth (2.5–2.7 GHz). The rectangular probes can also facilitate the construction of the I-NL. The I-NL and T-DN are simultaneously adopted to enhance the isolation (>40 dB). The decoupling structure is simple and compact. Furthermore, an equivalent circuit model is adopted to facilitate the illustration of decoupling principles. Rigorous design formulas are derived to help the design process. The experimental results show that the proposed antenna features wide operating bandwidth, high port isolation, and good radiation efficiency, making it a promising candidate for modern wireless communication systems.

2. Proposed Design Method

2.1. Structure of the Proposed Antenna

Figure 1 illustrates the 3D structure of the dual-polarized antenna with and without decoupling structures. In Figure 1a, the initial patch antenna is shown without any decoupling structures. It consists of a square metal patch that is suspended above the ground and fed by two rectangular probes. In Figure 1b, the antenna is shown with a C-shaped I-NL. The I-NL is also made of metal and connected to the feeding probes. This I-NL is used to create the first TZ at f_1. Figure 1c shows the antenna with both I-NL and a decoupling network (DN). The DN is constructed below the ground to create the second TZ at f_2. The T-DN has an inherent TZ at f_1, which enables independent control of two TZs. Impedance matching is realized by adjusting the length of the patch and the position of the feeding probes. Figure 2 shows the layout structure of the proposed dual-polarized patch antenna. The suspended patch is constructed using copper with a thickness of 1 mm. The Rogers4003 substrate with a permittivity of 3.55 and loss tangent of 0.0027 is adopted to construct the decoupling network. The detailed dimensions of the antenna and decoupling structures are listed in Table 1.

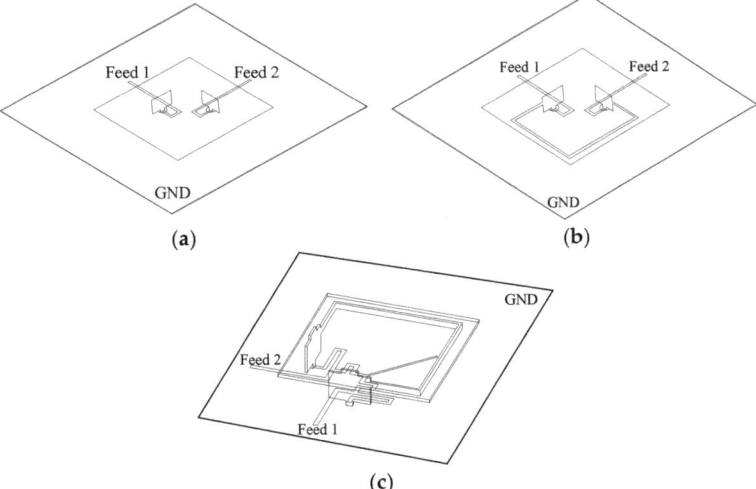

Figure 1. Structure of the proposed antenna. (**a**) Original patch; (**b**) with neutralization line; (**c**) with neutralization line and decoupling network.

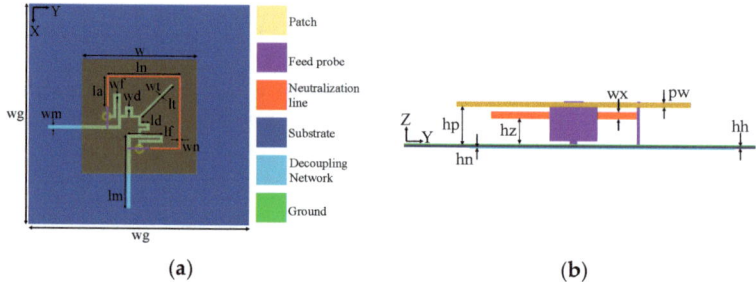

Figure 2. Layout of the proposed dual-polarized patch antenna (units: mm). (**a**) Top view; (**b**) side view.

Table 1. Dimensions of the proposed high-isolation dual-polarized patch antenna.

wg	w	wf	wd	wt	wn	hz
150	48.7	0.96	1.01	0.40	1.0	5.5
wm	hp	hn	wx	pw	hh	ln
1.13	8.0	0.035	1.5	1.0	0.508	30.9
lm	lf	ld	lt	la		
31.13	14.68	3.53	19.07	12.15		

2.2. Equivalent Circuit Model and Decoupling Mechanism

For further investigation, the equivalent circuit (EC) model of the high-isolation dual-polarized patch antenna (without a decoupling network) is proposed in Figure 3. This model can be subdivided into three parts: original patch antenna, initial coupling circuit (ICC), and I-NL. The radiating patch is equivalent to paralleled RLC circuits (R_1, L_1, and C_1). The feeding probe can be modeled by inductor L_2 and transmission line (e1). The initial coupling is represented by composite circuits R_2, L_3, and C_2. e_3 is used to adjust the phase effect of the coupling signal, which is mainly determined by patch dimensions. The I-NL (introduce additional coupling) is modeled by C_3, L_4, and e_4. Finally, the feeding lines of the path antenna are represented by e_2. The optimal parameters of this equivalent circuit model (corresponding to patch antenna with I-NL) are shown in Table 2. To validate the effectiveness of the EC model, we compared the S-parameters of the physical structure (simulated by SuperEM V2022) and the EC model (simulated by ADS2020), as shown in Figure 4. Specifically, Figure 4a depicts the S-parameter comparison of antennas without I-NL, while Figure 4b illustrates their comparison with I-NL. The results show that the phase and magnitude of the S-parameters are well-matched, indicating that the proposed EC model is accurate and reliable. As such, it can be utilized to expedite the optimization process of the proposed patch antenna design. Referring to Figure 3, let $[A_1, B_1; C_1, D_1]$ and $[A_2, B_2; C_2, D_2]$ denote the transmission matrices (TM) of ICC and I-NL, respectively. The TM of the resonance circuit with L_2 is denoted by $[A_3, B_3; C_3, D_3]$, and the TM with respect to the reference plane AA' is denoted by $[A_4, B_4; C_4, D_4]$. By applying network theory, the following equation can be derived.

$$\begin{bmatrix} A_1 & B_1 \\ C_1 & D_1 \end{bmatrix} = \begin{bmatrix} \cos e_3 & jz_3 \sin e_3 \\ jy_3 \sin e_3 & \cos e_3 \end{bmatrix} \begin{bmatrix} 1 & jw^3 L_3/w^2 L_3 C_2 - 1 \\ 0 & 1 \end{bmatrix} \begin{bmatrix} 1 & R_2 \\ 0 & 1 \end{bmatrix} \begin{bmatrix} \cos e_3 & jz_3 \sin e_3 \\ jy_3 \sin e_3 & \cos e_3 \end{bmatrix} \quad (1)$$

$$\begin{bmatrix} A_2 & B_2 \\ C_2 & D_2 \end{bmatrix} = \begin{bmatrix} \cos e_4 & jz_4 \sin e_4 \\ jy_4 \sin e_4 & \cos e_4 \end{bmatrix} \begin{bmatrix} 1 & jw^3 L_4/w^2 L_4 C_3 - 1 \\ 0 & 1 \end{bmatrix} \begin{bmatrix} \cos e_4 & jz_4 \sin e_4 \\ jy_4 \sin e_4 & \cos e_4 \end{bmatrix} \quad (2)$$

$$\begin{bmatrix} A_3 & B_3 \\ C_3 & D_3 \end{bmatrix} = \begin{bmatrix} 1 & 0 \\ \frac{R_1+jwL_1-w^2R_1L_1C_1}{jwR_1(L_1+L_2)-w^2L_1L_2-jw^3R_1L_1L_2C_1} & 1 \end{bmatrix} \quad (3)$$

$$\begin{bmatrix} A_4 & B_4 \\ C_4 & D_4 \end{bmatrix} = \begin{bmatrix} \cos e_1 & jz_1\sin e_1 \\ jy_1\sin e_1 & \cos e_1 \end{bmatrix} \begin{bmatrix} A_3 & B_3 \\ C_3 & D_3 \end{bmatrix} \begin{bmatrix} A_2 & B_2 \\ C_2 & D_2 \end{bmatrix} \begin{bmatrix} A_3 & B_3 \\ C_3 & D_3 \end{bmatrix} \begin{bmatrix} \cos e_1 & jz_1\sin e_1 \\ jy_1\sin e_1 & \cos e_1 \end{bmatrix} \quad (4)$$

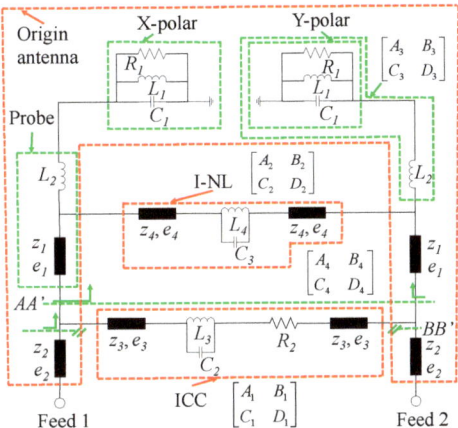

Figure 3. Equivalent circuit of the proposed patch antenna (without decoupling network).

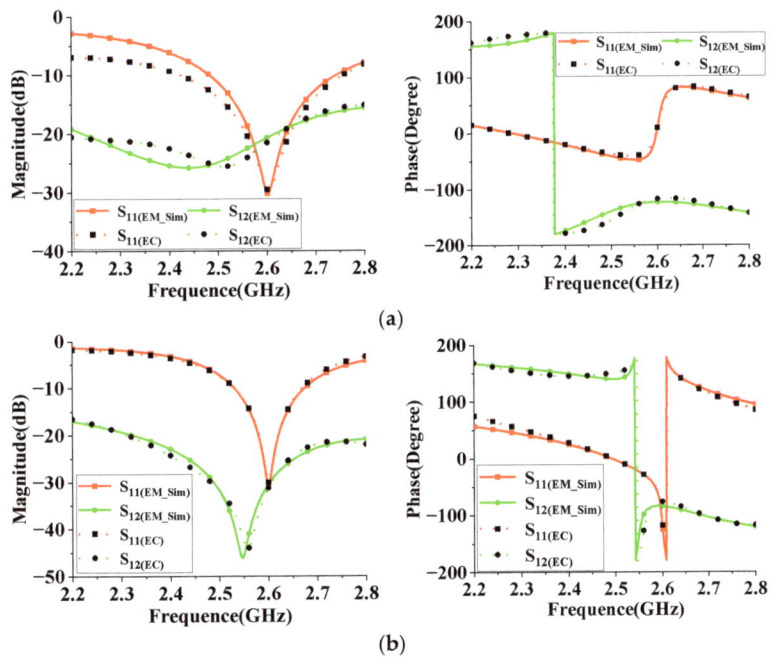

Figure 4. Comparison of S-parameters (magnitude and phase) of the EM structure and the equivalent circuit (EC) model. (**a**) Without I-NL; (**b**) with I-NL.

Table 2. Optimal parameters of the equivalent circuit model.

R_1	L_1	C_1	L_2	z_1	e_1
52.4 Ω	0.4 nH	9.4 pF	1.1 nH	53.1 Ω	179.4⁰
z_2	e_2	z_3	e_3	R_2	C_2
179.8 Ω	10.6⁰	85.4 Ω	49.8⁰	7.7 Ω	0.8 pF
L_3	z_4	e_4	L_4	C_3	
2.1 nH	15.3 Ω	10.6⁰	404.1 nH	0.5 pF	

Subsequently, the mutual admittance with reference to the plane BB' can be calculated as follows.

$$Y_{21}{}^B = Y_{21}{}^A + Y_{21}{}^C = -\frac{1}{B_4} - \frac{1}{B_1} \qquad (5)$$

As shown in (5), the I-NL can provide another mutual coupling to cancel out the original coupling. By adjusting the length/width and height of I-NL (C_3, L_2, L_4, z_4, and e_4), the first transmission zero can be created at f_1.

Figure 5 illustrates the schematic diagram of the proposed decoupling network, which comprises two sections of transmission lines (TLs) and a T-DN. As described in [12], the inserted TLs and T-DN serve to eliminate the real and imaginary parts of mutual admittance (by adjusting e_5 and z_6), respectively. A shunt quarter wavelength TL, evaluated at f_1, is positioned at the center of the T-DN to maintain the first TZ created by I-NL. This approach enables the independent control of the position of two TZs, resulting in deep and wideband decoupling.

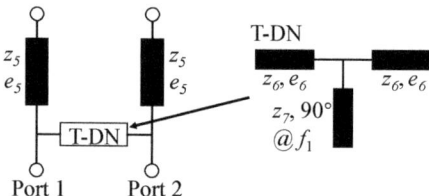

Figure 5. Schematic diagram of the proposed decoupling network.

The generation of two TZs using the proposed decoupling method is illustrated in Figure 6. The original antenna exhibits high mutual coupling (15–20 dB), which is significantly reduced after applying the I-NL, resulting in high isolation at f_1. However, the decoupling bandwidth is limited. To overcome this problem, a DN is then added, which generates another TZ and achieves wideband decoupling. The simulated in-band isolation is below 40 dB with two TZs.

The I-NL and T-DN are responsible for generating the first and second TZ, respectively. To further investigate this, the height of I-NL (hz) and the width of the microstrip line of T-DN (wd) are used for examination. Figure 7 illustrates the variation in S-parameters when these two parameters are changed. Excellent decoupling performance is attained with hz = 3.5 mm and wd = 1.05 mm. By adjusting hz, the first TZ at a lower frequency can be generated. By adjusting wd, the second TZ can be generated without affecting the first TZ. This demonstrates the independent control of the two TZs.

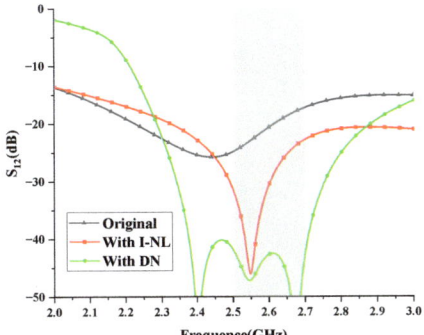

Figure 6. Generation of two TZs.

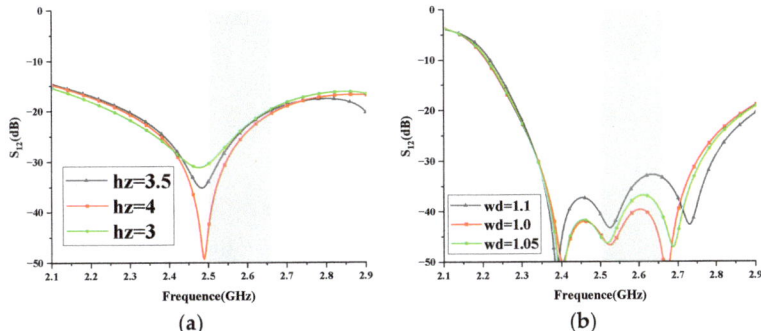

Figure 7. Key-parameter study. (**a**) hz; (**b**) wd.

3. Experimental Validation and Results

For verification, the proposed high-isolation dual-polarized patch antenna is designed, fabricated, and measured. Figure 8 shows the photographs of the fabricated antennas and the anechoic chamber. The suspended patch is supported by three plastic posts. The S-parameters are measured by the Keysight vector network analyzer E5071C, and radiation patterns are measured in an anechoic chamber. As shown in Figure 9, the measured reflection coefficient of the proposed antenna is below −10 dB from 2.5 to 2.7 GHz. High isolation (below 30 dB) is realized in the operating band by using the proposed decoupling method. Figure 10 shows the simulated and measured radiation patterns (yoz- and xoz-planes) of the proposed antenna. Good agreement of the simulated and measured results is observed. Figure 11 gives the measured total efficiency and realized antenna gain of the proposed antenna. High total efficiency (90%) and measured stable gain (9.6–10.3 dBi) is observed. Furthermore, the measured front-to-back ratio is about 23 dB.

Table 3 gives the performance comparison with other published works. As demonstrated, this design performs competitively compared to existing proposals, particularly in terms of realized gain, efficiency, and isolation performance.

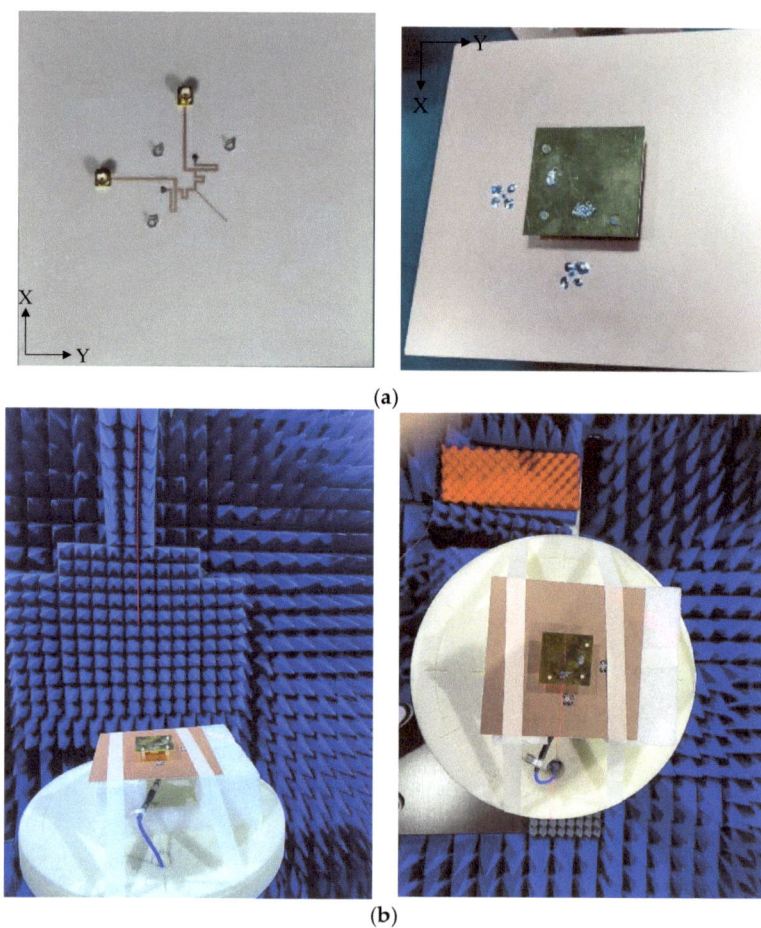

Figure 8. Photographs of the fabricated antenna and anechoic chamber. (**a**) The fabricated antenna; (**b**) the anechoic chamber.

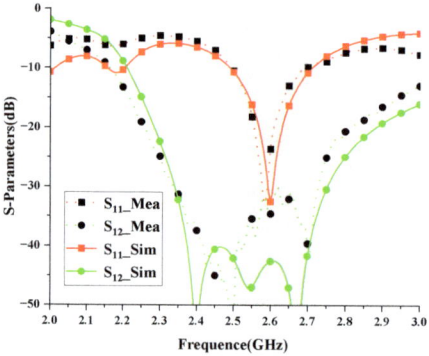

Figure 9. Simulated and measured S-parameters of the antenna.

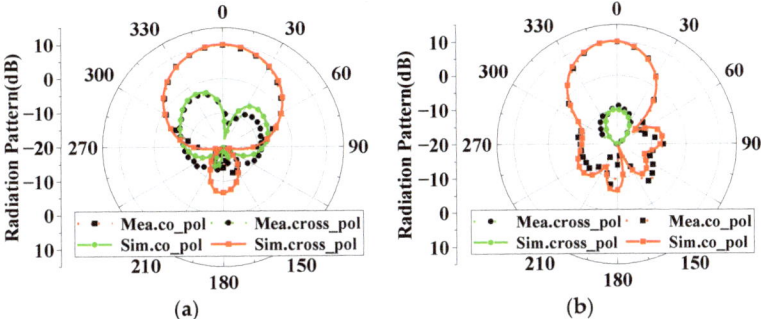

Figure 10. Radiation patterns. (**a**) xoz-plane; (**b**) yoz-plane.

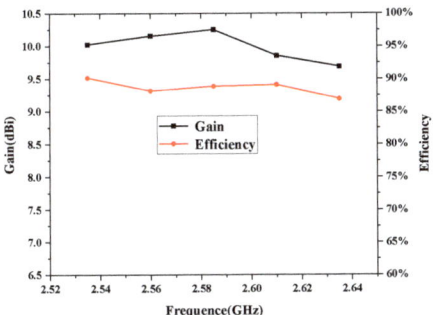

Figure 11. Measured total efficiency and realized gain of the proposed antenna.

Table 3. Performance comparison with other works.

Ref.	Method	Frequency (GHz)	Antenna Size (λ_0^3)	Isolation (dB)	Total Efficiency (%)	Average Gain (dBi)
[8]	Dual-feed technique	1.71–1.88	$0.55 \times 0.55 \times 0.11$	>30	N.A	<8
[9]	Complementary magneto-electric coupling feeding	1.53–2.95	$0.62 \times 0.62 \times 0.26$	>31	N.A	4
[10]	Introducing air bridge	1.6–2.3	N.A	>33	N.A	N.A
[11]	Using shorting vias and additional ground	7–12	$0.56 \times 0.56 \times 0.13$	>40	N.A	8.5
This work	Using I-NL and T-DN	2.5–2.7	$0.42 \times 0.42 \times 0.07$	>30	>85	10.0

4. Conclusions

In this paper, a novel high-isolation dual-polarized patch antenna with two transmission zeros has been proposed, designed, and demonstrated. To better reveal the decoupling principle, the equivalent circuit model of the proposed antenna is analyzed. Moreover, the decoupling condition of the two-layer decoupling structure is rigorously derived and equivalently represented by the two-port transmission matrix and Y-matrix. Finally, the isolation is improved by about 15–20 dB between two input ports. The proposed design features high port isolation, compact size, low cross-polarization, and high radiation performance.

Author Contributions: Conceptualization, F.L. and Y.-F.C.; Methodology, F.L.; Writing—original draft, F.L.; Data curation, Y.-F.C.; Formal analysis, Y.-F.C.; Writing—review and editing, Y.-F.C., G.W., and J.L.; Supervision, G.W. and J.L.; Project administration, G.W. and J.L. All authors have read and agreed to the published version of the manuscript.

Funding: This research was supported in part by the State Key Laboratory of Millimeter Waves under Grant K202316, in part by the National Natural Science Foundation of China under Grant 62201183, and in part by the Fundamental Research Funds for the Provincial Universities of Zhejiang under Grant GK229909299001-016 and GK229909299001-309.

Data Availability Statement: The data presented in this work are available within the article.

Acknowledgments: The authors would like to thank Faraday Dynamics, Ltd. [13]. for providing useful simulation software (SuperEM V2022) to assist the research in this paper.

Conflicts of Interest: The authors declare no conflict of interest.

References

1. Li, B.; Yin, Y.X.Z.; Hu, W.; Ding, Y.; Zhao, Y. Wideband dual-polarized patch antenna with low cross polarization and high isolation. *IEEE Antennas Wirel. Propag. Lett.* **2012**, *11*, 427–430.
2. Mirhadi, S. Single-layer, dual-port, and dual-mode antenna with high isolation for WBAN communications. *IEEE Antennas Wirel. Propag. Lett.* **2022**, *21*, 531–535. [CrossRef]
3. Lian, R.; Wang, Z.; Yin, Y.; Wu, J.; Song, X. Design of a low-profile dual-polarized stepped slot antenna array for base station. *IEEE Antennas Wirel. Propag. Lett.* **2016**, *15*, 362–365. [CrossRef]
4. Fang, Y.; Tang, M.; Zhang, Y.P. A decoupling structure for mutualcoupling suppression in stacked microstrip patch antenna array. *IEEE Antennas Wirel. Propagat. Lett.* **2022**, *21*, 1110–1114. [CrossRef]
5. Li, Y.; Chu, Q.-X. Dual-layer superstrate structure for decoupling of dual-polarized antenna arrays. *IEEE Antennas Wirel. Propag. Lett.* **2022**, *21*, 521–525. [CrossRef]
6. Qian, B.; Huang, X.; Chen, X.; Abdullah, M.; Zhao, L.; Kishk, A. Surrogate-assisted defected ground structure design for reducing mutual coupling in 2 × 2 microstrip antenna array. *IEEE Antennas Wirel. Propag. Lett.* **2022**, *21*, 351–355. [CrossRef]
7. Dai, X.-W.; Wang, Z.-Y.; Liang, C.-H.; Chen, X.; Wang, L.-T. Multi-291 band and dual-polarized omnidirectional antenna for 2G/3G/LTE appli-292 cation. *IEEE Antennas Wirel. Propag. Lett.* **2013**, *12*, 1492–1495. [CrossRef]
8. Wong, H.; Lau, K.-L.; Luk, K.-M. Design of dual-polarized L-probe patch antenna arrays with high isolation. *IEEE Trans. Antennas Propag.* **2004**, *52*, 45–52. [CrossRef]
9. Yu, H.-W.; Jiao, Y.-C. Complementary Magneto-Electric Coupling Feeding Methods for Low-Profile High-Isolation Dual-Polarized Microstrip Patch Antenna. In Proceedings of the 2019 International Symposium on Antennas and Propagation (ISAP), Xi'an, China, 27–30 October 2019; pp. 1–3.
10. Barba, M. A high-isolation, wideband and dual-linear polarization patch antenna. *IEEE Trans. Antennas Propag.* **2008**, *56*, 1472–1476. [CrossRef]
11. Wang, J.; Wang, W.; Liu, A.; Guo, M.; Wei, Z. Broadband metamaterial-based dual-polarized patch antenna with high isolation and low cross polarization. *IEEE Trans. Antennas Propag.* **2021**, *69*, 7941–7946. [CrossRef]
12. Cheng, Y.-F.; Cheng, K.-K.M. A Novel Dual-Band Decoupling and Matching Technique for Asymmetric Antenna Arrays. *IEEE Trans. Microw. Theory Tech.* **2018**, *66*, 2080–2089. [CrossRef]
13. *SuperEM, V2022*; Faraday Dynamics, Inc.: Hangzhou, China, 2022.

Disclaimer/Publisher's Note: The statements, opinions and data contained in all publications are solely those of the individual author(s) and contributor(s) and not of MDPI and/or the editor(s). MDPI and/or the editor(s) disclaim responsibility for any injury to people or property resulting from any ideas, methods, instructions or products referred to in the content.

Article

A Multiband Millimeter-Wave Rectangular Dielectric Resonator Antenna with Omnidirectional Radiation Using a Planar Feed

Tarek S. Abdou and Salam K. Khamas *

Communications Research Group, Department of Electronic and Electrical Engineering, University of Sheffield, Sheffield S1 3JD, UK; tsabdou1@sheffield.ac.uk
* Correspondence: s.khamas@sheffield.ac.uk

Abstract: In this study, a millimeter-wave (mmWave) dielectric resonator antenna (DRA) with an omnidirectional pattern is presented for the first time. A key feature of the proposed design is the utilization of a planar feed network to achieve omnidirectional radiation from a rectangular DRA, which has not been reported previously in the open literature. In addition, the proposed antenna offers multiband operation with different types of radiation patterns. The degenerate TE_{121}/TE_{211} modes were excited at 28.5 GHz with an overall internal electromagnetic field distribution that was similar to that of the $HEM_{21\delta}$ mode of a cylindrical DRA. The achieved omnidirectional bandwidth and gain were 1.9% and 4.3 dBi, respectively. Moreover, broadside radiation was achieved by exciting the TE_{111} fundamental mode at 17.5 GHz together with the resonance of the feeding ring-slot at 23 GHz. The triple-band operation offers a highly versatile antenna that can be utilized in on-body and off-body communications. Furthermore, the proposed design was validated through measurements, demonstrating good agreement with simulations.

Keywords: omnidirectional antenna; dielectric resonator antenna; multiband; mmWave communications

1. Introduction

Over recent years, the evolution of mmWave communication systems has led to more rigorous requirements for antenna designs such as high gain, wide and multiband operation, as well as pattern diversity. DRAs have the potential of addressing these requirements due to well-known advantages such as high radiation efficiency, wide impedance bandwidth, and design flexibility. Therefore, mmWave DRAs have received increased research interest with a focus on broadside radiation [1–4]. On the other hand, omnidirectional radiation is desired for 5G and Beyond 5G (B5G) communication systems to increase the coverage area in various applications such as on-body communications as well as device-to-device short-distance communications [5]. Therefore, several studies have been reported on the design of mmWave omnidirectional antennas [6,7]. However, an omnidirectional mmWave DRA has not been reported previously, which is in sharp contrast with the numerous published studies on the design of omnidirectional DRAs at lower frequencies with a focus on exciting specific transverse magnetic (TM) and transverse electric (TE) modes to achieve the required pattern.

For example, an omnidirectional cylindrical DRA was proposed by exciting the $TE_{01\delta}$ and $TM_{01\delta}$ resonance modes at 3.87 GHz and 4.02 GHz, respectively, using a central coaxial probe feed in [8]. Moreover, a multiband, multisense, circularly polarized hybrid patch/DRA omnidirectional antenna was reported by exciting the TM_{02} and TM_{011} resonance modes for the patch at 2 GHz and DRA at 2.6 GHz, respectively [9]. In a more recent study, a wideband filtering omnidirectional cylindrical DRA was presented using a hybrid feed that consisted of a coaxial probe and metallic disk to excite the $TM_{01\delta}$ and TM_{013} DRA resonance modes at 2.19 GHz and 3.37 GHz, respectively, in [10]. Further, a coaxial probe-fed omnidirectional hemispherical DRA was proposed by exciting the TM_{101} resonance mode at 3.7 GHz in [11]. Another probe-fed omnidirectional hemispherical DRA was designed for a wireless capsule endoscope system by exciting the $TM_{01\delta}$ and $TE_{01\delta}$ resonance modes for multipolarization at 2.45 GHz in [12]. Moreover, a probe-fed omnidirectional rectangular DRA with a square cross section was designed by exciting the quasi-TM_{011} mode at 2.4 GHz for linear and circular polarizations in [13]. Furthermore, innovative rectangular multifunction glass DRAs were reported with the capability of achieving linearly and circularly polarized omnidirectional radiation patterns by exciting the quasi-TM_{011} mode at 2.4 GHz in [14]. Similarly, the higher-order degenerate TE^x_{121} and TE^y_{211} modes were excited at 3.6 GHz, with equal amplitude and phase, to achieve omnidirectional radiation from a rectangular DRA in [15,16]. Subsequently, a multiband probe-fed omnidirectional rectangular DRA was proposed, where the TE^x_{121} and TE^y_{211} resonance modes were excited at 3.5 GHz together with the TE^x_{141} and TE^y_{411} resonance modes at 5.8 GHz, in [16].

It should be noted that in all the above-mentioned studies, omnidirectional radiation was attained using a centrally located coaxial feeding probe to excite the required resonance modes. On the other hand, owing to their capability of supporting various types of modes when placed above a ground plane, cylindrical DRAs have successfully been utilized recently with planar feed networks to achieve omnidirectional radiation patterns. For example, an omnidirectional cylindrical DRA with a planar feed of a shorted microstrip cross was demonstrated by exciting the $TM_{01\delta}$ and $TM_{011+\delta}$ resonance modes at ~2.4 GHz to achieve circular polarization diversity [17]. The first attempt to utilize a ring-slot aperture to feed an omnidirectional cylindrical DRA was proposed by exciting the $TM_{01\delta}$ fundamental resonance mode at 2.4 GHz in [18]. Furthermore, a pattern diversity cylindrical DRA was proposed using a meander line-loaded annular slot to excite the $TM_{01\delta}$ mode in combination with a differential strip to excite the $HEM_{11\delta}$ for omnidirectional and broadside radiations, respectively, at 2.4 GHz [19]. Moreover, four linear stubs were utilized to excite the $TM_{01\delta}$ and $TM_{011+\delta}$ resonance modes of an omnidirectional cylindrical DRA to achieve circular polarization at 5.8 GHz [20]. In a more recent study, arched-aperture feeding was employed in the design of a wideband omnidirectional cylindrical DRA by merging the bandwidth of the excited $TM_{01\delta}$ and $TM_{02\delta}$ resonance modes at ~5.8 GHz [21].

A rectangular DRA with a square cross section supports quasi-TM modes [22] that have been traditionally excited using a coaxial feeding probe to achieve omnidirectional radiation [13–16]. It is well-known that a coaxial feeding probe requires a hole to be drilled inside the DRA, which is impractical at mmWave frequencies due to the physically small DRA size. Furthermore, the probe's reactance can be large at millimeter-wave frequencies. Moreover, the power handling capacity of the probe is reduced at higher frequencies, leading to signal degradation and power dissipation [18,23]. Therefore, aperture–slot coupling is preferred to excite a DRA at higher operating frequencies as it provides a high level of isolation between the antenna and the planar feed network. On the other hand, compared to the cylindrical counterpart, a rectangular DRA offers an additional degree of design freedom together with simpler fabrication due to the planar sides. Therefore, an alternative noninvasive feeding approach needs to be utilized to design a mmWave omnidirectional rectangular DRA. Such a design is proposed in this study, where a ring-slot aperture is utilized to excite the required modes. In addition to the planar feed, the proposed antenna offers another advantage of multiband operation with two types of radiation patterns: broadside and omnidirectional. The first is achieved by exciting the fundamental TE_{111} mode at 17.5 GHz as well as a slot resonance at 23 GHz. An omnidirectional pattern is achieved by exciting the TE^x_{121} and TE^y_{211} higher-order degenerate modes. It should be noted that all the reported dual-band DRA designs radiate either broadside or omnidirectional patterns in both bands. As a result, the proposed DRA can be employed simultaneously for off-body and on-body applications, for example, by utilizing the broadside and omnidirectional patterns, respectively. A common problem with existing on-body antennas is the reduced radiation efficiency due to the impact of the human body, especially at mmWave frequencies. However, the utilization of on-body omnidirectional DRA can help in achieving more efficient on-body antennas, which necessitates the design of a planar feeding network.

This article is organized as follows: Section 2 presents the proposed DRA configuration. Section 3 investigates the excitable DRA modes at a frequency range of 20–30 GHz. Section 4 is focused on the design of the planar feed network. Section 5 presents an analysis of the performance of the on-body mmWave DRA. Section 6 presents the measured results that agree closely with the simulations and Section 7 is focused on the conclusions. All the simulations are implemented using CST microwave studio.

2. Antenna Configuration

The DRA was designed using a square cross section to facilitate the excitation of the required degenerate modes for omnidirectional radiation. In addition, alumina with a dielectric constant of $\varepsilon_d = 9.9$ and a loss tangent of $\tan\delta = 0.0001$ was chosen as the DRA's material. Figure 1 illustrates the utilized configuration in which the DRA was placed on a square ground plane with a size of $G_s = 12.5$ mm. The feed network also involved a square Rogers substrate, Ro4003, that was located at the lower side of the ground plane. The substrate had a thickness of $h_s = 0.308$ mm, dielectric constant of $\varepsilon_r = 3.5$, and loss tangent of 0.0027. Additionally, a 50 Ω microstrip feedline was printed on the substrate's lower surface with a respective length and width of $l_t = 6.25$ mm and $w_t = 0.3$ mm.

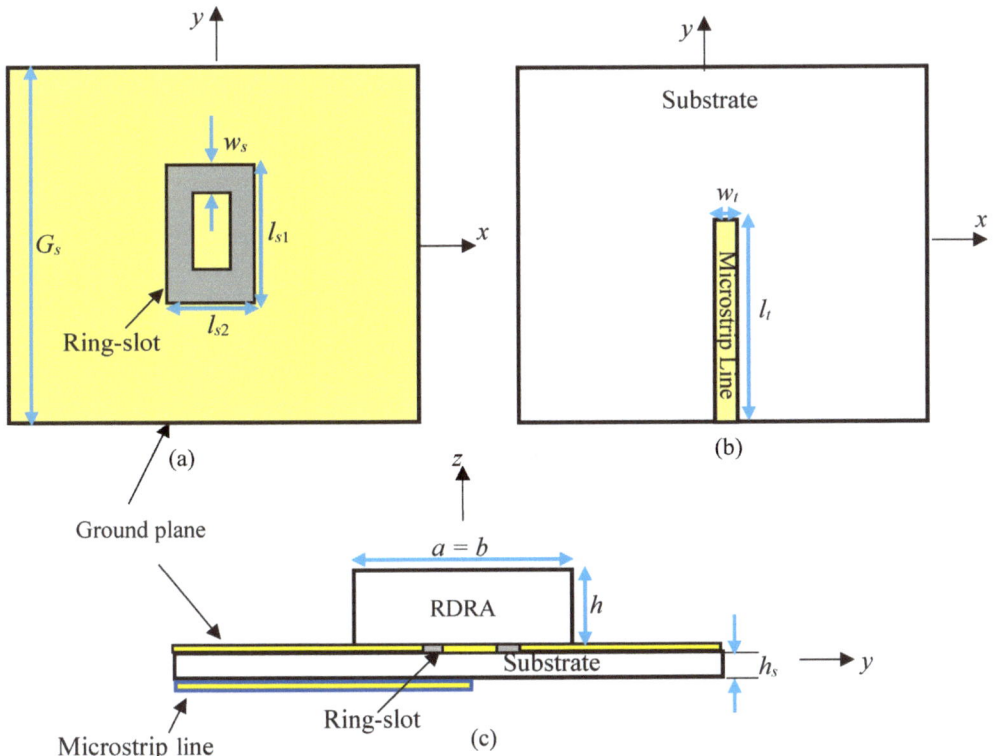

Figure 1. The proposed omnidirectional DRA: (**a**) ground plane with an etched ring-slot; (**b**) bottom view; (**c**) side view.

The design was evolved by noting that an electrically small vertical probe and a ring-slot that is etched in a metal ground plane represent the duals of a planar loop antenna with equivalent size. Therefore, the small ring-slot provided the same fields as an electrically short vertical probe and, hence, could be considered as an option to create the required planar feeding network that incorporated a rectangular ring-slot as a natural choice to feed a rectangular DRA, as can be observed from Figure 1. However, the ring-slot size may need to be increased depending on the field distribution of the required DRA mode. Furthermore, the utilized ring-slot consisted of x- and y-directed slot arms with side lengths of l_{s1} and l_{s2}. These slots behave as magnetic currents that excite the required magnetic fields inside the DRA. Since the ring-slot was positioned at the interface between the alumina DRA and the Rogers substrate, the circumference needed to be calculated in terms of the effective wavelength $\lambda_{\text{eff}} = \lambda_0 / \sqrt{\varepsilon_{\text{eff}}}$, where λ_0 is the free space wavelength and ε_{eff} is defined as [24]

$$\varepsilon_{eff} = \frac{\varepsilon_d \varepsilon_r (h + h_s)}{(\varepsilon_d h + \varepsilon_r h_s)} \quad (1)$$

In order to design an optimum feed, the supported DRA modes need to be identified over the frequency range of interest, as illustrated in the next section.

3. Supported Modes of the Proposed DRA

Based on the dielectric waveguide model (DWM) [25], the DRA dimensions were chosen to support the required degenerate modes for omnidirectional radiation at ~28.5 GHz when the DRA is located above a metal ground plane. Therefore, the DRA's length, width, and height were determined as $a = b = 3.8$ mm and $h = 1.7$ mm, respectively. These dimensions offer a compact DRA size that allows easy integration. The resonance frequencies of the TE^y_{mns} modes can be determined using the DWM as [25]:

$$\begin{aligned} k_x a &= m\pi - 2\tan^{-1}(k_x/(\varepsilon_d k_{x0})) \\ k_{x0} &= \left[(\varepsilon_d - 1)k_0^2 - k_x^2\right]^{\frac{1}{2}} \end{aligned} \quad (2)$$

$$\begin{aligned} k_y b &= n\pi - 2\tan^{-1}(k_y/k_{y0}) \\ k_{y0} &= \left[(\varepsilon_d - 1)k_0^2 - k_y^2\right]^{\frac{1}{2}} \end{aligned} \quad (3)$$

$$\begin{aligned} 2k_z h &= m\pi - 2\tan^{-1}(k_z/(\varepsilon_d k_{z0})) \\ k_{z0} &= \left[(\varepsilon_d - 1)k_0^2 - k_z^2\right]^{\frac{1}{2}} \end{aligned} \quad (4)$$

$$k_x^2 + k_y^2 + k_z^2 = \varepsilon_d k_0^2 \quad (5)$$

where k_0 is the free space wave number. Owing to the square cross section of the DRA, the resonance frequencies of the TE^x_{mns} and TE^y_{nms} modes are equal. Therefore, the required TE^x_{121} and TE^y_{211} higher-order modes can be simultaneously excited at the same frequency. This results in a total magnetic field distribution that is similar to that of the $\text{HEM}_{21\delta}$ mode of a cylindrical DRA that generates an omnidirectional pattern. Table 1 summarizes the supported resonance modes for the chosen rectangular DRA dimensions over a frequency range of 15–30 GHz based on the DWM.

Table 1. Resonance frequencies of the supported TE modes by the proposed DRA.

Frequency (GHz)	Resonance Mode
17.5	TE_{111}
28.5	$\text{TE}^x_{121}, \text{TE}^y_{211}$

The excitation of the required modes can be achieved by studying the magnetic field distributions of the supported resonance modes inside an isolated DRA which are illustrated in Figure 2. For example, from the TE^x_{121} mode's magnetic field distribution, it can be observed that the H-field is null when $y = 0.5b$, where b is the DRA size. Therefore, the utilization of a centrally located x-directed slot aperture will suppress this mode. Hence, the slot needs to be shifted along the y-axis to a strong H-field point to excite this mode effectively. Similarly, for the TE_{211}^y mode, in which the H-field is null at $x = 0.5b$, an off-set y-directed slot is needed at a strong H-field point for effective mode excitation. However, for omnidirectional radiation, the degenerate modes need to be excited simultaneously. Therefore, a ring-slot aperture was utilized, which involved y and x-directed slot arms that acted as magnetic current components exciting the aforementioned modes. Furthermore, the chosen DRA dimensions also support the fundamental broadside TE_{111} mode at 17.5 GHz and, hence, it would be beneficial if the same ring-slot aperture excites the fundamental TE_{111} mode as well as the TE_{211}^y and TE_{121}^x modes. As mentioned earlier, the interaction between these degenerate modes can result in a field distribution that is similar to the cylindrical $HEM_{21\delta}$ mode [22]. For a rectangular DRA, such a mode is defined as a quasi-$HEM_{21\delta}$ mode, which offers the required omnidirectional radiation pattern.

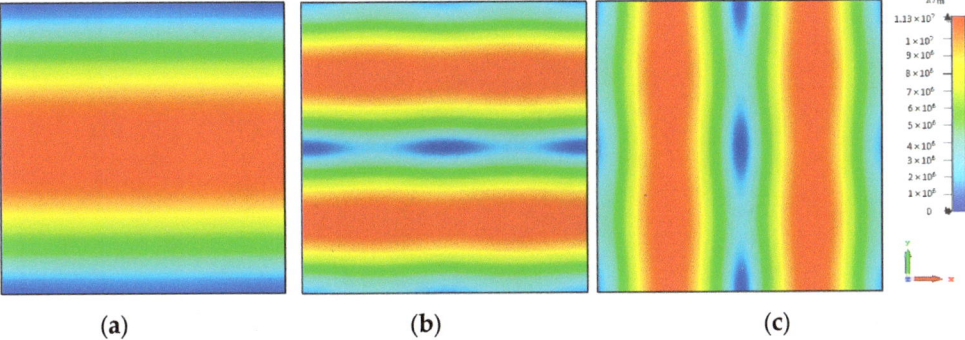

Figure 2. Magnetic field distributions inside the proposed isolated DRA at $z = 0$ when $a = b = 3.8$ mm and $h = 1.7$ mm; (**a**)TE_{111} mode at 17.5 GHz, (**b**) TE_{121}^x mode at 28.5 GHz, and (**c**) TE_{211}^y mode at 28.5 GHz.

Having identified the supported DRA resonance modes and understood the corresponding fields' distribution, the design of the required ring-slot needed to be implemented as described in the next section together with the achieved DRA performance.

4. Design of the Ring-Slot Feed

4.1. Square Ring-Slot Feed

For simplicity, the special case of a square ring-slot was considered first by setting $l_{s1} = l_{s2}$. It was important to ensure that the first slot's resonance, which had broadside radiation, was achieved at a frequency that was different from that of the required TE_{211}^y and TE_{121}^x DRA modes to avoid any interference between the different radiation patterns. Subsequently, the separated slot resonance could be suppressed or utilized as another operating frequency band, depending on the design requirements. The return losses are illustrated in Figure 3, where it can be noted that when $l_{s1} = l_{s2} = 2.1$ mm and $w_s = 0.5$ mm, the slot resonated at 27.7 GHz when the circumference was $\sim 1.1\lambda_{eff}$, which is too close to that of the degenerate modes. This was also combined with a broadside radiation pattern instead of the expected DRA's omnidirectional pattern, which was not observed initially at the expected frequency. Hence, the size of the slot was adjusted to avoid the coexistence of the DRA and ring-slot resonances at the same frequency. As demonstrated in Figure 3, by increasing the slot size, the DRA modes and slot's resonance frequencies could be separated as the latter was achieved at 24 GHz when $l_{s1} = l_{s2} = 2.5$ mm. Hence, the required omnidirectional and fundamental DRA modes were effectively excited at ~27.7 GHz and 16 GHz, respectively, using the same square ring-slot. In addition, the increased slot size meant the slot arms' positions could move closer to stronger H-field points of the corresponding

DRA's mode, which resulted in the effective excitation of the required modes. The excited degenerate modes provided an overall impedance bandwidth of 1.3%. Moreover, the TE_{111} broadside mode was excited with a bandwidth of 3%. Moreover, the resonance of the ring-slot was achieved with a bandwidth of 7.4%. It should be noted that the DRA modes were excited at resonance frequencies that were close to those listed in Table 1. Figure 4 presents the simulated magnetic field distribution inside the loaded DRA at 27.7 GHz, which was similar to that of the cylindrical $HEM_{21\delta}$ mode, and hence, an omnidirectional radiation pattern was achieved using the rectangular DRA.

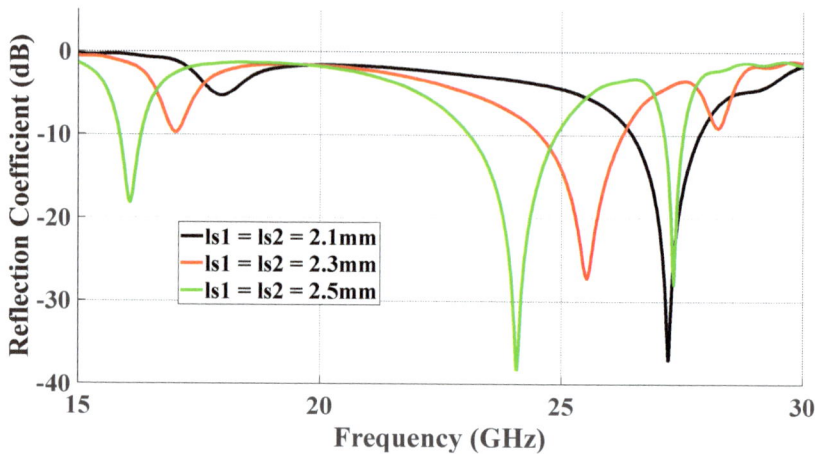

Figure 3. Return losses of the DRA using different sizes of the square-ring feeding slot.

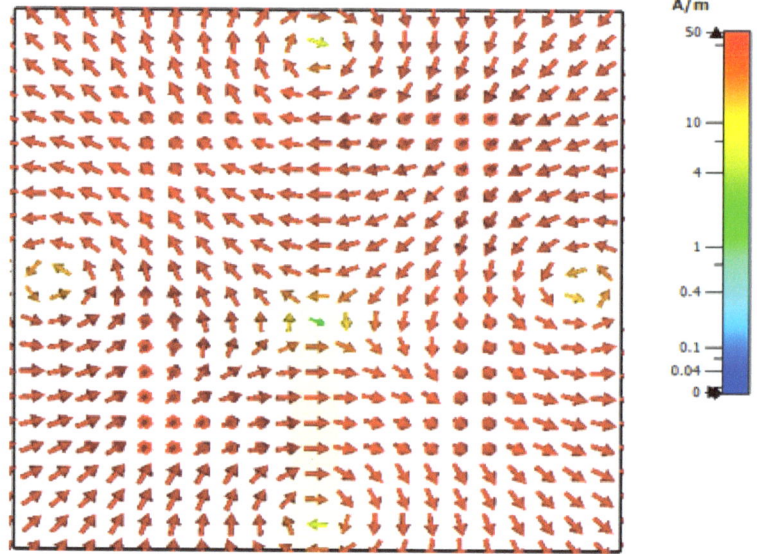

Figure 4. The *xy* plane view of the quasi-$HEM_{21\delta}$ internal magnetic field distribution of a square ring-slot-fed DRA at 27.7 GHz.

Figure 5 presents the achieved radiation patterns at the three resonance frequencies, where broadside radiation patterns were achieved at 16 GHz and 24 GHz due to the excitation of the fundamental DRA mode, TE$_{111}$, and the ring-slot's resonance, respectively. As mentioned earlier, the quasi-HEM$_{21\delta}$ mode was excited at 27.7 GHz when the size of the feeding square ring-slot was increased to 2.5 mm. As demonstrated in Figure 5c, omnidirectional radiation was attained with a maximum gain of 4.1 dBi at $\theta = 40°$. However, a slight asymmetry can be noted in the $\phi = 90°$ plane cut, which can be attributed to the asymmetrical feed point position compared to the traditionally used central coaxial probe that naturally enforces the fields' symmetry. Figure 5d presents the azimuthal patten at the $\theta = 40°$ plane, where it can be noted that an omnidirectional pattern was achieved with a modest out-of-roundness. Therefore, a triple-band operation was attained with two different types of radiation patterns using a planar feed network. The variation in the omnidirectional gain at the $\theta = 40°$ plane was investigated as illustrated in Figure 6, where it can be noted that a maximum variation of ~1.5 dB exists, which resulted in a pattern that was not perfectly omnidirectional. This can be explained in terms of the limitations imposed by the centrally located square ring-slot, since changing the slot's size is associated with a proportional shift in the position of each slot arm. This may result in having slot arms positioned at points with slightly different H-field strengths. It should be noted that this variation was slightly higher than that of 1.26 dB for an omnidirectional cylindrical DRA with a planar feed network [20]. An attempt to minimize the omnidirectional gain's variation is introduced in the next section. Figure 7 illustrates the realized gain at the main beam directions for the three bands. For the two broadside patterns, it can be observed that the realized gains of 6.5 dBi and 4.8 dBi were achieved at 16 GHz and 24 GHz, respectively. On the other hand, an omnidirectional gain of 4.1 dBi was achieved at 27.7 GHz at the main lobe direction of $\theta = 40°$. The simulated radiation efficiency is also illustrated in Figure 7, where it is evident that a high radiation efficiency of ~90% wsa achieved at all the operating frequency bands.

4.2. Rectangular Ring-Slot Feed

To minimize the azimuthal gain variations at $\theta = 40°$, a rectangular ring-slot aperture was considered as it offers the flexibility of changing the size of only one pair of the ring-slot arms at a time. The return losses are illustrated in Figure 8 when the longest slot arm's length, l_{s1}, was varied from 2.7 to 3.2 while $l_{s2} = 2$ mm. It should be noted these dimensions resulted in a slot circumference of ~$1.1\lambda_{eff}$ at 23 GHz. The results demonstrate that the proposed antenna configuration exhibited three operating frequency bands at 17.5 GHz, 23 GHz, and 28.5 GHz for the TE$_{111}$ DRA mode, slot resonance, and quasi-HEM$_{21\delta}$ mode, respectively, when $l_{s1} = 2.7$ mm. The achieved respective bandwidths for the three resonance modes were 17.3 GHz to 17.9 GHz, 22.1 GHz to 24 GHz, and 28.3 GHz to 28.8 GHz, which correspond to percentage bandwidths of 3.4%, 7.7%, and 1.9%. It can be noted that these bandwidths were wider than those achieved when a square ring-slot was utilized to excite the same DRA modes, which demonstrates the effectiveness of the rectangular ring-slot. It can also be observed from these results that the strongest impact of varying l_{s1} was on the slot's resonance frequency, which was expected owing to the change in the circumference of the rectangular ring-slot. On the contrary, smaller variations can be noted in the resonance frequencies of the excited DRA modes as they mainly depend on the DRA dimensions and permittivity. The combination of the reflection coefficient of square and rectangular slots is presented in Figure 9, where it is evident that wider matching bandwidths were achieved when a rectangular ring-slot was utilized.

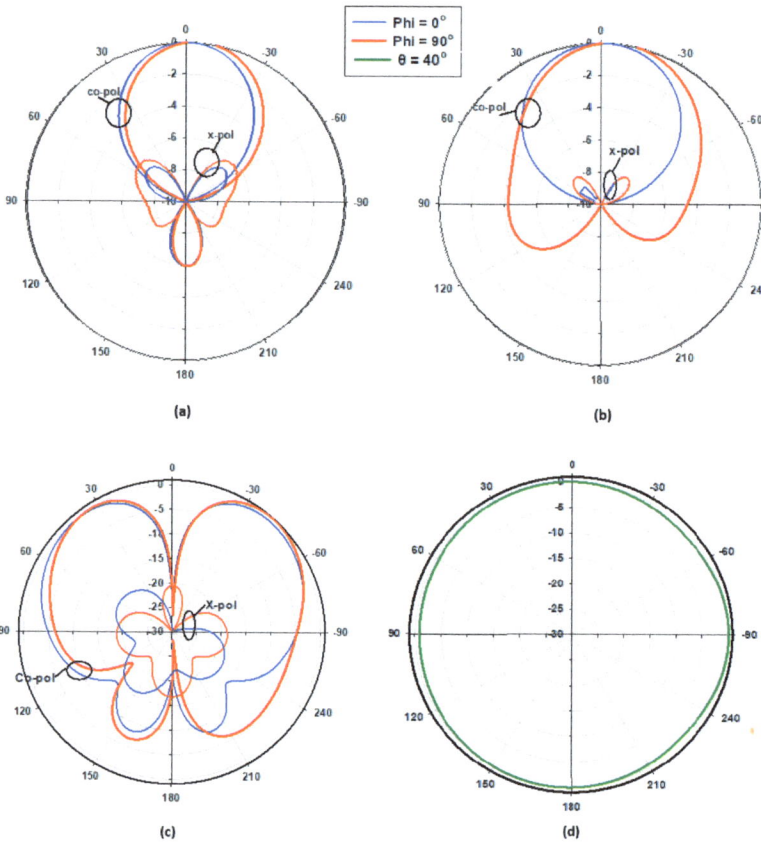

Figure 5. The radiation patterns of square ring-slot fed DRA with arm lengths of $l_{s1} = l_{s2} = 2.5$ mm, (a) 16 GHz, (b) 24 GHz, and (c,d) 27.7 GHz.

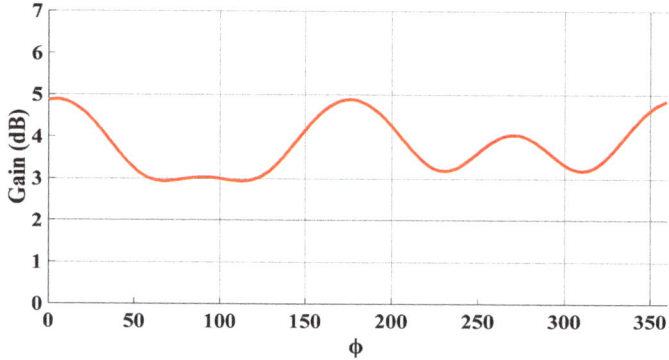

Figure 6. The omnidirectional gain variation at the θ = 40° azimuth plane when a square ring feeding slot is utilized.

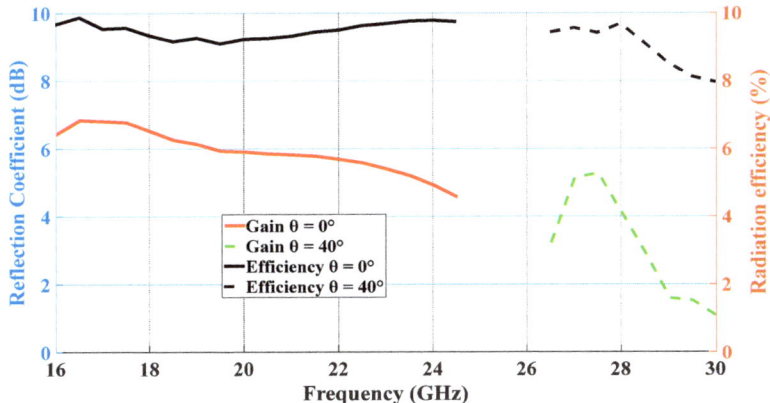

Figure 7. Realized gain and radiation efficiency of the square ring-slot-fed DRA.

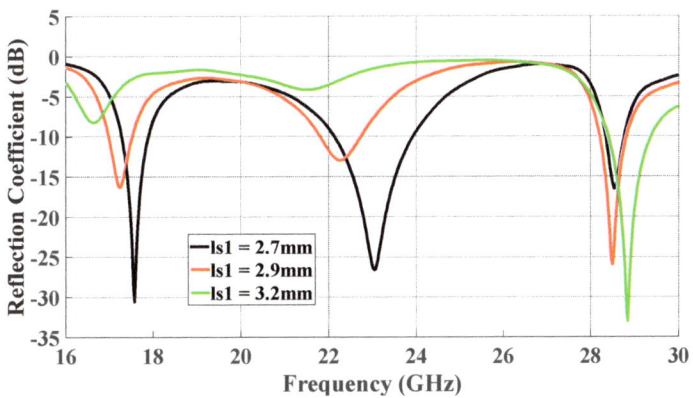

Figure 8. Simulated return losses of the rectangular ring-slot DRA when the longest slot arm's length, l_{s1}, was varied while $l_{s2} = 2$ mm.

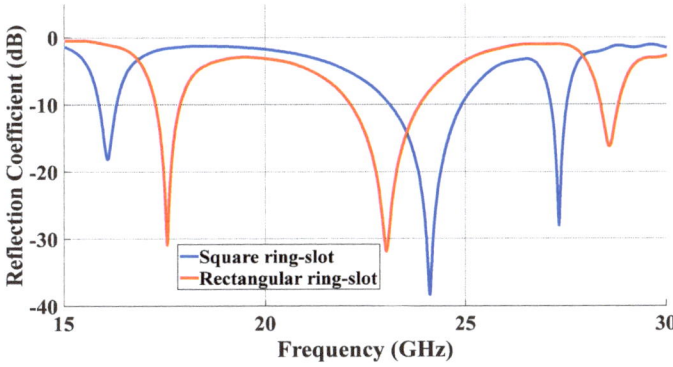

Figure 9. Simulated return losses of the square and rectangular ring-slots fed DRAs.

The simultaneous excitation of the degenerate modes was investigated by using a y-directed arm slot only with a length of l_{s1} and an offset of $0.5 l_{s2}$ from the x-axis, which excited the TE_{211} resonance mode at 30 GHz. Similarly, the TE_{121} resonance mode was individually excited at 30 GHz when an x-directed slot was utilized with a length of l_{s2} and an offset of $0.5l_{s1}$ from the y-axis. However, when the x- and y-directed linear slots were combined to create the rectangular ring-slot, resonance was achieved at a slightly lower frequency of 28.5 GHz with an overall field distribution that was similar to that of the cylindrical $HEM_{21\delta}$ mode, as demonstrated in Figure 10. The 3D radiation patterns for the three operating frequency bands are illustrated in Figure 11, where it can be noted that an omnidirectional pattern was achieved at 28.5 GHz with a maximum gain of 4.3 dBi.

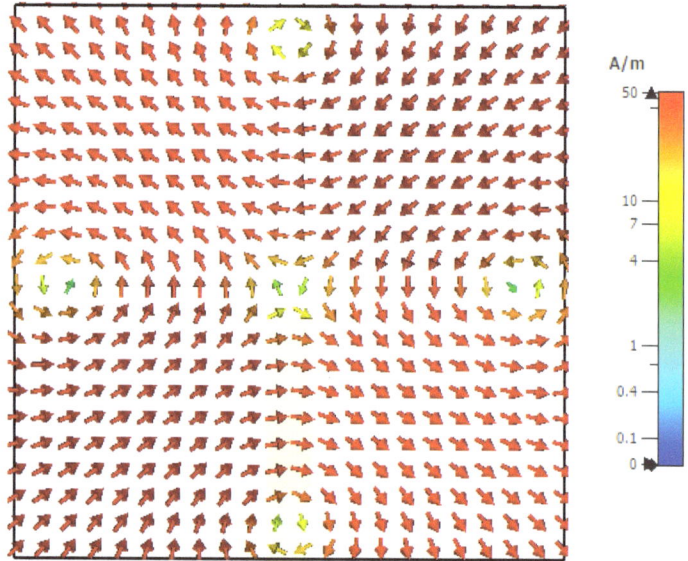

Figure 10. The xy plane view of the internal magnetic field distribution inside the rectangular ring-slot-fed DRA, which corresponds to the quasi-$HEM21\delta$ at 28.5 GHz.

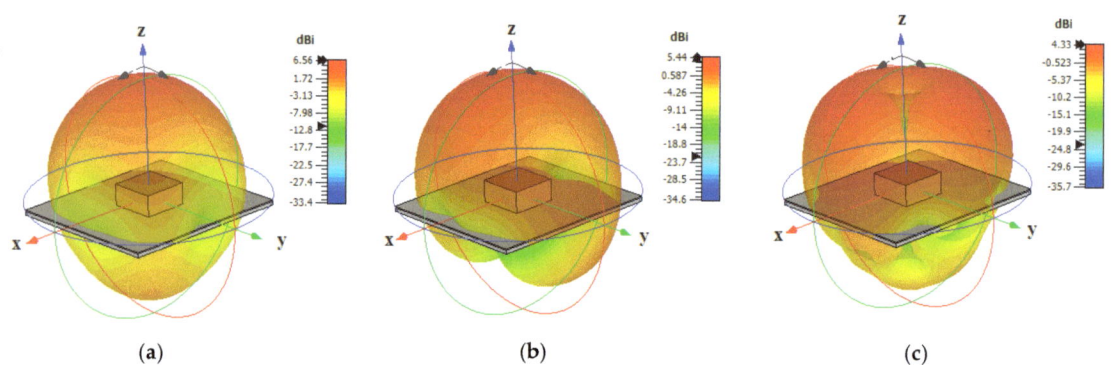

Figure 11. Three-dimensional radiation patterns at the three operating frequency bands: (**a**) 17.5 GHz, (**b**) 23 GHz, and (**c**) 28.5 GHz.

In addition, the electric and magnetic field distributions on the rectangular ring-slot are illustrated in Figures 12 and 13 at 17.5 GHz and 28.5 GHz, respectively. Figure 14 presents the variation in the return losses as a function of the ground plane size, where it can be observed that the slot's resonance was strongly dependent on the ground plane size [26]. As a result, there was also an impact on the excited DRA modes and the achieved resonance frequencies due to the variation in the performance of the feeding slot.

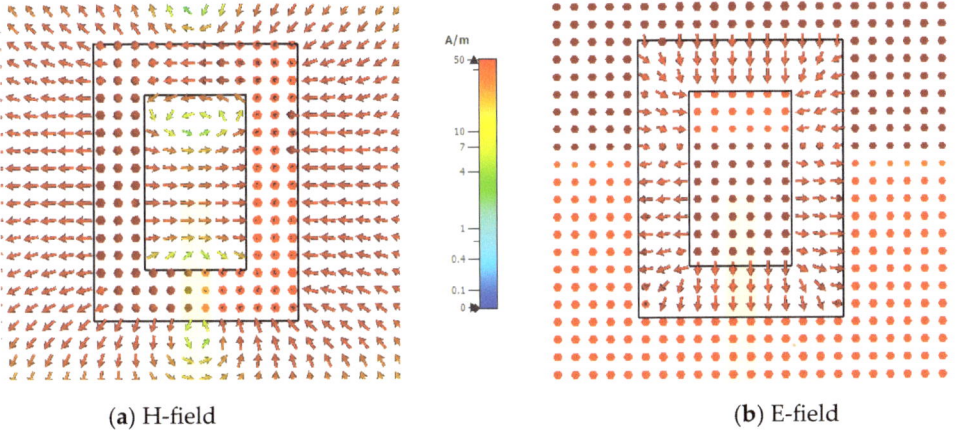

(a) H-field (b) E-field

Figure 12. The xy plane view of the electric and magnetic field distributions on the rectangular ring-slot feed structure at 17.5 GHz.

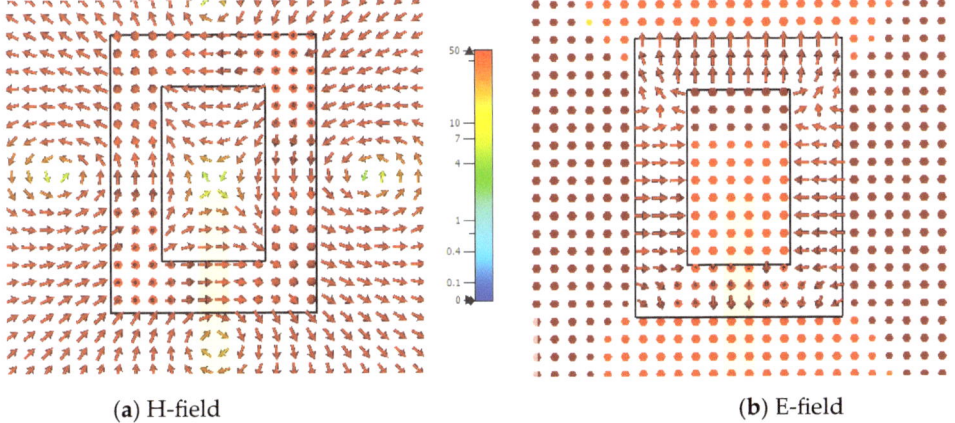

(a) H-field (b) E-field

Figure 13. The xy plane view of the electric and magnetic field distributions on the rectangular ring-slot feed at 28.5 GHz.

Another key parameter that was investigated was the sensitivity of the DRA performance to a range of alumina dielectric constants that have been used in the literature. The results of these investigations are presented in Figure 15, which demonstrates a stable DRA performance when ε_d was varied from 9.4 to 10.2.

The successful design of the omnidirectional DRA is presented in this section. In the next section, the performance of the DRA is investigated in proximity to a human body.

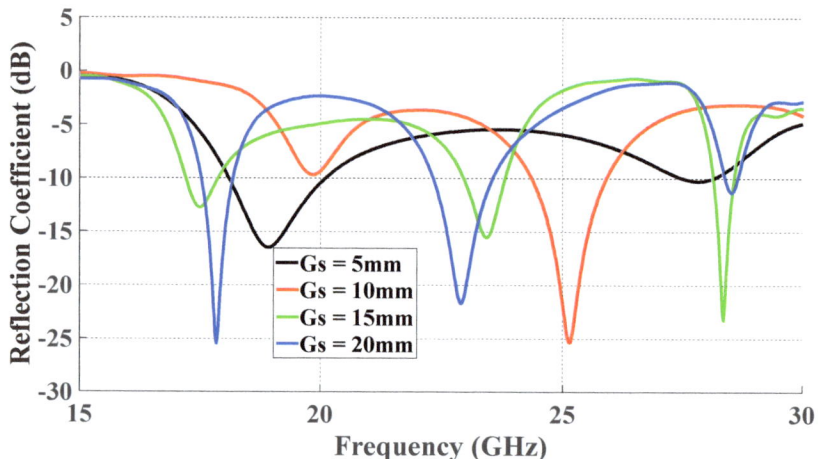

Figure 14. The variation in return losses as a function of the ground plane size.

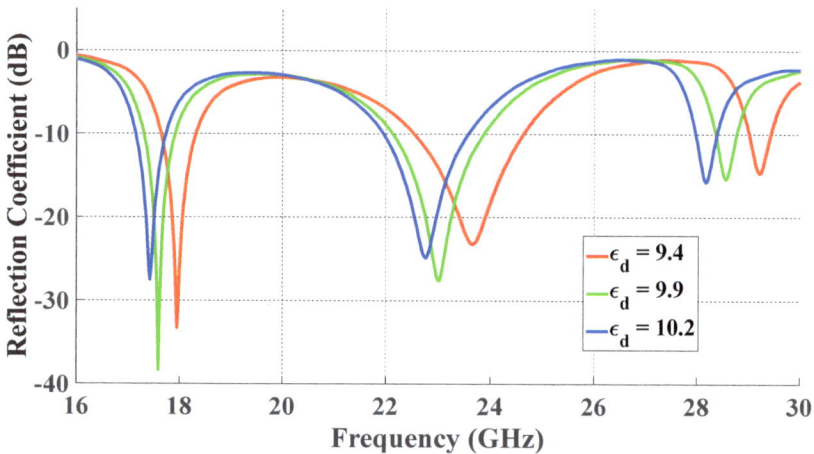

Figure 15. The variation in the return losses as a function of the dielectric constant of the DRA.

5. DRA Performance Next to a Human Body

For on-body applications, mmWave omnidirectional antennas are widely used. Therefore, it is important that the antenna's performance is assessed when the proposed antenna is placed next to the human body, as illustrated in Figure 16. In line with the literature, we assessed the DRA's performance next to three body areas: arm, chest, and stomach [27], where a three-layer phantom was utilized. The utilized parameters for the different tissue layers are illustrated in Table 2. The thicknesses of the three different body parts were based on those reported in [28]. The return losses when the DRA was placed next to arm, chest, and stomach are presented in Figure 17, where it can be noted that the presence of the ground plane minimized the impact of the human body on the resonance frequencies. In addition, the omnidirectional pattern was preserved in the presence of the chest, as demonstrated in Figure 18, which can also be attributed to the presence of the ground plane. However, reflections from the utilized phantom reduced the back lobes considerably and hence increased the omnidirectional realized gain from 4.33 dBi in free space to 5.8 dBi in the proximity of the human body tissue. On the other hand, the presence of the phantom reduced the radiation efficiency from 95% to 84%. However, this did not impact the gain as the increased directivity compensated for any loss due to the slightly reduced radiation efficiency.

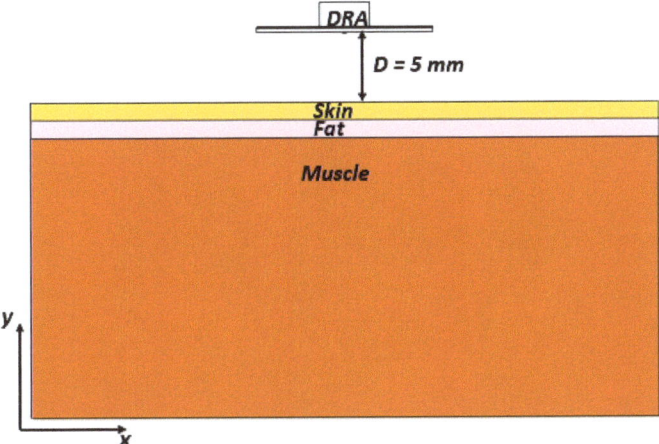

Figure 16. The proposed rectangular DRA in the proximity of a human body phantom.

Table 2. Human body tissue parameters at 28 GHz [29].

Tissue	Skin	Fat	Muscle
Relative permittivity	16.55	6.09	25.43
Loss tangent	0.2818	0.1454	0.242
Density (kg/m^3)	1109	911	1090

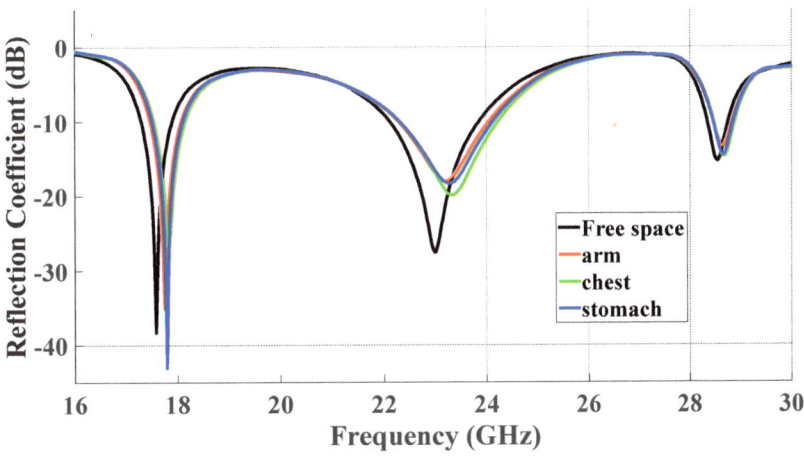

Figure 17. The variations in return losses when the antenna was placed on the arm, chest, and stomach.

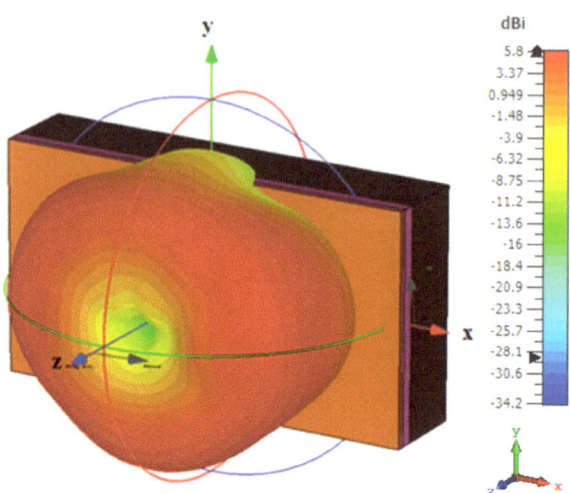

Figure 18. The 3D omnidirectional radiation pattern next to the equivalent tissue at 28.5 GHz when $d = 5$ mm.

The Specific Absorption Rate (SAR) indicates the safety threshold at which radio-frequency energy can be absorbed by human body tissue [29] The SAR must be assessed to ensure compliance with safety limits set by the Federal Communications Commission (FCC) and the International Commission for Non-Ionizing Radiation Protection (ICNIRP) standards. These standards define SAR thresholds of 1.6 and 2 W/kg for 1 g and 10 g tissues, respectively [29]. Unfortunately, the above guidelines do not offer dosimetric information or suggestions for mmWave frequencies [29,30] However, at 28 GHz, a 5 mm space is recommended between the antenna and the human body with input power levels of 15 dBm, 18 dBm, or 20 dBm at 28 GHz [30]. Subsequently, the proposed omnidirectional DRA was simulated next to a layered phantom, as demonstrated in Figure 16. The conducted SAR simulations confirmed that the radiation from the proposed antenna meets the safety requirements, as illustrated in Figures 19 and 20 for 1 g and 10 g tissues, respectively. It is worth noting that this SRA example is for the scenario of an antenna placed on the chest phantom.

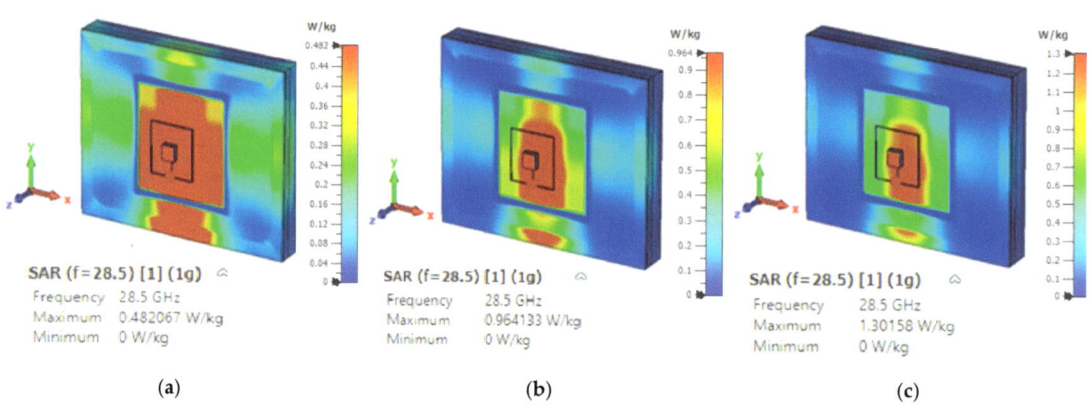

Figure 19. The SAR of the proposed DRA with various input power levels for a 1 g tissue: (**a**) 15 dBm, (**b**) 18 dBm, and (**c**) 20 dBm.

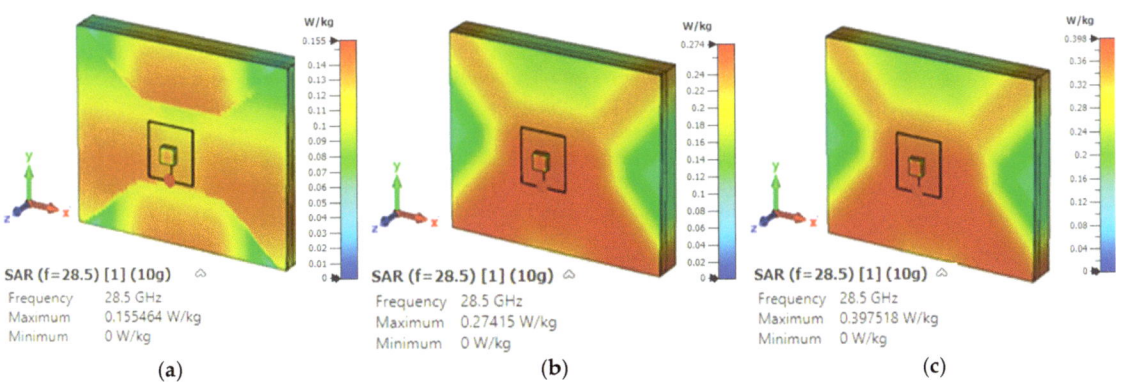

Figure 20. The SAR of the proposed DRA with various input power levels for a 10 g tissue: (a) 15 dBm, (b) 18 dBm, and (c) 20 dBm.

6. Measured Results

The alumina DRA and planar feed network incorporating a rectangular ring-slot were fabricated by T-ceramics [31] and Wrekin [32], respectively. At the mmWave frequency range, a precise alignment between the DRA and the feeding slot poses significant challenges. To overcome these challenges, a solution involving mapping out the DRA position on the ground plane was implemented during the fabrication stage [33]. The resulting fabricated feed network, which includes the outlined DRA position, is presented in Figure 21a. Following the outlining of the DRA position, ultrathin double-sided adhesive copper tape with a thickness of 0.08 mm was utilized to bond the antenna to the ground plane, ensuring secure assembly. The assembled DRA prototype is presented in Figure 21b, including the utilized ELF50-002 SMA connector that was attached using screws. In addition, the prototype was measured without experiencing any alignment or bonding issues. The implementation of this approach is critical in ensuring the mmWave measurements' accuracy, where even slight deviations can significantly affect the performance [34]. All measurements were carried out using the UKRI National Millimeter-Wave Facility [35], where an N5245B vector network analyzer (VNA) was employed to measure the return losses following a standard calibration procedure. Based on the analyzed data, the return losses were determined. On the other hand, an NSI-MI Technologies system was utilized in conducting the far-field measurements. By employing this specialized measurement system, various parameters, including the radiation pattern as a function of ϕ and θ, were accurately measured and visualized. To cover the elevation angle range of $\theta = -90°$ to $\theta = 90°$, the arm of the NSI-MI spherical system was set up to rotate across the upper hemisphere. Additionally, the gain of the antenna under test was determined with respect to a reference horn antenna.

As demonstrated in Figure 22, the measured and simulated return losses shared almost the same resonance frequencies of 17.5 GHz, 23 GHz, and 28.5 GHz for the TE_{111}, ring-slot, and quasi-$HEM_{21\delta}$ modes, respectively. In addition, the measured and simulated -10 dB impedance matching bandwidth of the lower band was 3.4%. In terms of the middle band that corresponds to the ring-slot resonance, the -10 dB impedance matching bandwidth was 1.8 GHz, demonstrating a good agreement between the measured and simulated percentage impedance bandwidths of 7.7% and 7.5%, respectively. However, a slight discrepancy can be noted in the omnidirectional mode's simulated and measured bandwidths of 1.9% and 3%, respectively. This discrepancy can be attributed to measurement uncertainties, including measurement setup as well as fabrication and calibration errors. In addition, the utilization of bulky SMA and fittings could have contributed to the discrepancy between simulated and measured results. It should be noted that the achieved impedance bandwidth of the omnidirectional mode was narrower than that of a probe-fed omnidirectional rectangular DRA. For example, impedance bandwidths of 22% were reported in [15] by merging the bandwidths of the DRA mode and that due to the feeding probe's resonance, which also offers an omnidirectional pattern. However, such a hybrid operation is not possible in the proposed configuration since the feeding ring-slot has broadside radiation, i.e., different from that of the excited DRA mode. Therefore, a feeding ring-slot with an omnidirectional pattern needs to be utilized for bandwidth enhancement. Alternatively, a dielectric coat layer [10], or concentric rectangular ring-slots, can be utilized to achieve a wider bandwidth.

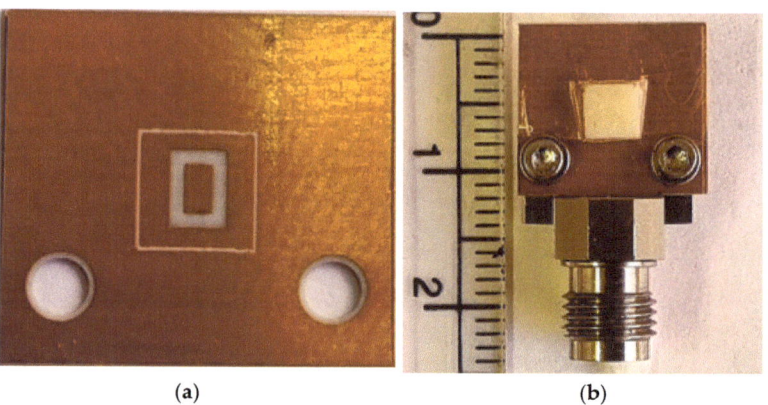

Figure 21. Prototype of the proposed antenna. (**a**) Ground plane with a rectangular ring-slot and outlined DRA position. (**b**) Assembled prototype.

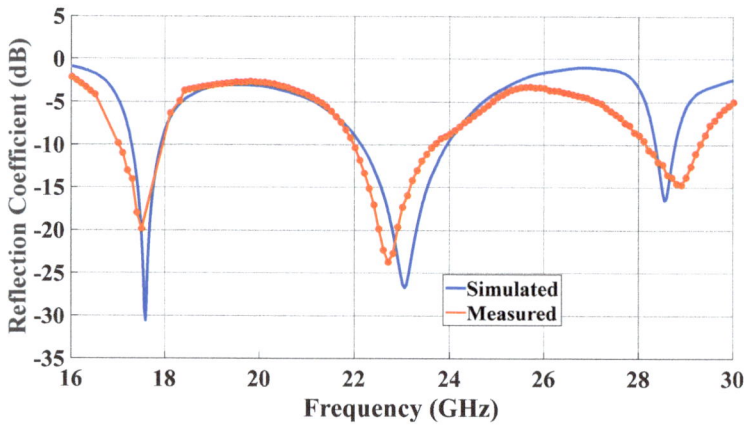

Figure 22. The measured and simulated return losses for the proposed DRA that is fed using a rectangular-ring-slot.

Figure 23 presents the measured and simulated radiation patterns at 17.5 GHz, where the TE_{111} broadside mode was excited. Close agreement can be observed between the simulated and measured broadside patterns. As mentioned earlier, the feeding slot's resonance was achieved at 23 GHz and the corresponding far field patterns are demonstrated in Figure 24, with reasonable agreement between the measurements and simulations. For example, the respective measured beamwidths were 88° and 108° in the E- and H- planes compared to 90° and 107° in the simulations. In addition, the simulated and measured omnidirectional radiation patterns are presented in Figure 25 for both the elevation and azimuth planes at 28.5 GHz. The results demonstrate close agreement between the simulated and measured radiation patterns, where an omnidirectional radiation pattern was achieved with a main beam direction at θ = 40°, as demonstrated in Figure 23a. The measured and simulated beamwidths of the omnidirectional patterns were 61.2° and 60.6°, respectively. However, a slight asymmetry can still be noted in the ϕ = 90° plane cut of Figure 25a, owing to the asymmetrical feed point position. An improved roundness of the azimuthal plane pattern can be observed in Figure 25b, which suggests that the rectangular ring-slot arms were placed at equally strong magnetic field points. Furthermore, the copolarized field component was considerably stronger than the cross-polarized component in all cases. The azimuthal plane gain variation presented in Figure 26, where it is evident that the variation was reduced considerably to ~0.85 dB, resulted in a more stable omnidirectional pattern with close agreement between the measurements and simulations. Additionally, the gain and radiation efficiency of the rectangular ring-slot-fed DRA are illustrated in Figure 27, where it can be noted that the maximum achieved gains were 6.56 dBi, 5.2 dBi, and 4.33 dBi for the TE_{111} mode, ring-slot resonance, and the quasi-$HEM_{21\delta}$ mode, respectively. Furthermore, a high radiation efficiency of ~90% was attained in the three operating bands.

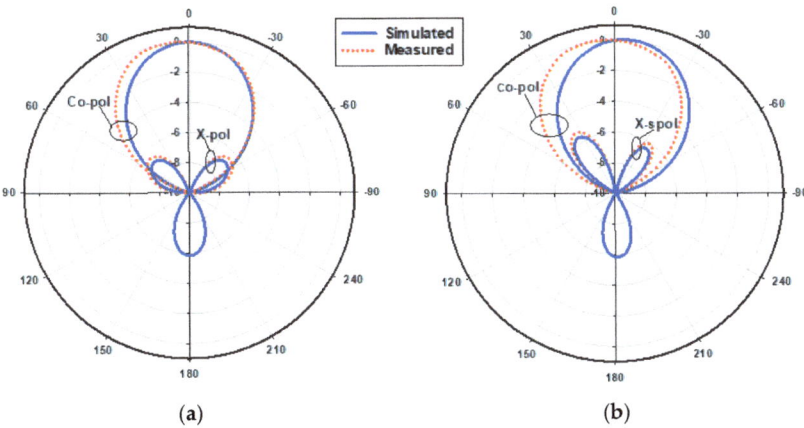

Figure 23. Normalized radiation patterns of the TE_{111} mode at 17.5 GHz: (**a**) E-plane; (**b**) H-plane.

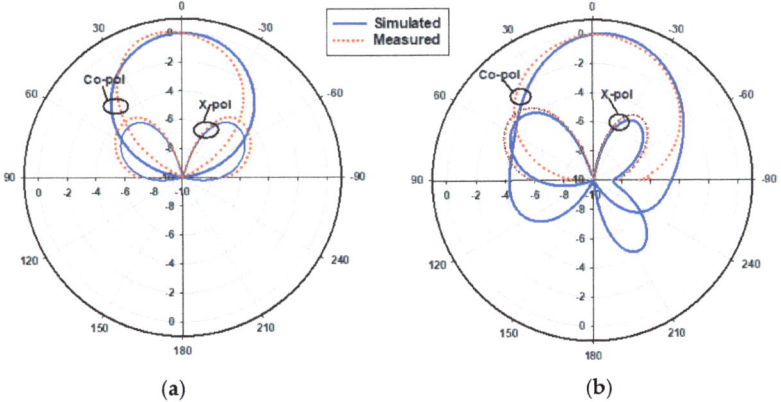

Figure 24. Normalized radiation patterns of the slot mode at 23 GHz: (**a**) E-plane; (**b**) H-plane.

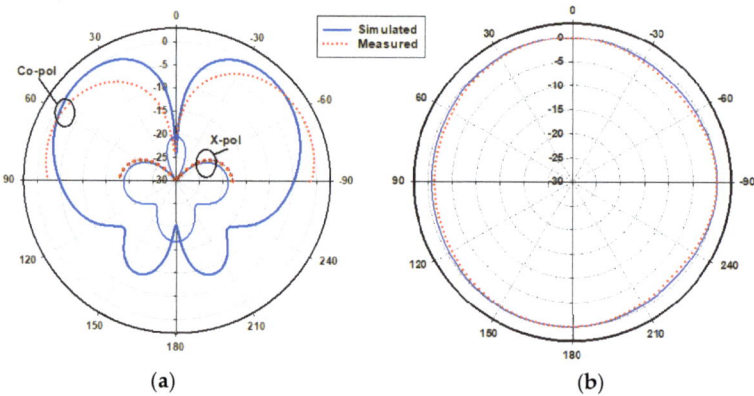

Figure 25. Normalized radiation patterns of the quasi-HEM$_{21\delta}$ mode at 28.5 GHz: (**a**) φ = 90° plane; (**b**) θ = 40° plane.

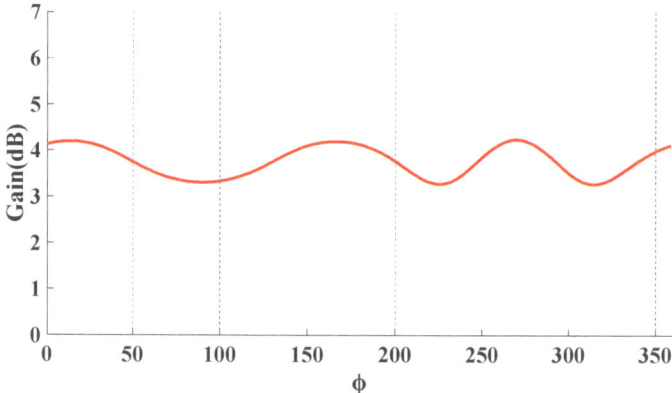

Figure 26. The azimuthal variation in the omnidirectional gain at the θ = 40° plane.

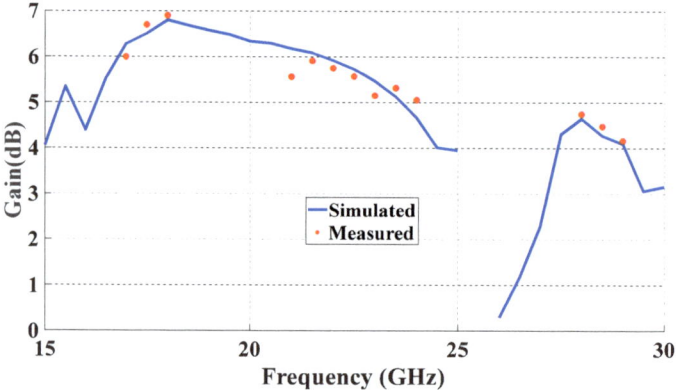

Figure 27. Measured and simulated gains at the main beam directions for the three resonance modes.

A comparative analysis of the ring-slot-fed DRA performance with respect to the reported on-body mmWave antenna designs is presented in Table 3. As mentioned earlier, there is no reported study on the on-body mmWave DRA in the open literature; hence, a comparison was made with respect to different antenna types that are available in the literature. The comparison was conducted with respect to the size, bandwidth, gain, and radiation efficiency. It is evident from Table 3 that the electrical size of the proposed antenna was smaller than most of the reported designs, except that of [36]. In addition, the utilized simple geometry resulted in simple and low-cost fabrication. On the other hand, a triple-band operation was achieved, which was also the case in [36]. owever, the individual bandwidths in the presented design were wider with higher gains compared to those in [36]. At the same time, the other antennas in Table 3 offer single-band operation, albeit with wider bandwidths. Furthermore, the proposed DRA outperformed the reported counterparts in terms of radiation efficiency.

Table 3. Comparison of the proposed on-body antenna's performance against cutting-edge mmWave counterparts.

Ref	Antenna Type	Frequency GHz	Size (λ^3) [1]	S_{11} Bandwidth (%)	On-Body Gain dBi	On-Body η_{rad} (%)
[36]	Slotted patch	28, 38, 61	$1.04 \times 1.02 \times 0.052$	3, 1, 1.5	8.1, 8.3, 7	54, 60, 58
[37]	Yagi array	60	$3.2 \times 1.6 \times 0.04$	15	9	41
[38]	Patch-like	60	$2.8 \times 2.1 \times 0.23$	16.3	12	63
[39]	Textile	28	$1.89 \times 0.87 \times 0.147$	33	6.6	53.5
[40]	Q Slot	60	$2.58 \times 2.8 \times 0.32$	12	8	56.68
[41]	Patch-like	60	$1.6 \times 1.02 \times 0.23$	-	5.4	62.2
This work	RDRA	17.5, 23, 28.5	$1.19 \times 1.19 \times 0.16$	3.4, 7.7, 1.9	7.3, 6.8, 5.8	90, 87, 84

[1] For [36] and this work, λ was calculated at the highest mentioned frequency point.

7. Conclusions

A multiband millimeter-wave rectangular DRA with different types of radiation patterns was demonstrated. A key achievement was the utilization of a planar feed network to excite an omnidirectional rectangular DRA instead of the traditionally used vertical coaxial probes. The dimensions of the feeding ring-slot were optimized to excite the required resonance modes, with improved performance in terms of the bandwidth and omnidirectional pattern quality. As a result, the quasi-$HEM_{21\delta}$ mode was excited for omnidirectional radiation. Moreover, broadside radiation was also achieved by exciting the fundamental TE_{111} mode and the ring-slot resonance mode. It should be noted that all the reported omnidirectional rectangular DRAs operate in the quasi-TM_{011} mode. Therefore, neither a planar feed network nor the excitation of the quasi-$HEM_{21\delta}$ mode were demonstrated earlier in the design of omnidirectional rectangular DRAs. Furthermore, an omnidirectional mmWave DRA of any shape has not been reported previously. The omnidirectional mode offers a gain of 4.33 dBi with a notably low azimuthal gain variation of 0.85 dB. The impact of different parts of the human body on the antenna performance was investigated and found to be marginal due to the presence of the ground plane. A comprehensive set of measurements was implemented with close agreement between the simulations and measurements. Overall, the proposed design offers considerable potential for a wide range of applications in the millimeter-wave frequency band. A comparison of the proposed DRA's performance against those of its earlier-reported counterparts showed that the DRA offers a smaller size and higher radiation efficiency, triple-band operation, and a low-cost, simple design.

Author Contributions: T.S.A.: simulation, manufacturing and measurements, writing; S.K.K.: supervision and writing. All authors have read and agreed to the published version of the manuscript.

Funding: This research received no external funding.

Institutional Review Board Statement: Not applicable.

Informed Consent Statement: Not applicable.

Data Availability Statement: Not applicable.

Acknowledgments: The authors would like to acknowledge the use of the National mmWave Facility and thank Steve Marsden for his support with the measurements.

Conflicts of Interest: The authors declare no conflict of interest.

References

1. Chowdhury, M.Z.; Shahjalal, M.; Ahmed, S.; Jang, Y.M. 6G wireless communication systems: Applications, requirements, technologies, challenges, and research directions. *IEEE Open J. Commun. Soc.* **2020**, *1*, 957–975. [CrossRef]
2. Abdulmajid, A.A.; Khalil, Y.; Khamas, S. Higher-order-mode circularly polarized two-layer rectangular dielectric resonator antenna. *IEEE Antennas Wirel. Propag. Lett.* **2018**, *17*, 1114–1117. [CrossRef]
3. Meher, P.R.; Behera, B.R.; Mishra, S.K.; Althuwayb, A.A. A chronological review of circularly polarized dielectric resonator antenna: Design and developments. *Int. J. RF Microw. Comput. -Aided Eng.* **2021**, *31*, e22589. [CrossRef]
4. Guha, D.; Banerjee, A.; Kumar, C.; Antar, Y.M. Higher order mode excitation for high-gain broadside radiation from cylindrical dielectric resonator antennas. *IEEE Trans. Antennas Propag.* **2011**, *60*, 71–77. [CrossRef]
5. Rappaport, T.S.; MacCartney, G.R.; Samimi, M.K.; Sun, S. Wideband millimeter-wave propagation measurements and channel models for future wireless communication system design. *IEEE Trans. Commun.* **2015**, *63*, 3029–3056. [CrossRef]
6. Liu, Y.; Yagoub, M.C. Compact, Broadband, and Omnidirectional Antenna Array for Millimeter-Wave Communication Systems. *J. Microw. Optoelectron. Electromagn. Appl.* **2021**, *20*, 297–306. [CrossRef]
7. Maurya, N.K.; Ammann, M.J.; Mcevoy, P. Series-fed Omnidirectional mm-Wave Dipole Array. *IEEE Trans. Antennas Propag.* **2023**, *71*, 1330–1336. [CrossRef]
8. Zou, L.; Abbott, D.; Fumeaux, C. Omnidirectional cylindrical dielectric resonator antenna with dual polarization. *IEEE Antennas Wirel. Propag. Lett.* **2012**, *11*, 515–518. [CrossRef]
9. Pan, Y.M.; Zheng, S.Y.; Li, W. Dual-band and dual-sense omnidirectional circularly polarized antenna. *IEEE Antennas Wirel. Propag. Lett.* **2014**, *13*, 706–709. [CrossRef]
10. Hu, P.F.; Pan, Y.M.; Leung, K.W.; Zhang, X.Y. Wide-/dual-band omnidirectional filtering dielectric resonator antennas. *IEEE Trans. Antennas Propag.* **2018**, *66*, 2622–2627. [CrossRef]
11. Fang, X.S.; Leung, K.W. Design of wideband omnidirectional two-layer transparent hemispherical dielectric resonator antenna. *IEEE Trans. Antennas Propag.* **2014**, *62*, 5353–5357. [CrossRef]
12. Lai, J.; Wang, J.; Zhao, K.; Jiang, H.; Chen, L.; Wu, Z.; Liu, J. Design of a dual-polarized omnidirectional dielectric resonator antenna for capsule endoscopy system. *IEEE Access* **2021**, *9*, 14779–14786. [CrossRef]
13. Pan, Y.M.; Leung, K.W.; Lu, K. Omnidirectional linearly and circularly polarized rectangular dielectric resonator antennas. *IEEE Trans. Antennas Propag.* **2011**, *60*, 751–759. [CrossRef]
14. Leung, K.W.; Pan, Y.M.; Fang, X.S.; Lim, E.H.; Luk, K.-M.; Chan, H.P. Dual-function radiating glass for antennas and light covers—Part I: Omnidirectional glass dielectric resonator antennas. *IEEE Trans. Antennas Propag.* **2012**, *61*, 578–586. [CrossRef]
15. Zou, M.; Pan, J. Investigation of resonant modes in wideband hybrid omnidirectional rectangular dielectric resonator antenna. *IEEE Trans. Antennas Propag.* **2015**, *63*, 3272–3275. [CrossRef]
16. Zou, M.; Pan, J.; Yang, D.; Xiong, G. Investigation of dual-band omnidirectional rectangular dielectric resonator antenna. *J. ElEctromagnEtic WavEs Appl.* **2016**, *30*, 1407–1416. [CrossRef]
17. Li, W.; Leung, K.W.; Yang, N. Omnidirectional dielectric resonator antenna with a planar feed for circular polarization diversity design. *IEEE Trans. Antennas Propag.* **2018**, *66*, 1189–1197. [CrossRef]
18. Yang, N.; Leung, K.W. Size reduction of omnidirectional cylindrical dielectric resonator antenna using a magnetic aperture source. *IEEE Trans. Antennas Propag.* **2019**, *68*, 3248–3253. [CrossRef]
19. Yang, N.; Leung, K.W. Compact cylindrical pattern-diversity dielectric resonator antenna. *IEEE Antennas Wirel. Propag. Lett.* **2019**, *19*, 19–23. [CrossRef]
20. Fang, X.S.; Weng, L.P.; Sun, Y.-X. Slots-coupled omnidirectional circularly polarized cylindrical glass dielectric resonator antenna for 5.8-GHz WLAN application. *IEEE Access* **2020**, *8*, 204718–204727. [CrossRef]
21. Fang, X.S.; Weng, L.P.; Fan, Z. Design of the Wideband and Low-Height Omnidirectional Cylindrical Dielectric Resonator Antenna Using Arced-Apertures Feeding. *IEEE Access* **2023**, *11*, 20128–20135. [CrossRef]
22. Pan, Y.M.; Leung, K.W.; Lu, K. Study of resonant modes in rectangular dielectric resonator antenna based on radar cross section. *IEEE Trans. Antennas Propag.* **2019**, *67*, 4200–4205. [CrossRef]
23. Keyrouz, S.; Caratelli, D. Dielectric resonator antennas: Basic concepts, design guidelines, and recent developments at millimeter-wave frequencies. *Int. J. Antennas Propag.* **2016**, *2016*, 6075680. [CrossRef]
24. Petosa, A.; Simons, N.; Siushansian, R.; Ittipiboon, A.; Cuhaci, M. Design and analysis of multisegment dielectric resonator antennas. *IEEE Trans. Antennas Propag.* **2000**, *48*, 738–742. [CrossRef]

25. Mongia, R.K.; Ittipiboon, A. Theoretical and experimental investigations on rectangular dielectric resonator antennas. *IEEE Trans. Antennas Propag.* **1997**, *45*, 1348–1356. [CrossRef]
26. Row, J.-S. The design of a squarer-ring slot antenna for circular polarization. *IEEE Trans. Antennas Propag.* **2005**, *53*, 1967–1972. [CrossRef]
27. Gao, G.; Wang, S.; Zhang, R.; Yang, C.; Hu, B. Flexible EBG-backed PIFA based on conductive textile and PDMS for wearable applications. *Microw. Opt. Technol. Lett.* **2020**, *62*, 1733–1741. [CrossRef]
28. Xu, R.; Zhu, H.; Yuan, J. Electric-field intrabody communication channel modeling with finite-element method. *IEEE Trans. Biomed. Eng.* **2010**, *58*, 705–712.
29. Hamed, T.; Maqsood, M. SAR calculation & temperature response of human body exposure to electromagnetic radiations at 28, 40 and 60 GHz mmWave frequencies. *Prog. Electromagn. Res. M* **2018**, *73*, 47–59.
30. Lak, A.; Adelpour, Z.; Oraizi, H.; Parhizgar, N. Design and SAR assessment of three compact 5G antenna arrays. *Sci. Rep.* **2021**, *11*, 21265. [CrossRef]
31. T-Ceram s.r.o. Available online: http://www.t-ceram.com/ (accessed on 3 January 2023).
32. Wrekin Circuits, Ltd. Available online: https://www.wrekin-circuits.co.uk/ (accessed on 6 December 2022).
33. Abdou, T.S.; Saad, R.; Khamas, S.K. A Circularly Polarized mmWave Dielectric-Resonator-Antenna Array for Off-Body Communications. *Appl. Sci.* **2023**, *13*, 2002. [CrossRef]
34. Alanazi, M.D.; Khamas, S.K. Wideband mm-wave hemispherical dielectric resonator antenna with simple alignment and assembly procedures. *Electronics* **2022**, *11*, 2917. [CrossRef]
35. UKRI National Millimetre Wave Facility. Available online: https://www.sheffield.ac.uk/mm-wave/ (accessed on 6 October 2022).
36. Ur-Rehman, M.; Adekanye, M.; Chattha, H.T. Tri-band millimetre-wave antenna for body-centric networks. *Nano Commun. Netw.* **2018**, *18*, 72–81. [CrossRef]
37. Chahat, N.; Zhadobov, M.; Le Coq, L.; Sauleau, R. Wearable endfire textile antenna for on-body communications at 60 GHz. *IEEE Antennas Wirel. Propag. Lett.* **2012**, *11*, 799–802. [CrossRef]
38. Ur-Rehman, M.; Malik, N.A.; Yang, X.; Abbasi, Q.H.; Zhang, Z.; Zhao, N. A low profile antenna for millimeter-wave body-centric applications. *IEEE Trans. Antennas Propag.* **2017**, *65*, 6329–6337. [CrossRef]
39. Wagih, M.; Weddell, A.S.; Beeby, S. Millimeter-wave textile antenna for on-body RF energy harvesting in future 5G networks. In Proceedings of 2019 IEEE Wireless Power Transfer Conference (WPTC), London, UK, 18–21 June 2019; pp. 245–248.
40. Khan, M.M.; Islam, K.; Alam Shovon, M.N.; Baz, M.; Masud, M. Design of a novel 60 GHz millimeter wave Q-slot antenna for body-centric communications. *Int. J. Antennas Propag.* **2021**, *2021*, 9795959. [CrossRef]
41. Rahman, H.A.; Khan, M.M.; Baz, M.; Masud, M.; AlZain, M.A. Novel compact design and investigation of a super wideband millimeter wave antenna for body-centric communications. *Int. J. Antennas Propag.* **2021**, *2021*, 8725263. [CrossRef]

Disclaimer/Publisher's Note: The statements, opinions and data contained in all publications are solely those of the individual author(s) and contributor(s) and not of MDPI and/or the editor(s). MDPI and/or the editor(s) disclaim responsibility for any injury to people or property resulting from any ideas, methods, instructions or products referred to in the content.

Article

Experimental Study on the Compatibility of PD Flexible UHF Antenna Sensor Substrate with SF6/N2

Xukun Hu [1,2], Guozhi Zhang [1,2,3,*], Guangyu Deng [1] and Xuyu Li [1]

[1] Hubei Engineering Research Center for Safety Monitoring of New Energy and Power Grid Equipment, Hubei University of Technology, Wuhan 430068, China; 2010221106@hbut.edu.cn (X.H.); 102110311@hbut.edu.cn (G.D.); 2010231403@hbut.edu.cn (X.L.)
[2] School of Electrical and Electronic Engineering, Hubei University of Technology, Wuhan 430068, China
[3] State Grid Electric Power Research Institute Wuhan Nanrui Co., Ltd., Wuhan 430074, China
* Correspondence: zhangguozhi@hbut.edu.cn

Abstract: The use of flexible, built-in, ultra-high-frequency (UHF) antenna sensors is an effective method to solve the weak high-frequency electromagnetic wave signal sensing of partial discharge (PD) inside gas-insulated switchgears (GISs), and the compatibility of flexible UHF antenna sensor substrate materials and SF6/N2 mixtures is the key to the realization of a flexible UHF antenna sensor inside a GIS. Based on this, this paper builds an experimental platform for the compatibility of a 30% SF6/70% N2 gas mixture and a PD flexible UHF antenna sensor substrate and conducts compatibility experiments between the 30% SF6/70% N2 gas mixture and PD flexible UHF antenna sensor substrate under different temperatures in combination with the actual operating temperature range of the GIS. In this article, a Fourier transform infrared spectrometer, scanning electron microscope and X-ray photoelectron spectrometer were used to test and analyze the gas composition, the surface morphology and the elemental change in the PD flexible UHF antenna sensor substrate, respectively. PET material will be slightly oxidized under the environment of a 30% SF6/70% N2 gas mixture at 110 °C, PI material will generate metal fluoride under the environment of a 30% SF6/70% N2 gas mixture and only PDMS material will remain stable under the environment of a 30% SF6/70% N2 gas mixture; therefore, it is appropriate to use PDMS substrate in the development of flexible UHF antenna sensors.

Keywords: flexible UHF antenna sensor; flexible base; DC gas-insulated switchgear (GIS); compatibility; SF6/N2 gas mixture

Citation: Hu, X.; Zhang, G.; Deng, G.; Li, X. Experimental Study on the Compatibility of PD Flexible UHF Antenna Sensor Substrate with SF6/N2. *Micromachines* **2023**, *14*, 1516. https://doi.org/10.3390/mi14081516

Academic Editor: Haejun Chung

Received: 19 June 2023
Revised: 22 July 2023
Accepted: 23 July 2023
Published: 28 July 2023

Copyright: © 2023 by the authors. Licensee MDPI, Basel, Switzerland. This article is an open access article distributed under the terms and conditions of the Creative Commons Attribution (CC BY) license (https://creativecommons.org/licenses/by/4.0/).

1. Introduction

SF_6 is widely used in insulation systems of electric equipment such as gas-insulated switchgears (GISs) because of its good insulating ability and arc-extinguishing ability. However, SF_6 has a large greenhouse effect, its decomposition products are toxic and corrosive and its cost is high. Therefore, in order to reduce the use of SF_6 gas, the most effective method is to develop the use of SF_6 gas mixtures. Current research shows that SF_6 mixed with N_2 has a good synergistic effect and its insulation performance is stronger than N_2 and more similar to SF_6 [1]. In recent years, the State Grid Corporation, in response to the national call for energy savings and emission reductions, organized work on the SF_6/N_2 gas mixture for GIS busbars and isolation and grounding switches, and they also determined the use of a 30% SF_6/70% N_2 gas mixture as an insulation medium for GIS busbars and isolation and grounding switches [2]. Partial discharge (PD) is the main cause of GIS insulation failure, and if not dealt with in a timely manner, it will likely lead to serious equipment accidents and grid accidents [3–5]. The ultra-high-frequency (UHF) method uses antenna sensors to sense the high-frequency electromagnetic wave signals radiated by PD, which has the advantages of strong anti-interference capability and high reliability, and it is widely used for the detection of GIS PD insulation defects [6,7]. At present, all newly commissioned GISs of 220 kV and above in China's power grid companies are installed with UHF detection systems or reserved with UHF detection interfaces. According to the installation position, UHF antenna sensors for GIS PD detection can be divided into two types: built-in and external [8–10]. External UHF antenna sensors will be affected by the serious attenuation of electromagnetic wave signals caused by outward PD leakage due to the GISs' own metal shell structure and the influence of the corona interference signal and mobile communication signal in the external environment, resulting in inadequacies in its sensitivity to the PD in the GIS, especially in intermittent PD signal sensing. Although built-in UHF antenna sensors can effectively overcome the shortcomings of external UHF antenna sensors, most of the current antenna sensors for UHF detection built into GISs are made of rigid materials such as those with FR-4 epoxy resin as the substrate [11–13], which cannot conform to the cylindrical metal structure housing of GISs and require the complex structural modification of the flange of the device itself. In addition, rigid substrates in built-in UHF antenna sensors also create the risk of problems such as the electric field distribution inside the device being affected [14].

For the creation of flexible, built-in GIS PD detection UHF antenna sensors, our group has developed a variety of UHF GIS PD detection antennas based on two PD flexible UHF antenna sensor substrates, namely thermosetting polyimide (PI) and polydimethylsiloxane (SU-8/PDMS). These flexible antennas have obvious advantages such as ultra-small sizes (maximum sizes of less than 60 mm), ultra-thinness (thicknesses of up to 0.2 mm), excellent radiation performances and stable performances in the bending radius deformation range of 150–500 mm [15–18].

Because the process of building a flexible UHF antenna sensor into a GIS puts it in direct contact with a SF_6/N_2 gas mixture, if the flexible UHF antenna sensor is incompatible with SF_6/N_2 inside the equipment, on the one hand, this may cause the PD flexible UHF antenna sensor substrate material to be corroded by the gas mixture, which will lead to problems such as the inaccurate detection of PD signals by the flexible antenna sensor; on the other hand, the flexible UHF antenna sensor may lead to the decomposition of the mixed gas, resulting in the degradation of the insulation performance of the mixed gas, and then causing more serious insulation accidents. At the same time, it is still unclear which built-in UHF antenna sensor substrate material has better compatibility with the SF_6/N_2 gas mixture, which is directly related to the safety of the long-term operation of built-in flexible UHF antenna sensors in GISs and the field implementability of the technology.

Based on this, an experimental study on the compatibility of PD flexible UHF antenna sensor substrates with SF_6/N_2 is proposed in this paper. Firstly, the experimental platform used to determine the compatibility of the SF_6/N_2 gas mixture and the flexible UHF antenna sensor substrate material was built, and thermal acceleration experiments were carried out with the SF_6/N_2 gas mixture and three commonly used PD flexible UHF antenna sensor substrate materials at different temperatures, according to the actual operating temperature of a GIS. The results are summarized and analyzed to enable us to select the most suitable PD flexible UHF antenna sensor substrate materials and provide basic reference data for the design and engineering application of GIS PD flexible built-in UHF antenna sensors.

2. Compatibility Experimental Platform and Experimental Materials

2.1. Compatibility Experimental Platform

The thermal acceleration experimental platform is shown in Figure 1. The experimental platform was mainly composed of a temperature control system and a switching system. The experimental temperature was regulated by the temperature control system at the beginning of the experiment, and the temperature inside the chamber was set using the internal detection system. The high-pressure sealed tank shell was made of stainless steel, and a stainless steel mesh holder was placed in the gas chamber for the PD flexible UHF antenna sensor substrate material, which was intended to be in full contact with the SF_6/N_2 gas mixture and to ensure uniform heating of the sample. The high-pressure sealed tank was cleaned with SF_6 gas several times before the experiment; the pressure value in the gas chamber was checked using a barometer during inflation; and the exhaust-gas-collection device was used to collect the excess gas and reduce the pollution to the environment.

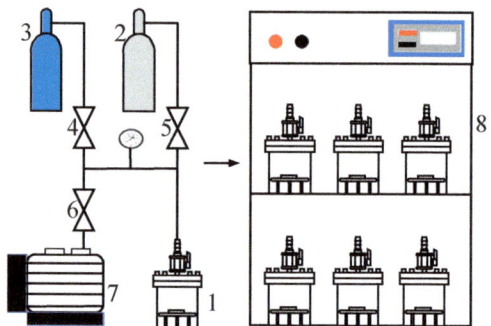

1. High pressure sealed tank, 2.N₂ gas, 3.SF₆ gas, 4.SF₆ Gas buffer valve, 5.N₂ Gas buffer valve, 6.Vacuum pump valves, 7.vacuum pump, 8.Constant temperature heating box

Figure 1. Experimental platform.

2.2. Experimental Materials

Due to the continuous development of flexible wearable devices, flexible antennas have become a research hotspot in recent years. Antennas with flexible materials as substrates have good deformation capabilities and can be co-profiled with complex structures, so the selection of suitable flexible substrates is the key to the realization of the development of flexible UHF antenna sensors. Common flexible dielectric substrates mainly include PI, polyethylene terephthalate (PET) and PDMS as PD flexible UHF antenna sensor substrate materials. These materials are bendable, have high tensile strength, are non-toxic, have high temperature resistance, have good insulation properties, have low dielectric constants and have good dimensional stability. It is these excellent properties that make these materials ideal for making flexible UHF antenna sensor substrates [19,20]. Therefore, in this paper, three PD flexible UHF antenna sensor substrate materials, PI, PET, and PDMS, were selected for the study to investigate their compatibility with SF_6/N_2. The dimensions of the experimental samples of the flexible materials used for accelerated thermal aging are shown in Figure 2.

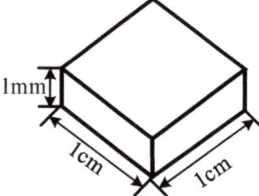

Figure 2. Sample of flexible material.

3. Experimental Method

According to GB/T11022-2020 [21], the maximum allowable temperature of the accessible parts of the equipment is 70 °C, and DL/T 617-2010 states that the maximum temperature rise inside the cavity is 40 K. Therefore, in order to investigate the compatibility of the SF_6/N_2 gas mixture with the PD flexible UHF antenna sensor substrate material under the actual operating temperature and the temperature limit of the equipment, the experimental temperatures were set to 40 °C, 70 °C and 110 °C, respectively. The temperature was controlled by using a constant temperature heating chamber. The experimental sample specimens were placed in the SF_6/N_2 gas mixture environment, and the thermal acceleration experiments were continuously carried out under three different temperatures for 7 days and 24 h per day.

According to GB/T11021-2014, "Electrical Insulation, Heat Resistance and Representation Method" and "Evaluation and Identification of Electrical Insulation System: 'IEC60505:2011'", according to the 6-degree rule, the compatibility of the set experimental temperature range and experimental time can better simulate the service life of gas-insulated equipment under normal working conditions.

$$\text{Equivalent formula}: t_2 = 2^{\frac{T_1-T_2}{6}} \times t_1 \qquad (1)$$

Among them, t_1 represents the duration of the thermal acceleration experiment, t_2 represents the equivalent duration of the GIS operating temperature, T_1 represents the experimental temperature and T_2 represents the GIS operating temperature. It is calculated that when the experimental duration is 7 days and the experimental temperature is 110 °C, the maximum ambient temperature of GIS operation is 40 °C, and the equivalent duration is calculated to be 62.34 years [22].

Experimental Steps

The experimental steps are shown in Figure 3. Before the experiment, the surface of the small experimental sample and the inner wall of the experimental gas tank were wiped with absolute ethanol, and after the sample and the tank were naturally air-dried, the experimental sample was put into the bottom bracket of the gas tank. The experimental gas tank had good air tightness, could withstand a maximum pressure of 0.8 MPa, and had a capacity of about 400 mL. Before the experiment, SF_6 was used to wash and vacuum the gas chamber, and the above operation was repeated 3 times to avoid impure gas affecting the results. The Dalton partial pressure law was used to fill the gas tank with 0.5 MPa 30% SF_6/70% N_2 gas, put the gas tank into a constant temperature heating box for the heating treatment for 7 days and set up room-temperature (20 °C) experimental control groups. After the experiment, the composition of the mixed gas was analyzed using a Fourier transform infrared spectrometer (FTIR), and a scanning electron microscope (SEM) and X-ray photoelectron spectroscopy (XPS) were used to test and analyze the surface morphology and elemental changes in the PD flexible UHF antenna sensor.

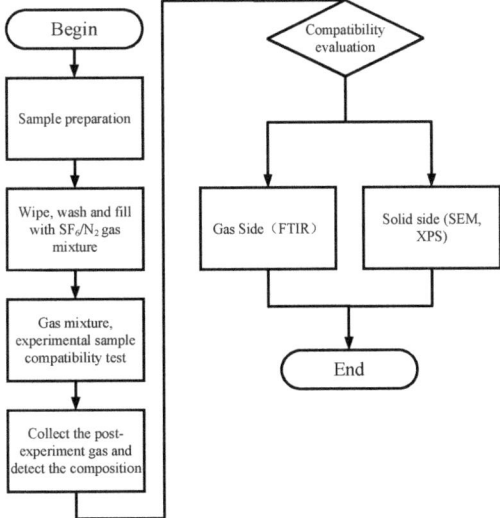

Figure 3. Flowchart of experimental steps.

4. Experimental Results and Analysis

4.1. Analysis of the Influence of PD Flexible UHF Antenna Sensor Substrate on SF_6 Gas Itself

An FTIR is an infrared spectrometer developed based on the principle of the Fourier transform of infrared light after interference, which is mainly composed of an infrared light source, diaphragm, interferometer (beam splitter, moving mirror and fixed mirror), sample chamber, detector and various infrared mirrors, lasers, control circuit boards and power supplies. Fourier infrared spectroscopy can process a signal via Fourier transform to obtain an infrared absorption spectrum containing absorbance with a wavenumber. At the same time, since the chemical vibrations of different functional groups have different absorption characteristics of infrared light with different wavenumbers, the qualitative and quantitative analysis of SF_6/N_2 properties can be realized by comparing the FTIR infrared absorption spectra before and after the experiment.

After the 7-day thermal acceleration experiment of 40 °C, 70 °C and 110 °C, respectively, on the base materials of the three PD flexible UHF antenna sensor substrates of PI, PET and PDMS, the mixed gas after the gas collection experiment was used for FTIR detection, and the FTIR detection results of the mixed gas before and after the experiment of the base material of PI, PET and PDMS for three PD flexible UHF antenna sensors are shown in Figures 4–6.

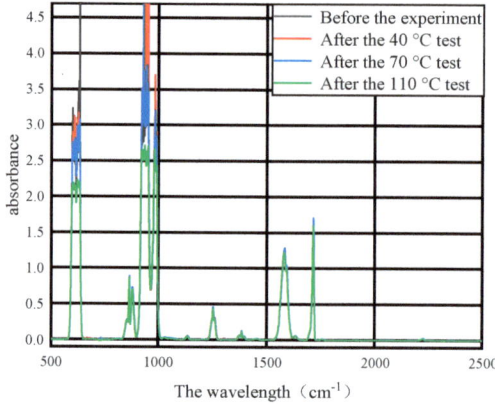

Figure 4. Infrared spectra of gas after PI and SF_6/N_2 tests at different temperatures.

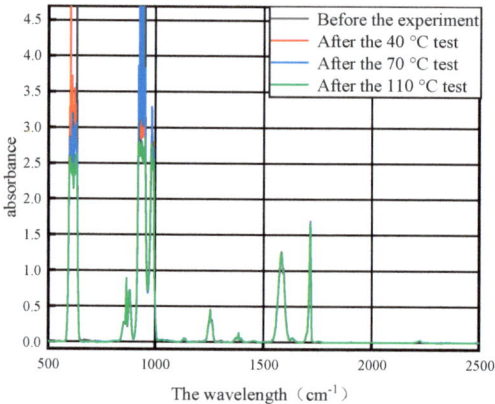

Figure 5. Infrared spectra of gas after PET and SF_6/N_2 tests at different temperatures.

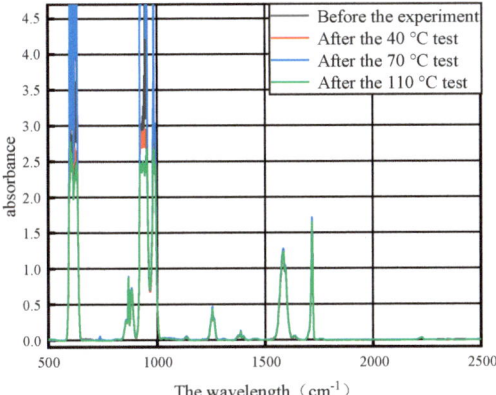

Figure 6. Infrared spectra of gas after PDMS and SF_6/N_2 tests at different temperatures.

It can be seen from Figures 4–6 that the SF_6/N_2 mixture in the control group peaks in the infrared spectrum with the SF_6/N_2 mixture after the experiment, which indicates that the composition of the mixed gas does not change after the experiment, and the base materials of the three PD flexible UHF antenna sensors of PI, PET and PDMS will not affect the SF_6/N_2 mixture within the GIS operating temperature range.

4.2. Analysis of the Influence of SF_6/N_2 Mixture on PD Flexible UHF Antenna Sensor Substrate

If the SF_6/N_2 mixture reacts with the PD flexible UHF antenna sensor substrate material, it may not only cause the SF_6/N_2 mixture to decompose but also cause corrosion to the PD flexible UHF antenna sensor substrate material and introduce new uncertain risk factors.

4.2.1. PD Flexible UHF Antenna Sensor Substrate Topography Detection

In order to explore the changes in the surface morphologies of PD flexible UHF antenna sensor substrate materials, scanning electron microscopy was used to magnify the surface of PD flexible UHF antenna sensor substrate materials by 300 times and 2500 times before and after the experiment [23]. Since PDMS has low adhesion and cannot be stably fixed in the SEM sample stage, only the surface topographies of the PI and PET PD flexible UHF antenna sensor substrates were analyzed, and the results are shown in Figures 7 and 8, respectively.

It can be seen from Figures 7 and 8 that in the SF_6/N_2 mixed gas environment, when the magnification is 300 times, the surfaces of the PI and PET materials are smooth at three temperatures, and there is no obvious corrosion phenomenon. When the magnification is 2500 times, because the test points of the PI and PET materials are different at different temperatures, the surface morphologies of the PI and PET materials are different at different temperatures. The PI and PET materials have relatively few surface impurities in the mixed gas environment of 40 °C SF_6/N_2, and the PI and PET materials have comparatively many surface impurities in the mixed gas environment of 70 °C and 110 °C SF_6/N_2, among which, the surface impurities of the PI materials and PET materials show the shape of small islands in the mixed gas environment of 110 °C SF_6/N_2. Since the surfaces of the PI and PET materials show impure morphologies in three different temperature environments of 40 °C, 70 °C and 110 °C, and there is no obvious corrosion phenomenon, it can be concluded that neither PI nor PET reacts with SF_6/N_2 mixed gas.

4.2.2. PD Flexible UHF Antenna Sensor Substrate Material Surface Element Detection

In order to analyze the possible changes in the substrate material surface of the PD flexible UHF antenna sensor, XPS analysis of the possible elements (C1s and F1s) on the sample surface was carried out. The detected absorption peaks were analyzed with Multipak software, and Shirley-type fitting subtraction was used to fit the peaks with the

Gauss algorithm, where the charge calibration element was C1s (284.8 eV) [24,25]. The full spectra of elements on the substrates of three kinds of PD flexible UHF antenna sensors (PI, PET and PDMS) before the experiment are shown in Figure 9; from the figure, it can be seen that the substrates of three kinds of PD flexible UHF antenna sensors, PI, PET and PDMS, mainly contain C, O and S elements and contain trace amounts of F elements.

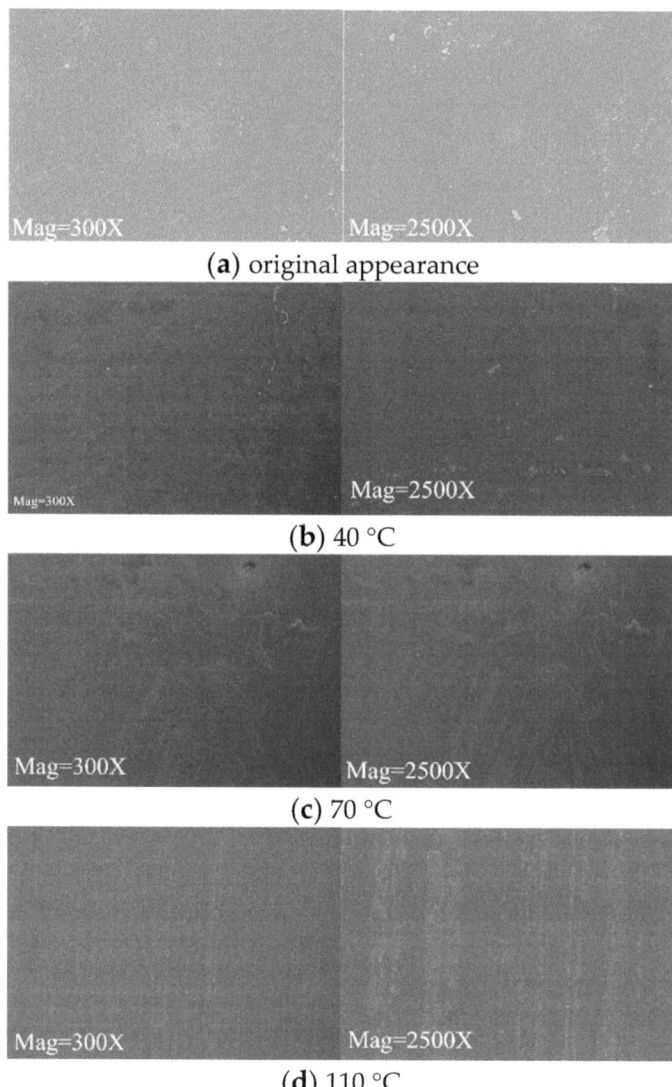

Figure 7. SEM of PI material.

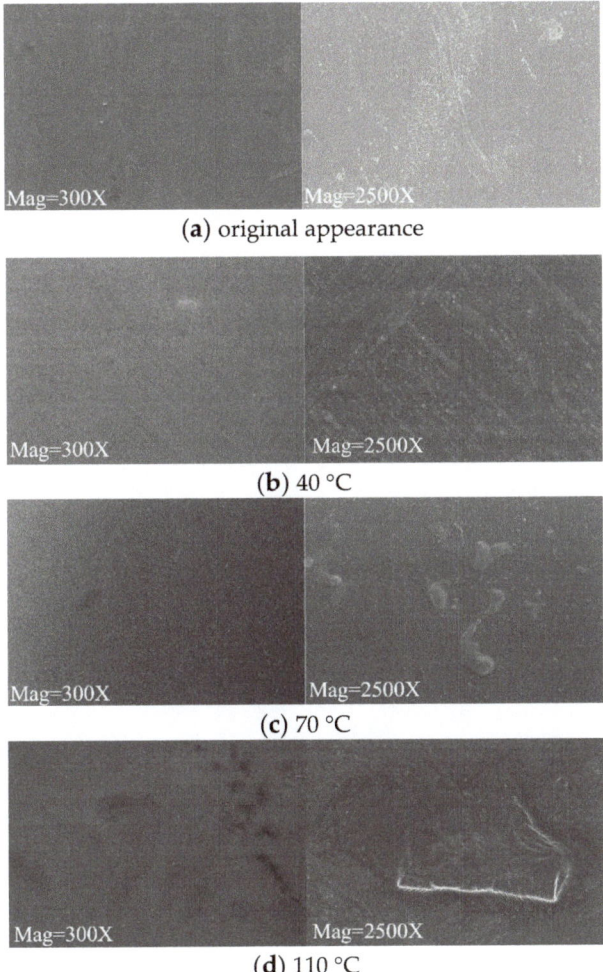

Figure 8. SEM of PET material.

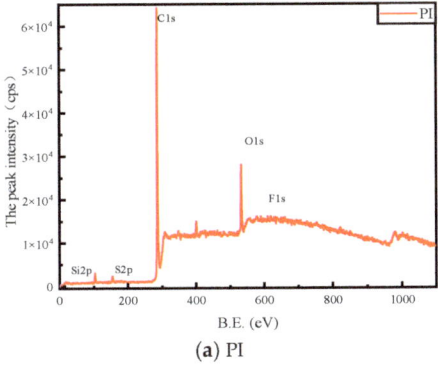

(**a**) PI

Figure 9. *Cont.*

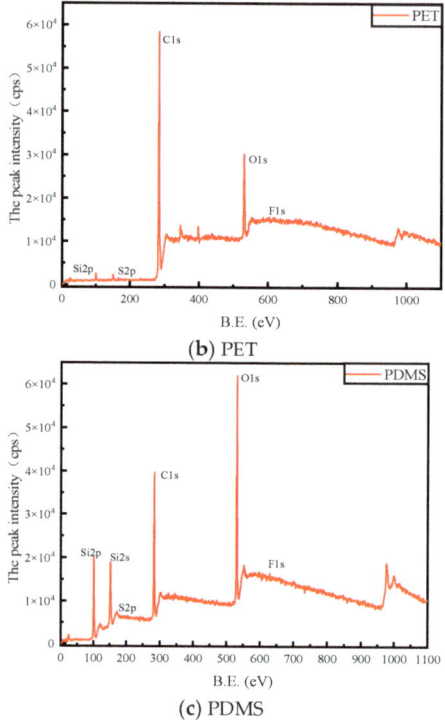

Figure 9. Photoelectron spectra of the full spectrum of elements before PI, PET and PDMS material experiments.

Figure 10 shows the high-resolution photoelectron spectra of C1 and F1 elements before and after experiments on three kinds of PD flexible UHF antenna sensor substrates: PI, PET and PDMS. It can be seen from Figure 10a,b that the C1s of the PI and PET materials before and after the experiment detected characteristic peaks of C-C, C-H bonds and C=O, O-C-O bonds at 284.8 eV and 288.5 eV. It can be seen from Figure 10c that the C1s of the PDMS materials before and after the experiment only detected the characteristic peaks of C-C and C-H bonds at 284.8 eV. It can also be seen from Figure 10 that the content of C1s of PI, PET and PDMS did not change significantly with the increase in the experimental temperature in the SF_6/N_2 mixed gas environment at different temperatures of 40 °C, 70 °C and 110 °C. At the same time, by observing Figure 10b, it is not difficult to see that the absorption peaks of C=O and O-C-O bonds of PET materials in the 110 °C SF_6/N_2 mixed gas environment are higher than those of C=O and O-C-O bonds in the 40 °C and 70 °C SF_6/N_2 mixed environment. At the same time, the characteristic peaks (C=O and O-C-O bond) of C1s at the binding energy of 288.5 eV also increased with the increase in the experimental temperature, indicating that PET surface oxidation occurred under the action of the SF_6/N_2 gas mixture.

In addition, it can be seen from Figure 10a that when PI is in the 70 °C and 110 °C SF_6/N_2 mixed gas environments, F1s detects metal-F bonds and C-F bonds at the binding energy of 684.5 eV and 688.15 eV, respectively, indicating that the PI material reacts with the SF_6/N_2 mixed gas. Metal fluoride is formed, and the newly generated metal fluorine compounds are adsorbed on the surface of the PI material, resulting in the accumulation of element F on the surface of the PI material. At the same time, the detection of C-F bonds on the surface of the PI material indicates that the PI material reacts with components in the mixed gas, so that the F element replaces the H element in the PI material, and a new fluorine-containing compound is generated. At the same time, the newly generated fluorine-containing compound is adsorbed on the surface of the PI material, resulting in the accumulation of the F element on the surface of the PI material. At the same time, it can also be seen from Figure 10a that the peak intensity of F1s did not change significantly with the increase in the experimental temperature. As can be seen from Figure 10b, F1s also detected metal-F and C-F bonds at the binding energy of 684.5 eV and 688.15 eV. However, with the increase in the experimental temperature, the peak strength of the metal-F bond and C-F bond gradually increased, indicating that more fluoride accumulated on the surface of the PET material with the increase in temperature. As can be seen from Figure 10c, when PDMS was under three different-temperature SF_6/N_2 gas mixtures, F1s did not detect characteristic peaks, indicating that PDMS would not react with the SF_6/N_2 gas mixtures. Through the above analysis, it can be concluded that the PDMS material has good compatibility with SF_6/N_2 gas mixtures.

The results show that PDMS has good compatibility with SF_6/N_2 gas mixtures.

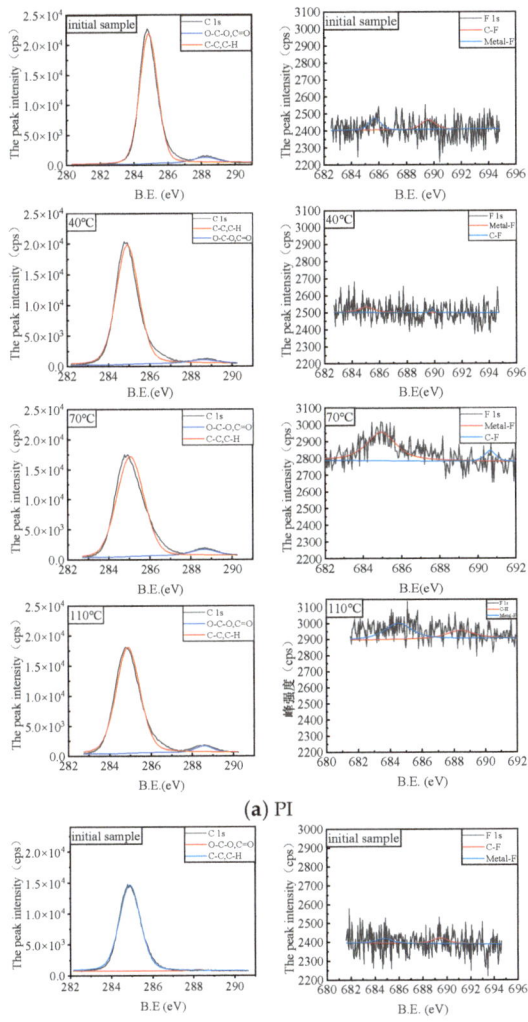

Figure 10. *Cont.*

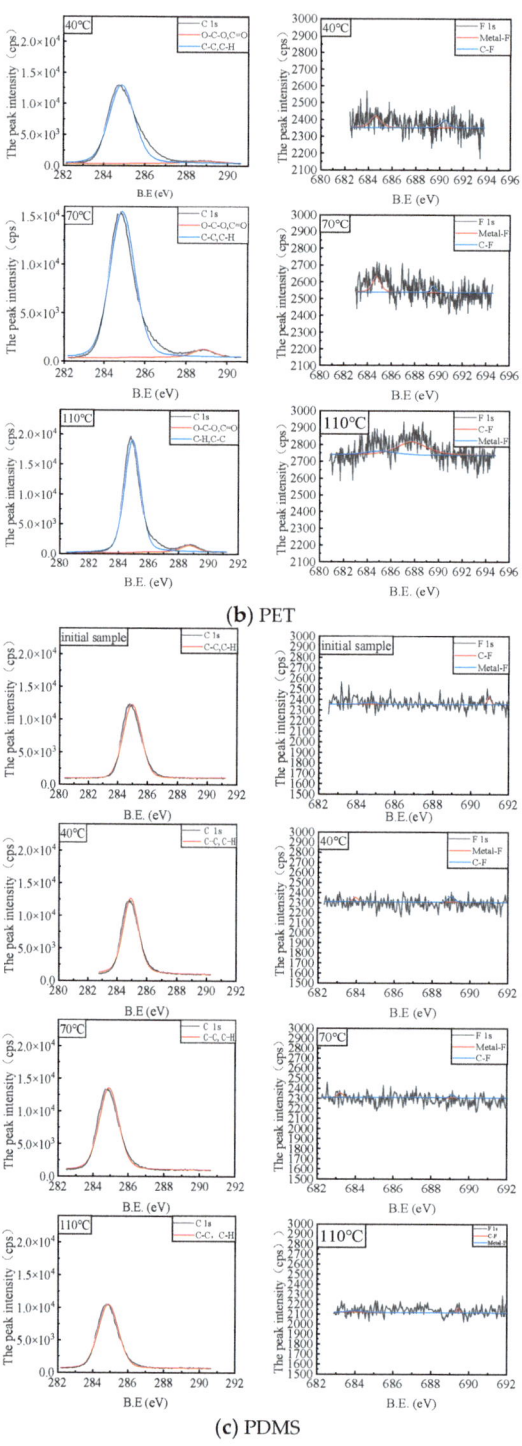

Figure 10. Fine spectra of surface elements of PI, PET and PDMS before and after the experiment.

5. UHF Antenna Sensor Based on PDMS

This research group developed a flexible Hilbert antenna and a flexible planar bicone antenna based on PDMS. In reference [26], the flexible Hilbert antenna designed by our research group could reach 73.6% coverage at frequencies of 422–800 MHz, 1.23–1.58 GHz, 1.65–1.7 GHz and 1.79–3 GHz without bending, and bending did not affect the antenna bandwidth. When the curvature of the antenna was low, the cross polarization component of the antenna was less in the low frequency band, showing a better polarization effect, while the influence of the curvature was more obvious in the high frequency band, and some curvature would increase the cross polarization ratio. When the bending radius of the antenna was 350 mm, the peak gain of the antenna reached 1.34 dB. In the PD detection band, the radiation efficiency of the antenna with different bending radii was 30~50%. The flexible planar bipyramidal antenna designed by the research group in the literature [27] was divided into a finite lateral branch plane bipyramidal antenna and an infinite lateral branch plane bipyramidal antenna. The return loss characteristics of the two antennas were less than −10 dB in the 0.3–0.5 GHz, 0.64–1.25 GHz and 1.4–3.0 GHz bands. Meanwhile, the two planar double-cone antennas had good bandwidth and gain under different bending radii of 0 mm, 150 mm, 350 mm and 500 mm. The infinite lateral branch plane bipyramidal antenna could achieve a gain of 5.38 at 1.37 GHz. The finite lateral branch-plane bipyramidal antenna had stable radiation performance at 0.3–3 GHz, and the infinite lateral branch-plane bipyramidal antenna had stable radiation performance at 0.5 GHz–3 GHz. Both antennas could effectively detect UHF PD signals.

6. Conclusions

In order to investigate the compatibility of the substrate material of a GIS PD flexible UHF antenna sensor with SF_6/N_2 mixed gas, this paper built an experimental platform to determine the compatibility of SF_6/N_2 gas mixtures with substrates of PD flexible UHF antenna sensors to study the compatibility of SF_6/N_2 gas mixtures with the flexible antenna materials at different temperatures; through the analysis of the experimental results, the following conclusions are drawn:

(1) The composition of the SF_6/N_2 mixture was almost the same before and after the compatibility experiment, and the substrate materials of PI, PET and PDMS had no influence on the composition of the SF_6/N_2 mixture.

(2) PET slightly oxidized at 110 °C 30% SF_6/70% N_2, and the metal fluoride formation of PI was observed at 30% SF_6/70% N_2.

(3) The PDMS substrate is suitable for the development of flexible UHF antenna sensors.

Author Contributions: G.Z. proposed the research direction and supervised the writing of the paper. X.H., G.D. and X.L. designed the thermal acceleration experiment, tested experimental data, completed the experiment, and wrote the manuscript. X.H. and G.Z. revised the manuscript. All authors have read and agreed to the published version of the manuscript.

Funding: This project was supported by the National Natural Science Foundation of China (52107144) and the Educational Commission Program of Hubei Province of China (Q20211401).

Data Availability Statement: The data presented in the article is original and has not been inappropriately selected, manipulated, enhanced or fabricated by us.

Conflicts of Interest: The authors declare that they have no known competing financial interests or personal relationships that could have appeared to influence the work reported in this paper.

References

1. Wang, Y. *Study on Gap Breakdown and Surface Flashover Characteristics of SF6/N2 Mixture Gas*; Shenyang University of Technology: Shenyang, China, 2021; pp. 6–8.
2. Ji, Y.; Zhang, M.; Wang, C.; Bi, J.; Mei, K.; Feng, Y.; Gong, Y.; Huang, Y.; Wang, H. Detection of Decomposition Products of SF6/N2 mixture gas. *High Volt. Electr. Equip.* **2020**, *56*, 97–102.
3. Song, H.; Dai, J.; Li, Z.; Luo, L.; Sheng, G.; Jiang, X. An assessment method of partial discharge severity for GIS in service. *Proc. CSEE* **2019**, *39*, 1231–1240.

4. Zhang, X.X.; Meng, F.S.; Ren, J.B.; Tang, J.; Yang, B. Simulation on the B-doped single-walled carbon nanotubes detecting the partial discharge of SF6. *High Volt. Technol.* **2011**, *37*, 1689–1694.
5. Li, J.; Han, X.; Liu, Z.; Yao, X. A novel GIS partial discharge detection sensor with integrated optical and UHF methods. *IEEE Trans. Power Deliv.* **2018**, *33*, 2047–2049. [CrossRef]
6. Zhang, X.; Zhang, J.; Xiao, S. Design of external ultra-high frequency partial discharge sensor for large transformer. *High Volt. Eng.* **2019**, *45*, 499–504.
7. Qin, J.; Wang, C.C.; Shao, W. Applying UHF to partial discharge on-line monitoring of electric power apparatus. *Power Syst. Technol.* **1997**, *21*, 33–36.
8. Tang, J.; Wei, G.; Sui, X.X. Research on the dipole antenna sensor with broadband for partial discharge detection in GIS. *High Volt. Eng.* **2004**, *30*, 29–31.
9. Wang, L.; Zheng, S.; Li, C. Distribution of electric field strength and spectral characteristic of UHF signal of partial discharge inside GIS at resin sprue of metal ring. *Power Syst. Technol.* **2014**, *38*, 3843–3849.
10. Lu, Q.; Zheng, S.; Li, X.; Wang, L.; Tang, Z.; Zhan, H. Study on propagation characteristics of UHF signal via hole of GIS mental flange and development of external radiating antenna. *Power Syst. Technol.* **2013**, *37*, 2303–2309.
11. Tan, Q.; Tang, J.; Zeng, F.P. Design of fourfold-band micro-strip monopole antenna for partial discharge detection in gas insulated switch-gear. *Trans. China Electrotech. Soc.* **2016**, *31*, 127–144.
12. Bao, L.; Li, J.; Xue, W.; Zhang, J.; Cheng, C. Application of genetic algorithms in optimization of partial discharge ultra-high frequency fractal Hilbert antenna. *High Volt. Eng.* **2015**, *41*, 3959–3966. [CrossRef]
13. Zhou, W.J.; Liu, Y.S.; Li, P.F.; Yu, J.H. Modified Vivaldi antenna applied to detect partial discharge in electrical equipment based on ultra-high frequency method. *Trans. China Electrotech. Soc.* **2017**, *32*, 259–267.
14. Ji, S.C.; Wang, Y.Y.; Li, J.H. Review of UHF antenna for detecting partial discharge in GIS. *High Volt. Appar.* **2015**, *51*, 163–172.
15. Zhang, G.; Chen, K.; Li, X.; Wang, K.; Fang, R. Flexible Built-in Miniaturized Archimedes Helical Antenna Sensor for GIS PD Detection. *High Volt. Technol.* **2022**, *48*, 2244–2254.
16. Zhang, G.; Han, J.; Liu, J.; Chen, K.; Zhang, S. GIS PD detection antenna body and balun coplanar flexible miniaturized UHF antenna sensor. *Trans. China Electrotech. Soc.* **2023**, *38*, 1064–1075.
17. Zhang, G.; Zhang, S.; Zhang, X.; Chen, K.; Han, J.; Liu, J. Research on a new type of GIS Partial Discharge flexible built-in Archimedes spiral antenna. *High Volt. Appar.* **2022**, 1–10. Available online: https://kns.cnki.net/kcms/detail/61.1127.TM.20220130.1459.002.html (accessed on 1 July 2023).
18. Trajkovikj, J.; Zürcher, J.F.; Skrivervik, A.K. A Robust Casing for Flexible W-BAN Antennas. *IEEE Antennas Propag. Mag.* **2013**, *55*, 287–297. [CrossRef]
19. Lin, C.P.; Chang, C.H.; Cheng, Y.T.; Jou, C.F. Development of a flexible SU-8/PDMS-based antenna. *IEEE Antennas Wirel. Propag. Lett.* **2011**, *10*, 1108–1111.
20. Yuan, R. *Study on the Compatibility of Environmental Protection INSULATING Gas C4F7N/CO2 with Epoxy Resin*; Wuhan University: Wuhan, China, 2020; pp. 5–19.
21. GB/T11022-2020; Common Specifications for High-Voltage Switchgear and Controlgear Standards. Standards Press of China: Beijing, China, 2023. Available online: https://openstd.samr.gov.cn/bzgk/gb/std_list?p.p1=0&p.p90=circulation_date&p.p91=desc&p.p2=GB/T11022-2020 (accessed on 1 July 2023).
22. Muniz-Miranda, M.; Muniz-Miranda, F.; Caporali, S. SERS and DFT study of copper surfaces coated with corrosion inhibitor. *Beilstein J. Nanotechnol.* **2014**, *5*, 2489–2497. [CrossRef] [PubMed]
23. She, C.; Tang, J.; Cai, R.; Li, H.; Li, L.; Yao, Q.; Zeng, F.; Li, C. Compatibility of C5F10O with common-used sealing materials: An experimental study. *AIP Adv.* **2021**, *11*, 065200. [CrossRef]
24. Zhang, X.; Wu, P.; Cheng, L.; Liang, S. Compatibility and Interaction Mechanism between EPDM Rubber and a SF6 Alternative Gas-C4F7N/CO$_2$/O$_2$. *ACS Omega* **2021**, *6*, 13293–13299. [CrossRef]
25. Yuan, R.; Li, H.; Zhou, W.; Zheng, Z.; Yu, J. Study of Compatibility between Epoxy Resin and C4F7N/CO$_2$ Based on Thermal Ageing. *IEEE Access* **2020**, *8*, 119544–119553. [CrossRef]
26. Tian, J.; Zhang, G.; Ming, C.; He, L.; Liu, Y.; Liu, J.; Zhang, X. Design of a Flexible UHF Hilbert Antenna for Partial Discharge Detection in Gas-Insulated Switchgear. *IEEE Antennas Wirel. Propag. Lett.* **2022**, *22*, 794–798. [CrossRef]
27. Zhang, G.; Tian, J.; Zhang, X.; Liu, J.; Lu, C. A Flexible Planarized Biconical Antenna for Partial Discharge Detection in Gas-Insulated Switchgear. *IEEE Antennas Wirel. Propag. Lett.* **2022**, *21*, 2432–2436. [CrossRef]

Disclaimer/Publisher's Note: The statements, opinions and data contained in all publications are solely those of the individual author(s) and contributor(s) and not of MDPI and/or the editor(s). MDPI and/or the editor(s) disclaim responsibility for any injury to people or property resulting from any ideas, methods, instructions or products referred to in the content.

Article

Design of a Ka-Band Five-Bit MEMS Delay with a Coplanar Waveguide Loaded U-Shaped Slit

Yongxin Zhan [1,2,3,4], Yu Chen [1,2,3,4], Honglei Guo [1,2,3,4], Qiannan Wu [2,3,4,5,*] and Mengwei Li [1,2,3,4,6]

1 School of Instrument and Electronics, North University of China, Taiyuan 030051, China; lmwprew@163.com (M.L.)
2 Academy for Advanced Interdisciplinary Research, North University of China, Taiyuan 030051, China
3 Center for Microsystem Intergration, North University of China, Taiyuan 030051, China
4 School of Instrument and Intelligent Future Technology, North University of China, Taiyuan 030051, China
5 School of Semiconductors and Physics, North University of China, Taiyuan 030051, China
6 Key Laboratory of Dynamic Measurement Technology, North University of China, Taiyuan 030051, China
* Correspondence: qiannanwoo@nuc.edu.cn

Abstract: This paper designs a five-bit microelectromechanical system (MEMS) time delay consisting of a single-pole six-throw (SP6T) RF switch and a coplanar waveguide (CPW) microstrip line. The focus is on the switch upper electrode design, power divider design, transmission line corner compensation structure design, CPW loading U-shaped slit structure design, and system simulation. The switch adopts a triangular upper electrode structure to reduce the cantilever beam equivalent elastic coefficient and the closed contact area to achieve low drive voltage and high isolation. The SP6T RF MEMS switch uses a disc-type power divider to achieve consistent RF performance across the output ports. When designed by loading U-shaped slit on transmission lines and step-compensated tangents at corners, the system loss is reduced, and the delay amount is improved. In addition, the overall size of the device is 2.1 mm × 2.4 mm × 0.5 mm, simulation results show that the device has a delay amount of 0–60 ps in the frequency range of 26.5–40 GHz, the delay accuracy at the center frequency is better than 0.63 ps, the delay error in the whole frequency band is less than 22.2%, the maximum insertion loss is 3.69 dB, and the input–output return rejection is better than 21.54 dB.

Keywords: SP6T RF MEMS switch; triangular upper electrode; CPW loading gaps; high delay accuracy; low insertion loss; high-frequency band

Citation: Zhan, Y.; Chen, Y.; Guo, H.; Wu, Q.; Li, M. Design of a Ka-Band Five-Bit MEMS Delay with a Coplanar Waveguide Loaded U-Shaped Slit. *Micromachines* **2023**, *14*, 1508. https://doi.org/10.3390/mi14081508

Academic Editor: Haejun Chung

Received: 20 June 2023
Revised: 18 July 2023
Accepted: 25 July 2023
Published: 27 July 2023

Copyright: © 2023 by the authors. Licensee MDPI, Basel, Switzerland. This article is an open access article distributed under the terms and conditions of the Creative Commons Attribution (CC BY) license (https://creativecommons.org/licenses/by/4.0/).

1. Introduction

Currently, the wireless communications market is highly competitive, and tiny, low-cost, reconfigurable RF modules are a major research hotspot. Delayers are the essential components of phased array antennas, mainly used in radar systems [1]. Digital delay techniques require the sampling of signals, normally requiring analog-to-digital converters (ADC) [2], therefore leading to relatively high power consumption, quantization noise, and sampling confusion, in addition to the inclusion of ADCs causing problems such as clock injection, nonlinearity, and bandwidth limitations, which can be avoided by using analog delay circuits [3]. Conventional analog delayers use PIN diodes as signal conduction and disconnection elements, but such transistor switching introduces significant losses, up to 8–9 dB at 35 GHz [4]. Instead, it becomes a performance-limiting component of the system. Compared with the delayers discussed above, delayers based on RF MEMS technology have the advantages of low power consumption, superior linearity, small size, low cost, and easy monolithic integration [5]. Therefore, in recent years, RF MEMS delayers have been extensively studied by both domestic and international scholars.

There are several classifications of MEMS delayers, such as distributed transmission line delay [6], reflective delay [7], and switched linear delay (TTDL) [8]. However, published literature and already-released TTDL products typically suffer from low device operating bands, low accuracy, and high loss. In 2007, Chu et al. of the University of Southern California proposed a path-sharing true-delay structure [9], with a design bandwidth of 1–15 GHz, a delay resolution of 15 ps, and a maximum delay of 225 ps. The chip adopts an 8-layer CMOS process, and its size is 3.1 mm × 3.2 mm. In 2013, Park et al. of Koryo University designed and fabricated a 15–40 GHz three-dimensional CMOS real-time delay (TTD) circuit with a maximum group time delay of 40 ps but with insertion loss as steep as 14 dB [10]. In 2019, a novel structure to achieve broadband true delay was proposed and implemented [11]. There is a maximum delay of 109.3 ps in the 8–18 GHz band and an average insertion loss of 18.2–22.5 dB in the expected band with an error of less than 4 ps.

This paper proposes an RF MEMS delay device with high delay accuracy and low loss in the Ka-band. The MEMS delay adopts the structure design of a single-blade multi-throw switch equipped with a linear time delay unit, which is characterized by a moderate chip area and an easily machinable structure. In addition, conventional switch delayers often use a cascade structure [12,13], with a minimum of 2n (n delay bits) switches driven for delay operation, and only two switches need to be driven to activate a delay state using the structure of this paper.

The paper is organized as follows. First, the development of a single-blade multi-throw switch (SPMT) is reviewed, and the designs of the upper electrode of the switch and the power divider are introduced in detail. Finally, the SP6T switch model designed in this paper is proposed, which has superior RF performance at DC-40 GHz. Then, a new delayed microwave line structure is presented, with innovations in transmission line corners and coplanar waveguide centerline for better delay performance. Finally, the proposed Ka-band five-bit MEMS delay, obtained by integrating two SP6T RF MEMS switches and a modern CPW microstrip, achieves elevated delay accuracy and low transmission loss.

2. Model Design

2.1. SP6T RF MEMS Switch Design

The SP6T RF MEMS switch is used as the core device to control the signal transmission path of the delay, and its RF performance has a great impact on the overall performance of the delay. The (SPMT) [14–16], developed so far, mainly has PIN diode class, RF coaxial class, and MEMS class. The listed PIN diode SPMT switches have the problems of narrower frequency band and larger size; the volume of RF coaxial SPMT switch developed by RF connector is also large; the MEMS SPMT switches still have the problems of poor isolation and insertion loss performance on the characteristics of small size and wide frequency band. To design the SP6T RF MEMS switch with good RF performance in Ka-band, this paper will optimize the upper electrode structure and power divider structure of the switch, respectively.

2.1.1. Switch Upper Electrode Design

The proposed SP6T RF MEMS switch uses a series contact RF MEMS switch with a cantilever beam structure for the upper electrode of the switch, where one end of the upper electrode is fixed to the signal line by an anchor point and the remaining end is suspended above the contact. By applying a voltage to the driving electrode below the cantilever beam, the upper electrode is displaced by the electrostatic force and pulled down to contact with the contact, and the signal is conducted [17]. From the theory related to RF MEMS, it is known that the driving voltage for switch conduction [18] is:

$$V = \sqrt{\frac{8k}{27\varepsilon_0 A}g_0^3} \tag{1}$$

$$k = \frac{Ewt^3}{4l^3} \tag{2}$$

where k refers to the elastic coefficient of the upper electrode of the switch; E means Young's modulus of the material; w is the width of the upper electrode; l indicates the length of the upper electrode; t denotes the thickness of the upper electrode; A denotes the positive actuation region of the upper and lower electrodes of the switch; g_0 denotes the initial spacing between the upper and lower electrodes; ε_0 denotes the relative dielectric constant of air. The above equation shows that the driving voltage is proportional to the elastic coefficient and inversely proportional to the driving area.

Most switches currently use a rectangular, straight-plate upper electrode [19]. Since it is more effective to reduce the cantilever beam elasticity coefficient than to increase the local driving area in reducing the driving voltage, a triangular upper electrode structure is used in this paper. The total length of the triangular upper electrode cantilever beam is 106 μm, consisting of a 40 μm rectangle, a 60 μm triangle, and a 6 μm rectangular block. Due to the skinning effect, 2 μm is chosen in this paper when designing the thickness of the upper electrode and CPW, which can ensure that the thickness of the metal is greater than 2 times the skinning depth at frequencies above 10 GHz, and reduce the metal resistance at low frequencies. The specific structure parameters are shown in Figure 1.

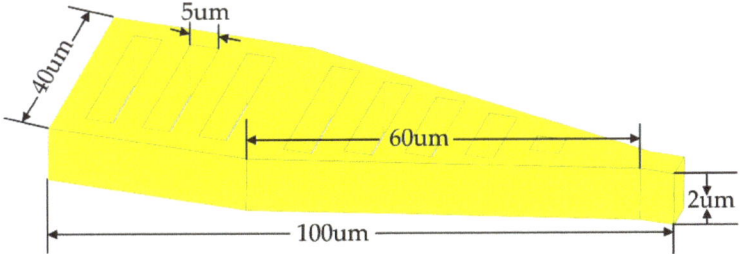

Figure 1. Triangular upper electrode structure and its specific parameters.

The proposed triangular upper electrode RF MEMS switch reduces the driving voltage in three main ways. First, by reducing the width of the cantilever beam to obtain a lower elasticity coefficient; second, by opening several openings in the upper electrode plate to reduce the air damping of the switch pull-down [20]; third, the hollowed-out triangular design significantly eliminates most of the switch mass when compared to the rectangular shape, thus reducing the weight of the switch. In addition, the opening of the hole in the upper pole plate facilitates the release of the sacrificial layer during process processing. The contact area between the triangular upper electrode structure and the transmission line is significantly smaller than the transmission line width compared to the straight upper electrode. A smaller contact area will reduce metal adhesion and provide better isolation performance.

To verify that the optimized upper electrode structure is easy to pull down, the COMSOL software was used to apply equivalent pressure to both upper electrode structures, and the simulation results are shown in Figure 2. The displacement of the straight upper electrode is 0.8 μm, and that of the displacement of the triangular upper electrode is 2.64 μm. The results show that the triangular upper electrode structure is easier to pull down than the straight one. Using the above equations, the equivalent elastic coefficient of the cantilever beam was calculated to be 5.32 N/m, and the driving voltage of the triangular upper electrode was 13 v.

To verify that the triangular upper electrode structure can improve the switching isolation, ANSOFT HFSS software was used to simulate and compare the switches of the two structures, the results of which are shown in Figure 3. It can be seen that the triangular upper electrode structure effectively improves the switching isolation in the Ka-band, with an improvement of 11.86 dB at 40 GHz and better isolation of 31.46 dB in the full frequency band.

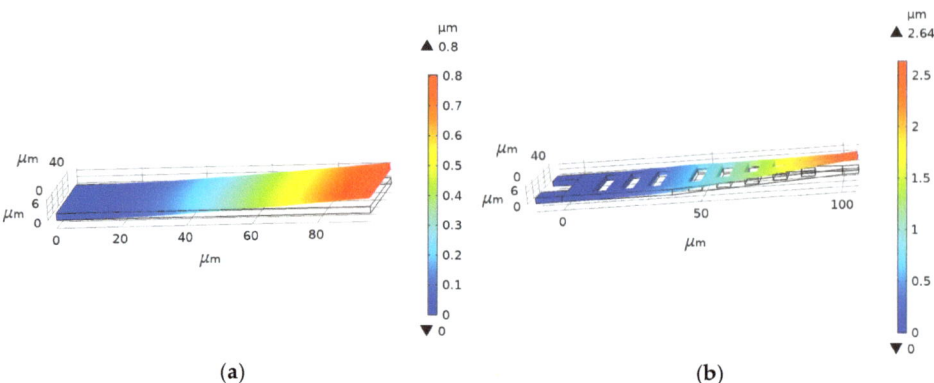

(a)   (b)

Figure 2. Schematic diagram of displacement of different structures under the same pressure (**a**) Schematic diagram of the displacement of a straight-plate structure; (**b**) Schematic diagram of the displacement of a triangular structure.

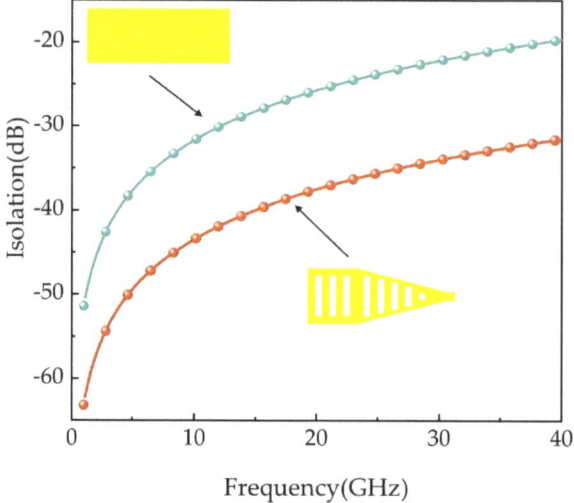

Figure 3. Isolation performance of different upper electrode structures.

2.1.2. Switch Power Divider Design

The function of the power divider (power splitter) is to proportionally distribute the input signal into the end branches [21], and its performance has an important impact on the performance of the SP6T RF MEMS switch in terms of insertion loss and signal splitting. The design expects to design a power divider structure to make the RF performance of each port of the SP6T RF MEMS switch consistent, so three power divider design models are proposed in this paper, as shown in Figure 4. Figure 4a is a general six-out-of-one power divider, whose structure is designed to divide into six branches of 100 μm length directly at the signal shunt. Figure 4b is a circular power divider, whose structure is designed to connect a 10 μm wide ring at the signal shunt and set a 75 μm block in the center of the circle as a ground wire. Figure 4c shows the disc-type power divider, whose structure is designed to set a 100 μm radius disc at the signal shunt.

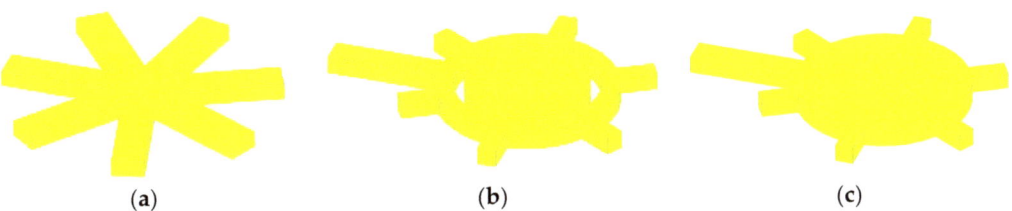

Figure 4. (a) General six-out-of-one power divider; (b) Circular power divider; (c) Disc-type power divider.

Next, the three power dividers and switches are cascaded, and their insertion loss performance was simulated, and the results are shown in Figure 5. A previous article verified that the closer the SPMT switch output is to the input, the stronger the electric field strength at its port, resulting in greater signal loss [22]. A look at the data in the three result plots shows that the closer the channel is to the input, the worse the insertion loss results for channel conduction, which is in line with the above theory. Because of the symmetrical geometry of the power divider, the insertion loss performance of Port 1 is similar to that of Port 6, Port 2 is similar to that of Port 4, and Port 3 is similar to that of Port 5.

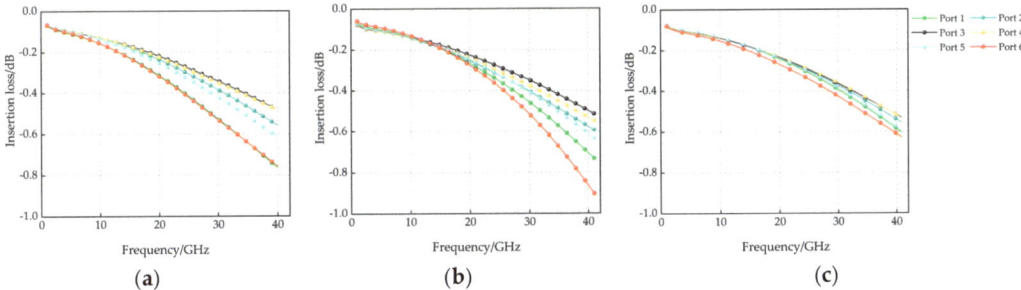

Figure 5. Insertion loss simulation results for different power divider structures: (a) General six-out-of-one power divider; (b) Circular power divider; (c) Disc-type power divider.

From the perspective of signal loss, the insertion loss of the disc-type power splitter structure is the smallest at 40 GHz, and the maximum loss is only 0.62 dB. From the perspective of port performance consistency, the difference in insertion loss performance between the ports of the general six-out-of-one power divider at 40 GHz is 0.28 dB, the difference in insertion loss performance between the ports of circular-type power splitters is 0.39 dB, and the difference in insertion loss performance between the ports of disc-type power splitters is only 0.094 dB. On balance, the disc-type power splitter design best meets the RF performance requirements of this paper.

The SP6T RF MEMS switch obtained by integrating the single-throw switch and the disc power divider is shown in Figure 6. The overall size of the switch is less than 0.6 mm × 0.6 mm, and the insertion loss of each port is less than 0.62 dB in Ka-band, with excellent overall RF performance.

2.2. Delay Structure Design

2.2.1. CPW Transmission Line Corner Design

MEMS delayers are used to select different lengths of transmission lines through switches to achieve a multi-bit delay function. The formula for calculating the delay amount of the same microwave transmission line structure is as follows [23]:

$$\Delta\varphi = \frac{2\pi f \sqrt{\varepsilon_{eff}}}{c}(l_d - l_r) \tag{3}$$

c is the free-space velocity; ε_{eff} is the relative permittivity of the CPW equivalent; f is the operating frequency of the delay; l_d is the delay transmission line length; l_r is the reference line length. Equation (3) shows that the amount of delay is proportional to the length of the transmission line. To reasonably use the limited area of the substrate at the same time, as often as possible to achieve a larger delay, here is the use of zigzag microstrip line wiring, as shown in Figure 7a. Due to the CPW discontinuity at the corner, it is very easy to produce slot line mode, which is usually suppressed by erecting an air bridge to connect the ground lines on both sides [24].

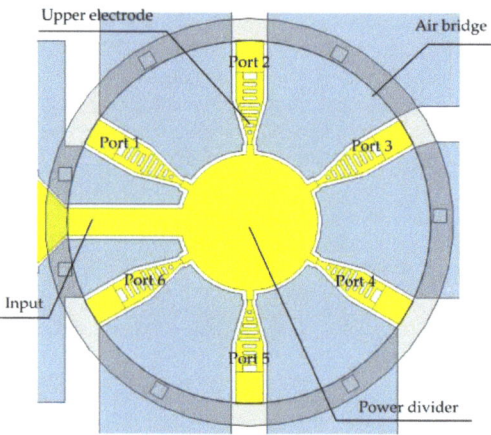

Figure 6. SP6T RF MEMS switch model.

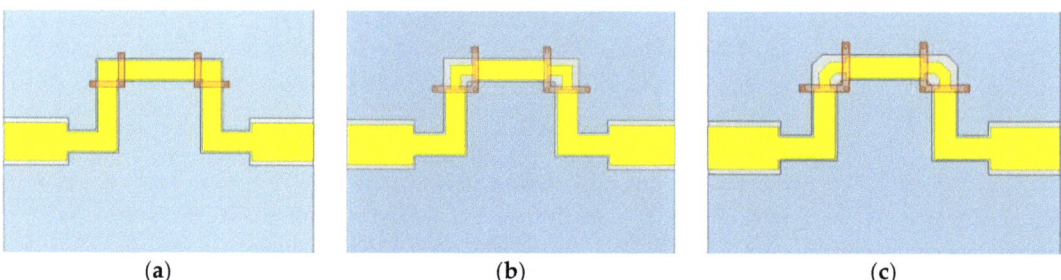

Figure 7. (a) Zigzag microstrip line layout; (b) Step compensation design; (c) Step compensation cut-angle design.

To further reduce the transmission loss, the geometric structure at the corner of the microstrip line is optimally designed in this paper. As shown in Figure 7b,c, the step compensation design and step compensation tangent design are carried out at the corners, respectively.

The simulation results of the transmission characteristics of different designs at the corner of the curved microstrip line are shown in Figure 8. From the simulation results, it seems that the RF performance of the design with step-compensated corner-cutting at the Ka-band is significantly improved, the insertion loss is improved by 0.39 dB at 40 GHz, and the return loss is improved by 5.16 dB at 40 GHz. Therefore, the design of step-compensated corner-cutting at the corner of the curved microstrip line is used.

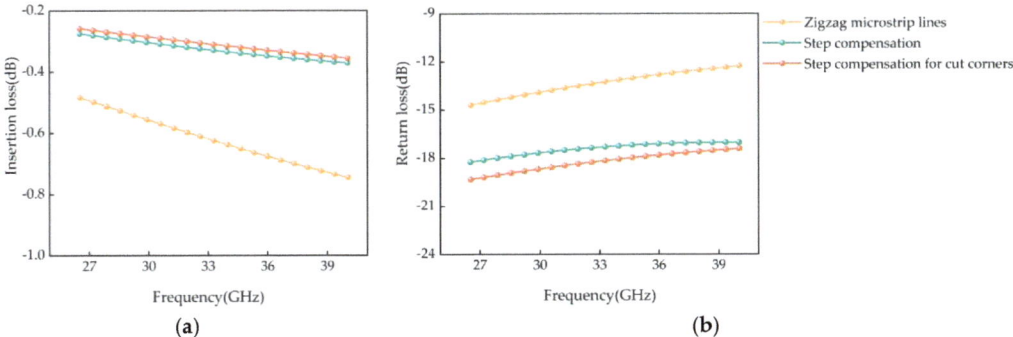

Figure 8. (a) Insertion loss simulation results for different corner designs; (b) Simulation results of echo resistance in different corner designs.

For the corner-cutting design, there is an effect of the corner-cutting width d on the transmission characteristics of the zigzag microstrip line. The simulation results of the RF performance of the zigzag microstrip line with different cut-angle widths are shown in Figure 9, where w = 30 μm is the corner transmission line width. From Figure 9, it can be seen that different corner-cutting methods have a considerable impact on the standing wave and delay of the microstrip line. From the simulation results, with the increase of the length of the corner-cutting line (d = 0.2–1.4 W), the delay performance of the transmission line deteriorates progressively, and the amount of the delay decreases by nearly 0.39 ps; at the same time, the VSWR obtained when the length of the corner-cutting line is d = w is the best, and it is optimized by 0.12 relative to that of d = 0.2 w. Therefore, as a compromise, the width of the tangent angle with d = w is chosen.

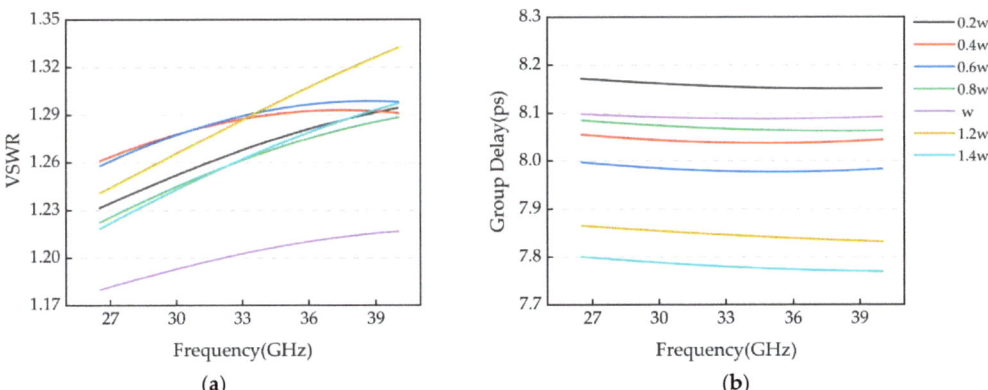

Figure 9. (a) Effect of angle of tangency on standing waves; (b) Effect of angle of tangency on the amount of delay.

2.2.2. Coplanar Waveguide Loading Gap Design

In this paper, a coplanar waveguide structure is used for signal transmission, as shown in Figure 10a, and its equivalent circuit is shown in Figure 10b [25], where the equations for the unit capacitance C_t and unit inductance L_t in the circuit are Equations (4) and (5), respectively, where c is the velocity in free space, Z_0 is the characteristic impedance of the unloaded transmission line, and ε_{eff} is the effective dielectric constant of the CPW.

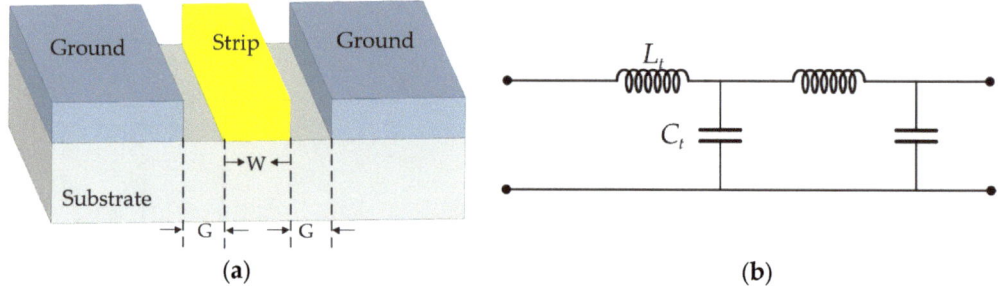

Figure 10. (a) Structure of Coplanar Waveguide; (b) CPW equivalent circuit.

The coplanar waveguide structure achieves the delay by controlling the phase velocity v of the transmission line, as shown in Equation (6), where the magnitude of the phase velocity depends on the unit capacitance C_t and the unit inductance L_t.

$$C_t = \frac{\sqrt{\varepsilon_{eff}}}{cZ_0} \quad (4)$$

$$L_t = C_t Z_0^2 \quad (5)$$

$$v = \sqrt{\frac{1}{L_t\left(\frac{C_t}{G}\right)}} \quad (6)$$

To increase the delay amount without increasing the length of the delay line, this paper improves the unit inductance of the delay line by replacing the common transmission line with a coplanar waveguide loading gap, thus increasing the phase speed and increasing the delay amount. A CPW discontinuous transmission structure was previously proposed by Tang et al. [26], in which two rectangular slots are hollowed out in the central signal line of the CPW, which forms another coplanar waveguide in the central guide member. According to transmission line theory, this structure is equivalent to a short circuit along the propagation direction and can be equated to a parallel inductor. The results of the article show that this CPW discontinuous structure design can not only widen the operating band but also keep the insertion loss low while obtaining a large phase shift. Therefore, this paper proposes a structure that adds a U-shaped slit to the center guide of the coplanar waveguide to equivalently form three coplanar waveguides on the center guide, thus achieving the purpose of introducing a larger equivalent inductance L_2. The specific structure of the coplanar waveguide loading gap is shown in Figure 11a, and the equivalent circuit is shown in Figure 11b.

The RF performance simulations of the normal CPW structure and the loaded U-shaped slit CPW structure were performed using ANSOFT HFSS software, and the comparison results are shown in Figure 12. As shown in Figure 12a, the characteristic impedance of the CPW center guide is simulated before and after loading the U-shaped slit, and the result is that the introduction of the surface U-shaped slit has a small impact on the characteristic impedance of the transmission line, which only changes 0.11 Ω at 40 GHz. As the characteristic impedance of RF devices is generally 50 Ω, the final determination of the model signal input and output port width is 120 μm, the center guide and both sides of the ground spacing is 16 μm. As can be seen in Figure 12b,c, the RF performance of the common CPW structure deteriorates sharply at the high operating frequency band of 39.78 GHz. Comparatively, the CPW-loaded U-shaped slit outperforms the normal CPW structure in terms of insertion loss performance and delay performance in the full frequency band 26.5–40 GHz. In terms of insertion loss performance, the CPW-loaded U-shaped slit structure reduces 0.22 dB on average over the full frequency band. In terms of delay performance, the CPW-loaded U-shaped slit structure not only increases the delay by an average of 0.78 ps in the whole frequency band but also provides better delay flatness. In summary, the CPW load gap not only does not have a greater impact on the characteristic impedance of the transmission structure but also reduces the insertion loss, improves the delay amount and delay flatness, and is more suitable for the delay transmission structure in this paper.

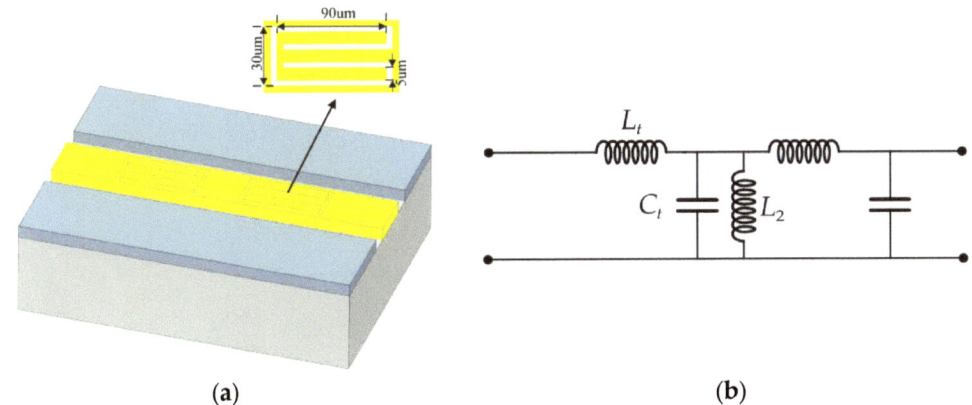

Figure 11. (**a**) Coplanar waveguide loading gap structure; (**b**) CPW loads a gap equivalent circuit.

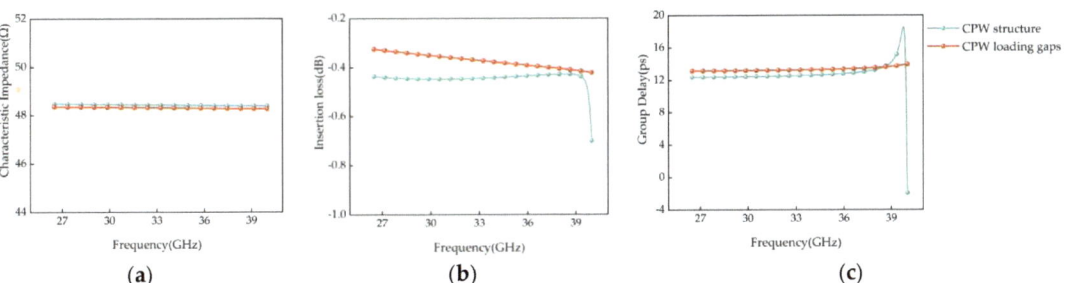

Figure 12. (**a**) characteristic Impedance simulation comparison; (**b**) Insertion loss simulation comparison; (**c**) Simulation comparison of time delay.

2.3. Five-Bit Delay Model Design

Determined by the above structure design, a five-bit MEMS delayer for an integrated modeling simulation model is shown in Figure 13, with an overall size of 2.1 mm × 2.4 mm × 0.5 mm. The model delay center frequency is 30 GHz, and the simulation results of the RF performance are shown in Figure 14 and Table 1. The simulation results of the five-bit MEMS delayers show that the error of the delay amount at the center frequency of 30 GHz is less than 0.63 ps, and the return loss is better than 28 dB. In the whole operating band (26.5–40 GHz), the delay error of all states is within 22.2%, the return rejection is better than 21.54 dB, and the insertion loss is less than 3.69 dB. The results show that the overall performance of the device is superior, although the delay accuracy and delay error increase with the increase of delay amount when each delay unit works individually in the five-bit MEMS delay.

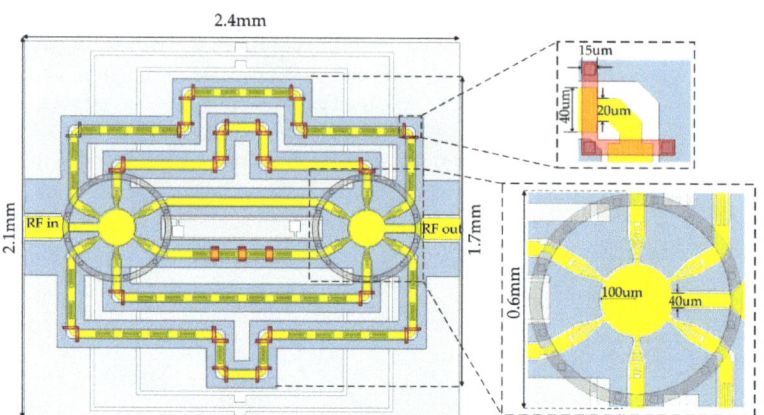

Figure 13. Model diagram of a Ka-band five-bit MEMS delay.

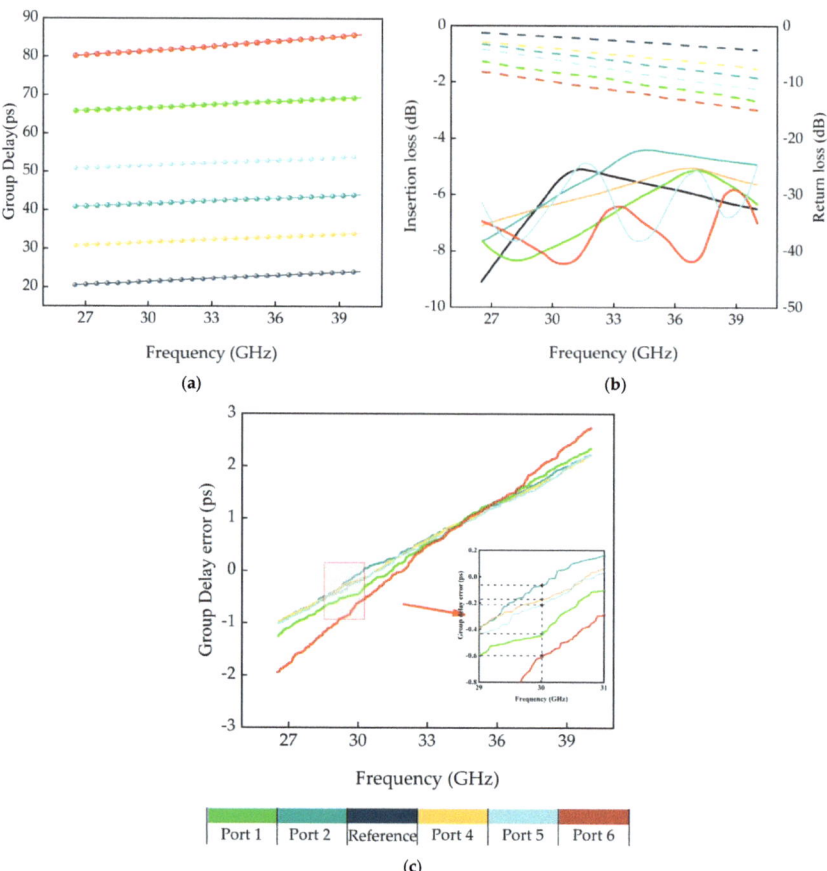

Figure 14. (**a**) Device delay simulation results; (**b**) Simulation results of device insertion loss and return resistance; (**c**) Delay error.

Table 1. Simulation performance of a five-bit delay.

State (bit)	Target (ps)	Delay (ps) Center	Delay (ps) Subtraction	Delay (ps) Accuracy	Error	Return Loss (dB)
0	0		21.50			−28.68
1st	10	31.44	9.94	±2.22	−0.06	−31.74
1st	10	31.44	9.94	±2.22	−0.06	−31.74
2nd	20	41.33	19.83	±2.21	−0.17	−30.60
3rd	30	51.29	29.79	±2.20	−0.21	−39.19
4th	45	61.07	44.57	±2.34	−0.43	−40.71
5th	60	80.9	59.40	±2.54	−0.6	−31.20

3. Comparison and Discussion

Table 2 shows the performance comparison of the five-bit MEMS delay designed in this paper and the multi-bit delay studied by various institutions in recent years. As shown in Table 2, the main process used in the current multi-digit delay design is the CMOS technology. Comparison of this paper with other published results shows that the multi-bit delay timers fabricated by the pseudomorphic HEMT (PHEMT) process and CMOS process have the advantages of small device size but suffer from the problems of low bandwidth and high insertion loss. Compared to the five-bit MEMS delay designed in this paper, which is close to the multi-delay made by the CMOS technology in terms of device size, and in terms of RF performance has the advantages of a high-frequency band, low loss, and high accuracy, and the device has a high potential for application in Ka-band.

Table 2. Comparison of delay in recent years.

Ref	Frequency (GHz)	Bit	Delay (ps)	Max Error (ps)	Insertion Loss dB	Technology	Size mm
2001 [27]	0–40	3	86	3	4.3	MEMS	5 × 6
2007 [9]	1–15	4	255	15	-	CMOS	3.1 × 3.2
2013 [10]	15–40	3	40	5	14	CMOS	1.1 × 0.9
2018 [28]	6–18	8	255	29	-	CMOS	2.0 × 0.6
2019 [11]	8–18	3	109.3	4	22.5	CMOS	0.9 × 2.1
2021 [29]	8–18	5	120	3.9	20.5	CMOS	1.2 × 2.7
2022 [30]	0–0.8	4	3800	4	-	CMOS	1.45 × 1.37
2023 [31]	6–18	3	106	10	15	PHEMT	2.7 × 0.73
This work	26.5–40	5	60	2.54	3.69	MEMS	2.1 × 2.4

4. Conclusions

A five-bit MEMS delay device integrated by a symmetric SP6T RF MEMS switch and a CPW microstrip line is proposed. The device is innovatively designed with a triangular upper electrode structure, disc-type power divider, a step-compensated tangent structure at the corner, and a coplanar waveguide loaded U-shaped slit structure to achieve 0–60 ps delay on an area of 2.1 mm × 2.4 mm. Simulation results show this design has high delay accuracy at the 30 GHz frequency point and superior return loss and insertion loss performance in the frequency band of 26.5–40 GHz. Compared with other reported delayers, the optimization method proposed in this paper can provide a new idea for the design of delayers operating in high-frequency bands. However, this paper's MEMS five-bit delay research is only for theoretical modeling, which can be used in practice in the radio frequency system, but also needs to be modeled through the MEMS process technology to process the finished product after measurement debugging. Therefore, future work in this paper will focus on device process research.

Author Contributions: Conceptualization, Y.Z. and Q.W.; methodology, Y.Z.; software, Y.Z.; validation, Y.Z., Y.C. and H.G.; formal analysis, Y.Z.; investigation, Y.Z.; resources, Q.W.; data curation, Y.C.; writing—original draft preparation, Y.Z.; writing—review and editing, Y.Z.; visualization, Y.Z.; supervision, Y.C.; project administration, Q.W.; funding acquisition, M.L. All authors have read and agreed to the published version of the manuscript.

Funding: This research was funded by Equipment Development Department New Product Project, Shanxi Province Postgraduate Education Reform Project, "Double First-Class" disciplines National first-class curriculum construction, and National Future Technical College Construction Project, grant number 2019XW0010, 11012103, 11012133, 11013168, and 11013169.

Data Availability Statement: Not applicable.

Conflicts of Interest: The authors declare no conflict of interest.

References

1. Zhu, Z.Q.; Zhang, W.Q. Broadband cell digital array radar delay method. *Mod. Radar* **2016**, *38*, 27–29+33.
2. Zhang, J.P.; Li, J.X.; Sun, H.B. Study on the application of broadband phased-array antenna with real-time delayers classification. *Mod. Radar* **2010**, *32*, 75–78.
3. Huang, Z.H.; Liu, B.; Zhang, J.C.; Liu, M.; Meng, Q.D. A low-power CMOS programmable analog time-delay circuit. *Microelectronics* **2019**, *49*, 225–229+236.
4. Rebeiz, G.M.; Tan, G.L.; Hayden, J.S. RF MEMS phase shifters: Design and applications. *IEEE Microw. Mag.* **2002**, *3*, 72–81. [CrossRef]
5. Yao, J.J. RF MEMS from a device perspective. *J. Micromech. Mi-Croeng.* **2000**, *10*, R9–R38. [CrossRef]
6. Barker, N.S.; Rebeiz, G.M. Optimization of distributed MEMS transmission-line phase shifters-U-band and W-band designs. *IEEE Trans. Microw. Theory Tech.* **2000**, *48*, 1957–1966.
7. Malczewski, A.; Eshelman, S.; Pillans, B.; Ehmke, J.; Goldsmith, C.L. X-band RF MEMS phase shifters for phased array applications. *IEEE Microw. Guid. Wave Lett.* **1999**, *9*, 517–519. [CrossRef]
8. Hacker, J.B.; Mihailovich, R.E.; Moonil, K.; DeNatale, J.F. A Ka-band 3-bit RF MEMS true-time-delay network. *IEEE Trans. Microw. Theory Tech.* **2003**, *51*, 305–308. [CrossRef]
9. Chu, T.S.; Roderick, J.; Hashemi, H. An Integrated Ultra-Wideband Timed Array Receiver in 0.13 μm CMOS Using a Path-Sharing True Time Delay Architecture. *JSSC* **2007**, *42*, 2834–2850.
10. Park, S.; Jeon, S. A 15–40 GHz CMOS True-Time Delay Circuit for UWB Multi-Antenna Systems. *IEEE Microw. Wirel. Compon. Lett.* **2013**, *23*, 149–151. [CrossRef]
11. Ghazizadeh, M.H.; Medi, A. Novel Trombone Topology for Wideband True-Time-Delay Implementation. *IEEE Trans. Microw. Theory Tech.* **2019**, *99*, 1–11. [CrossRef]
12. Liu, R.L.; Bao, J.F.; Huang, Y.L.; Li, X.Y. Design of four-position phase shifter based on microelectromechanical system switch. *J. Terahertz Sci. Electron. Inf. Technol.* **2016**, *14*, 127–130.
13. Yang, Y.; Chan, K.Y.; Ramer, R. Design of 600 GHz 3-bit delay-line phase shifter using RF NEMS series switches. In Proceedings of the 2011 IEEE International Symposium on Antennas and Propagation (APSURSI), Spokane, WA, USA, 3–8 July 2011; pp. 3287–3290. [CrossRef]
14. Gong, L.Y.; Wang, X.S. A compact structure of PIN diode single-pole nine-throw switch. *Electron. Compon. Mater.* **2016**, *35*, 69–72.
15. Yang, H.H.; Achref, Y.; Hosein, Z.; Blondy, P.; Gabriel, M.R. Symmetric and compact single-pole multiple-throw (SP7T, SP11T) RF MEMS switches. *J. Microelectromech. Syst.* **2015**, *24*, 685–695. [CrossRef]
16. Liu, B.; Yang, Q.; Wang, Y.J. Design of JPT45-6-S24-1 type single-blade six-throw RF coaxial switch. *Public Commun. Sci. Technol.* **2017**, *9*, 60–63.
17. Hou, Z.H.; Liu, Z.W.; Hu, G.W.; Liu, L.T.; Li, Z.G. Research on the design and fabrication of series capacitive RF MEMS switches. *Chin. J. Sens. Actuators* **2008**, *4*, 660–663.
18. Ou, S.J.; Zhang, G.J.; Wang, S.Y.; Dai, L.P.; Zhong, Z.J. A comparative study of low drive voltage RF MEMS cantilever beam switches. *Electron. Eng. Prod. World* **2020**, *27*, 81–84.
19. Chen, G.H.; Wu, Q.X.; Yu, Y.Y.; Lou, Z.X. Research on RF MEMS switches. *China Instrum.* **2008**, *8*, 37–39.
20. Guo, Q.; Li, Y.; Gao, S.; Zhang, P.Y. Simulation and analysis of capacitive micromechanical ultrasound transducer using sacrificial layer technology. *Chin. J. Biomed. Eng.* **2018**, *37*, 705–713.
21. Gao, Y.S.; Li, Q.; Wu, Q.N.; Chen, H.; Li, M.W. Optimization and design of power dividers in RF MEMS attenuators. *J. North Univ. China: Nat. Sci. Ed.* **2020**, *41*, 85–90.
22. Fan, L.N.; Wu, Q.N.; Wang, S.S.; Han, L.L.; Hou, W.; Li, M.W. Research and design of a snowflake-type MEMS single-blade five-throw switch. *Microelectronics* **2021**, *51*, 533–538.
23. Yue, C.J. Research on high performance microwave broadband delay lines. Master's Thesis, University of Electronic Science and Technology of China, Chengdu, China, 16 August 2018.

24. Beilenhoff, K.; Heinrich, W.; Hartnagel, H.L. The Scattering Behaviour of Air Bridges in Coplanar MMICs. In Proceedings of the 21st European Microwave Conference, Stuttgart, Germany, 9–12 September 1991; pp. 1131–1135.
25. Barker, S.; Rebeiz, G.M. Distributed MEMS true-time delay phase shifters and wide-band switches. *IEEE Trans. Microw. Theory Tech.* **1998**, *46*, 1881–1890. [CrossRef]
26. Tang, K.; Wu, Q.; Yang, G.H.; He, X.J.; Fu, J.H.; Li, L.W. RF characteristics investigation of MEMS phase shifter with CPW discontinuities. In Proceedings of the 2008 International Conference on Microwave and Millimeter Wave Technology, Nanjing, China, 13 June 2008; pp. 1393–1396.
27. Kim, M.; Hacker, J.B.; Mihailovich, R.E.; DeNatale, J.F. A DC-to-40 GHz four-bit RF MEMS true-time delay network. *IEEE Microw. Wirel. Compon. Lett.* **2001**, *11*, 56–58. [CrossRef]
28. Jeong, J.C.; Yom, I.B.; Kim, J.D.; Lee, W.Y.; Lee, C.H. A 6-18-GHz GaAs Multifunction Chip With 8-bit True Time Delay and 7—bit Amplitude Control. *IEEE Trans. Microw. Theory Tech.* **2018**, *66*, 2220–2230. [CrossRef]
29. Zhang, M.; Xu, Q. Design of 8 GHz-18 GHz CMOS Passive Delay Line with Low Insertion Loss. *Chin. J. Electron Devices* **2021**, *5*, 1041–1046.
30. Lin, C.C.; Puglisi, C.; Boljanovic, V.; Yan, H.; Ghaderi, E.; Gaddis, J.; Xu, Q.Y.; Poolakkal, S.; Cabric, D.; Gupta, S. Multi-Mode Spatial Signal Processor with Rainbow-Like Fast Beam Training and Wideband Communications Using True—Time—Delay Arrays. *JSSC* **2022**, *57*, 3348–3360. [CrossRef]
31. Hao, D.N.; Zhang, W. GaAs bidirectional true time delay chip design. *J. Xidian Univ.* **2023**, *1*, 102–108+128.

Disclaimer/Publisher's Note: The statements, opinions and data contained in all publications are solely those of the individual author(s) and contributor(s) and not of MDPI and/or the editor(s). MDPI and/or the editor(s) disclaim responsibility for any injury to people or property resulting from any ideas, methods, instructions or products referred to in the content.

Article

Ultra-Wideband and Narrowband Switchable, Bi-Functional Metamaterial Absorber Based on Vanadium Dioxide

Xiaoyan Wang [1,2,3,4,*], Yanfei Liu [2,3,4,5], Yilin Jia [2,3,4,5], Ningning Su [2,3,4,5] and Qiannan Wu [2,3,4,5,*]

1. School of Information and Communication Engineering, North University of China, Taiyuan 030051, China
2. Center for Microsystem Integration, North University of China, Taiyuan 030051, China; 15373166906@163.com (Y.L.); 13384612312@163.com (Y.J.); suningning826032@163.com (N.S.)
3. School of Instrument and Intelligent Future Technology, North University of China, Taiyuan 030051, China
4. Academy for Advanced Interdisciplinary Research, North University of China, Taiyuan 030051, China
5. School of Semiconductors and Physics, North University of China, Taiyuan 030051, China
* Correspondence: wangxiaoyan@nuc.edu.cn (X.W.); qiannanwoo@nuc.edu.cn (Q.W.)

Abstract: A switchable ultra-wideband THz absorber based on vanadium dioxide was proposed, which consists of a lowermost gold layer, a PMI dielectric layer, and an insulating and surface vanadium dioxide layer. Based on the phase transition properties of vanadium dioxide, switching performance between ultra-broadband and narrowband can achieve a near-perfect absorption. The constructed model was simulated and analyzed using finite element analysis. Simulations show that the absorption frequency of vanadium dioxide above 90% is between 3.8 THz and 15.6 THz when the vanadium dioxide is in the metallic state. The broadband absorber has an absorption bandwidth of 11.8 THz, is insensitive to TE and TM polarization, and has universal incidence angle insensitivity. When vanadium dioxide is in the insulating state, the narrowband absorber has a Q value as high as 1111 at a frequency of 13.89 THz when the absorption is more excellent than 99%. The absorber proposed in this paper has favorable symmetry properties, excellent TE and TM wave insensitivity, overall incidence angle stability, and the advantages of its small size, ultra-widebands and narrowbands, and elevated Q values. The designed absorber has promising applications in multifunctional devices, electromagnetic cloaking, and optoelectronic switches.

Keywords: terahertz; metamaterial; perfect absorber; vanadium dioxide

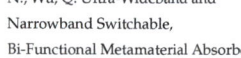

Citation: Wang, X.; Liu, Y.; Jia, Y.; Su, N.; Wu, Q. Ultra-Wideband and Narrowband Switchable, Bi-Functional Metamaterial Absorber Based on Vanadium Dioxide. *Micromachines* 2023, 14, 1381. https://doi.org/10.3390/mi14071381

Academic Editor: Haejun Chung

Received: 20 June 2023
Revised: 3 July 2023
Accepted: 4 July 2023
Published: 6 July 2023

Copyright: © 2023 by the authors. Licensee MDPI, Basel, Switzerland. This article is an open access article distributed under the terms and conditions of the Creative Commons Attribution (CC BY) license (https://creativecommons.org/licenses/by/4.0/).

1. Introduction

Metamaterials are a class of artificially fabricated microstructured materials with unique electromagnetic properties that cannot be achieved with natural materials. They are widely used in communication [1], imaging [2], stealth [3], sensing [4], and other fields. With the development of technology, the science and technology of terahertz have also led to the rapid growth of metamaterial devices. To facilitate the development of THz technologies, various metamaterial-based functional devices have been proposed, such as filters [5,6], polarization converters [7,8], modulators [9,10], antennas [11], and perfect absorbers [12,13]. Among these devices, metamaterial perfect absorbers have been a popular research topic due to their wide range of applications in solar energy, cloaking technology, etc. Most THz absorbers, however, have a single function. They cannot be dynamically tuned, etc., and actively tunable metamaterial absorbers are more suitable for complex electromagnetic applications in practical applications.

Metamaterials can be used in perfect absorbers based on metamaterials, which was first proposed by LANDY [14]. People have gradually introduced some active materials in the devices in order to be able to design tunable absorbers, such as vanadium dioxide [15,16], graphene [17–21], molybdenum disulphide [22–24], strontium titanate oxide [25,26], indium antimonide [27–30], etc. Among these phase-changing materials, vanadium dioxide can undergo a phase transition from an insulating state at room temperature to a stable high-loss metallic state at higher temperatures, and the optical and electrical properties can be significantly modified during the phase transition, which is reversible. As the conductivity of vanadium dioxide increases steadily when heated from room temperature to higher temperatures, vanadium dioxide can vary in the range of 20 S/m to 200,000 S/m [31–33]. Based on the fact that vanadium dioxide phase transitions require modest temperatures and can be accomplished at room temperature with few limitations in experimental testing and practical applications, vanadium dioxide is gradually being applied with these properties, and the optical properties of vanadium dioxide films can be altered by heating or cooling the films to temperatures close to the phase transition temperature. The phase-shift properties of vanadium dioxide are well suited to the tuning needs of metamaterial absorbers and are gradually being used more and more by researchers to prepare tunable metamaterial absorbers based on the phase-shift material vanadium dioxide [34,35]. In the last decade or so, metamaterial absorbers have been used in the preparation of virtual narrowband absorbers [36–38], broadband absorbers [39–48], and broadband and narrowband tunable absorbers.

For example, in 2021, Chunyu Zhang et al. proposed a dual-modulated broadband terahertz absorber based on vanadium dioxide and graphene, which could achieve more than 90% absorption from 1.04 THz to 5.51 THz with a broadband absorption of 4.07 THz [49]. In the same year, Zhipeng Zheng et al. proposed a terahertz perfect absorber based on an ultra-broadband flexible active switch with more than 90% absorption intensity at 8.5–11 THz, and it was switchable to a narrowband absorber by changing the vanadium dioxide conductivity [50]. In 2022, H Peng et al. proposed a broadband terahertz tunable multiple absorber based on phase-shift materials with a bandwidth of 5.5 THz in the range of 4.5–10 THz where the absorption intensity exceeds 90% [51]. In the same year, Pengyu Zhang et al. proposed an ultra-broadband tunable THz metamaterial absorber based on a double layer of vanadium dioxide in a square ring array with an absorption intensity in the 1.63 THz range and an absorption intensity exceeding 90% in the 12.39 THz range. The absorption bandwidth was 10.76 THz [52]. In 2023, Niujunhao et al. proposed a switchable bi-functional metamaterial based on vanadium dioxide for broadband absorption and broadband polarization in the terahertz band, with absorption intensity exceeding 90% in the range of 3.3–5.62 THz [53]. Peng Gao et al. also proposed a broadband terahertz polarization converter based on the phase transition properties of vanadium dioxide in the same year, with a bandwidth of 2.87 THz in the range of 2.71–5.58 THz where the absorption exceeds 90% [54]. In 2021, Zhangbo Li et al. proposed an "Ultra-narrow-band metamaterial perfect absorber based on surface lattice resonance in a WS_2 nanodisk. Array", which has a sensitivity of 1067 nm/RIU when used as a narrowband absorber [55]. In 2022, Xianglong Wu et al. proposed a "High performance dual-control tunable absorber with switching function and high sensitivity based on. Dirac semi-metallic film and vanadium oxide", which has a sensitivity of 462 GHz/RIU as a narrowband absorber [56]. Although there have been numerous studies of broadband absorbers for several years, terahertz absorbers that combine broadband absorption and narrowband absorption with a wider band and a simple structure have occasionally been reported. Moreover, currently reported broadband absorbers and narrowband switchable absorbers suffer from issues such as insufficient absorption bandwidth to meet practical applications and low Q values when switching to narrowband absorbers.

This paper proposes a dual-function THz metamaterial absorber based on vanadium dioxide for ultra-broadband and ultra-narrowband switching. The absorber size is considerably smaller than all current absorbers, making it more suitable for practical applications. When vanadium dioxide is in the metallic state, the absorber behaves as a broadband absorber with more than 90% absorption in the 3.8–15.8 THz range. When vanadium dioxide is in the insulating state, the absorber can be switched to a narrowband absorber with elevated Q values. The absorption at 13.89 THz is 99.99% with a Q value of 1111. Due to the extreme symmetry of the designed absorber, the absorber also has the characteristics of polarization insensitivity and insensitivity within a wide incidence angle of 50°, which considerably reduces the limitations of the absorber in a practical application. The ultra-wide and ultra-narrow bifunctional absorbers designed in this paper can provide current research ideas for versatile and tunable devices in the terahertz and its infrared bands, with promising applications in terahertz imaging, detection, and sensing.

2. Design and Simulation

The structure diagram of our proposed vanadium-dioxide-based switchable absorber unit cell is shown in Figure 1, which consists of four layers: the lowermost structure is 0.2 µm of thick gold, which acts as a reflecting mirror to guarantee the complete reflection for the impinging terahertz wave, thereby suppressing the transmission; the second layer is a PMI (polymethacryl imide) layer with a relative permittivity of 1.1 [57]; the third layer is an insulating layer (Topas (cyclic olefin copolymer)) with a relative permittivity of 1.96, which is assumed to be lossless [58]; and the uppermost layer is a vanadium dioxide layer with a thickness of 200 nm. Figure 1b shows a top view of the unit cell structure with period p. The uppermost layer structure consists of a cross-like structure and four L-shaped dart-like structures. The optimal geometrical parameters were determined by analyzing the effect of the geometrical parameters on the absorption broadening, as shown in Table 1.

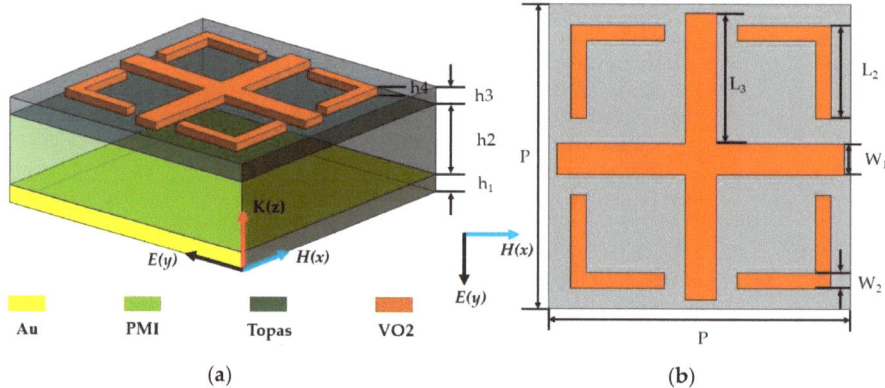

Figure 1. (a) 3D schematic of a terahertz metamaterial absorber; (b) top view of the x–y plane.

Table 1. Parameters of the designed absorber.

Parameter	P	h_1	h_2	h_3	h_4	W_1	W_2	L_2	L_3
Value/µm	22	0.2	6.55	0.15	0.2	2.8	1.5	7.4	9.1

In this paper, we numerically simulate the absorption properties of a designed switchable metamaterial absorber using the CST software. When a THz wave is vertically incident on the surface of a metamaterial absorber, the electric field of the incident electromagnetic wave is polarized in the x-direction, and the magnetic field is polarized in the y-direction. The structure of the absorbing metamaterial receiver cell extends indefinitely in the x–y plane. The Drude model describes the dielectric constant of VO_2 in the THz range, where the dielectric constant of VO_2 can be expressed as [59]:

$$\varepsilon(\omega)_{VO_2} = \varepsilon_\infty - \frac{\omega_p^2}{\omega(\omega + i\gamma_2)}, \quad (1)$$

where $\varepsilon_\infty = 12$ is the permittivity at infinite frequency for vanadium dioxide, $\gamma_2 = 5.75 \times 10^{13}$ rad/s is the collision frequency, and ω_{p^2} is the plasmon frequency which depends on the conductivity σ. The relation between them may be stated as follows:

$$\omega_{p^2}^2 = \frac{\sigma}{\sigma_0}\omega_{p0}^2, \qquad (2)$$

with $\sigma_0 = 3 \times 10^5$ S/m and $\omega_{p^2}^2 = 1.4 \times 10^5$ S/m, the conductivity σ can vary with the phase transition of VO$_2$. When the temperature changes, VO$_2$ can switch back and forth between the insulating and metallic states [60], and the change in conductivity is reversible with the change in temperature. The variation in the conductivity of vanadium dioxide with ambient temperature is shown in Figure 2. Vanadium dioxide is insulated at room temperature and has an electrical conductivity of 20 S/m. When the temperature gradually increases, vanadium dioxide reaches the phase transition state, and the rice dioxide is in a metallic state with an electrical conductivity of 2×10^5 S/m; moreover, the phase transition state of vanadium dioxide is reversible. When the ambient temperature is lowered, the metallic state can be transformed into an insulating state. The variation in the conductivity of vanadium dioxide at different temperatures is mainly due to the effect of temperature on the permittivity. Figure 3 shows the real and imaginary parts of the permittivity at different conductivities of vanadium dioxide. As shown in Figure 3, the imaginary part of the permittivity of vanadium dioxide is considerably larger than the real part, and the rate of change in conductivity at the transition is also larger than the real part. In practice, the following two methods can be used to make the vanadium dioxide phase transition. The first method is to heat the metal at the bottom of the absorber and convert the vanadium dioxide from an insulating to a metallic state by heat transfer, at which point the conductivity gradually increases. When the heating source is removed, the temperature gradually decreases, and the conductivity of vanadium dioxide gradually decreases; the metallic titanium transitions to an insulating state. The second method is to add a metallic patch to the vanadium dioxide layer and transfer heat to the vanadium dioxide by applying a voltage at the two ends of the metal, thus changing it from an insulating to a metallic state. When the voltage at both ends is removed, the temperature gradually decreases, and the vanadium dioxide transitions from a metallic to an insulating state. In addition, there are chemical mixing methods, etc., which can enable the phase transition of vanadium dioxide from the insulating to the metallic regime. With the above method, we can change the state of vanadium dioxide in practical applications so that the absorber designed in this paper can switch between ultra-widebands and ultra-narrowbands, enabling multi-functional device applications.

In this paper, we use the finite element theory to obtain ultra-narrowband and ultra-wideband absorption spectra of the proposed THz metamaterial absorber in the insulating and metallic states of vanadium dioxide. In the simulations, the absorption rate can be expressed as follows.

$$A = 1 - R - T, \qquad (3)$$

In Equation (3) where A, T, and R above are the absorptance, transmittance, and reflectance of the absorber, respectively. $R = |S_{11}|^2$ and $T = |S_{21}|^2$. $|S_{11}|$ and $|S_{21}|$ represent the reflection and transmission coefficients of the metamaterial absorber, respectively. In this paper, the structure has an underlying metal thickness of 0.2 µm, which is larger than the skin depth of THz waves at the target frequency, and the transmittance $T(\omega)$ is close to 0, which is simplified as $A = 1 - R$ [61].

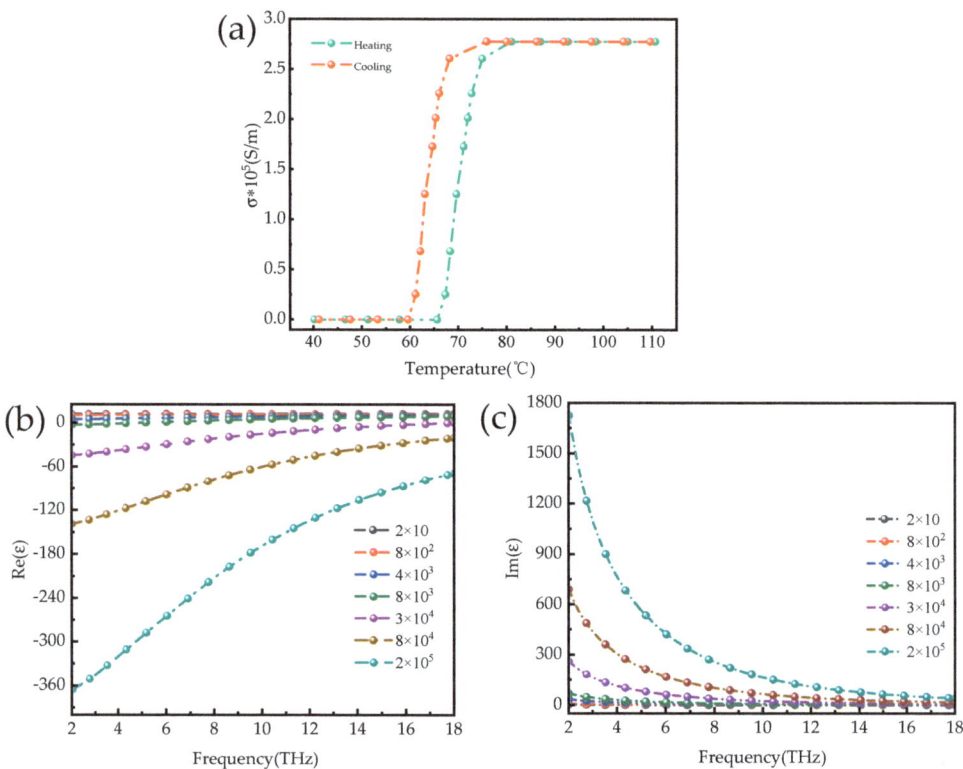

Figure 2. (**a**) The curve of VO$_2$ conductivity as a function of ambient temperature; (**b**,**c**) real and imaginary parts of the relative permittivity of VO$_2$ at different electrical conductivities.

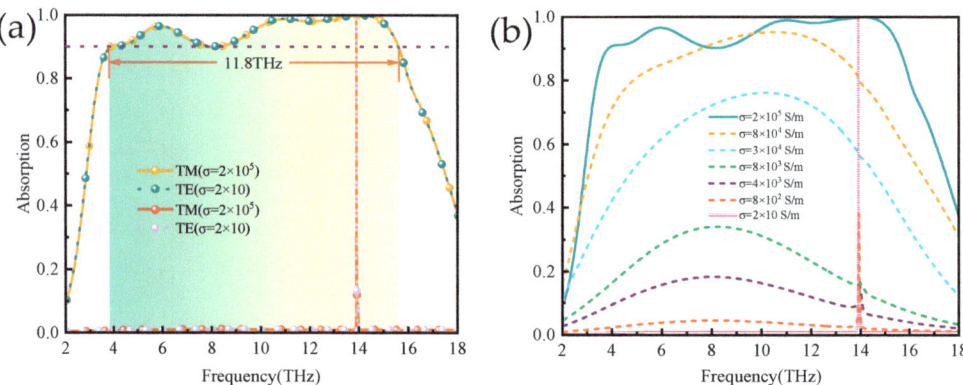

Figure 3. (**a**) Broadband absorption curve of VO$_2$ in metallic phase (green line) and narrowband absorption curve of VO$_2$ in insulating phase (pink line); (**b**) plots of absorptivity for absorbers with different conductivities.

3. Results and Discussion

Through simulation, we calculated the absorption spectra of the absorber at different polarization modes (TE, TM) and different conductivities, respectively, as shown in Figure 3.

From Figure 3a, when the VO$_2$ is in the metal phase with a conductivity of 2×10^5 S/m, the absorption bandwidth of more than 90% absorption is observed to be 11.8 THz in the frequency range from 3.8 THz to 15.6 THz, with a central frequency of around 9 THz, and the absorber has the advantage of being polarization insensitive. When VO$_2$ undergoes a phase transition from a metallic to an insulating state under the influence of temperature, the absorber switches from a broadband absorber to a narrowband absorber with elevated Q values. The absorber achieves an absorption of >99.99% at the electromagnetic frequency f = 13.89 THz with a quality factor of 1111.

The absorption spectra of different conductivities are shown in Figure 3b. The figures show the switching function of the absorber in the broadbands and narrowbands as the conductivity of vanadium dioxide varies from 2×10 S/m to 2×10^5 S/m. Upon heating, the vanadium dioxide transitions from an insulating to a metallic state while the absorber forms a conventional metal–dielectric–metal structure. Broadband absorbers enable switching between total reflection and perfect absorption. The absorber designed in this paper implements a switchable function between broadband and narrowband, which can also be referred to as the on–off function of a THz absorber.

To further explain the absorption principle of the designed absorber, the impedance matching theory is used. The relative impedance formulae are shown in Equations (4) and (5) [62,63].

$$A = 1 - R = 1 - \left| \frac{Z - Z_0}{Z + Z_0} \right|^2 = 1 - \left| \frac{Z_r - 1}{Z_r + 1} \right|^2, \tag{4}$$

$$Z_r = \pm \sqrt{\frac{(1 + S_{11})^2 - S_{21}^2}{(1 - S_{11})^2 - S_{21}^2}}, \tag{5}$$

where S_{11}, S_{21}, Z, and Z_0 are the S-parameters, effective impedance and free space impedance of the proposed absorber, respectively, and $Zr = Z/Z_0$ represents the relative impedance. Formulas (4) and (5) can be obtained to make the metamaterial absorber x absorption rate reach the maximum. At this time, in free space impedance matching, Z and Z_0 represent the equivalent impedance and free space impedance of the absorber, while Z_r represents the relative impedance ($Z_r = Z/Z_0 = 1$); Figure 4 shows the real and imaginary parts of the relative impedance of the metallic and insulating absorber under TE polarization. In Figure 4a, when VO$_2$ is in the metallic state, the real part is close to 1, and the imaginary part is close to 0 in the frequency range of 3.8~15.6 THz. In this frequency range, the absorber's impedance and the free's impedance are already adequately matched. When the electromagnetic wave is incident to the absorber, the reflected wave is approximately 0, and most of the energy is lost in the insulating layer, thus achieving perfect absorption.

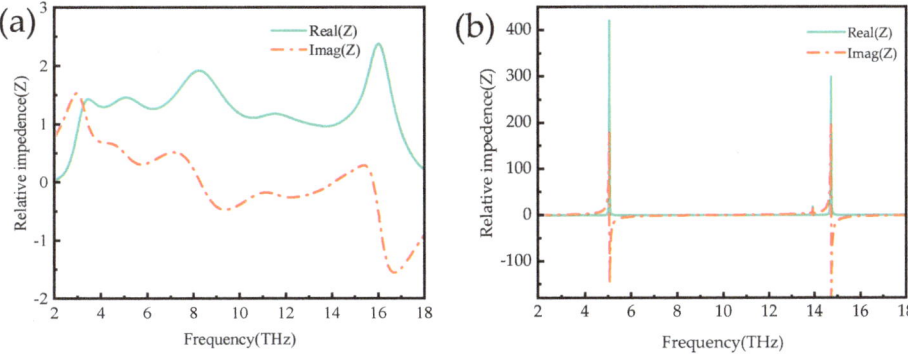

Figure 4. (a) Normalized impedance of the structure at $\sigma = 2 \times 10^5$ S/m. (b) Normalized impedance of the structure at $\sigma = 2 \times 10$ S/m.

In order to further explain the specific mechanism of the proposed absorber in wideband absorption, Figure 5 shows the wideband absorption field distribution of metal vanadium dioxide absorbent at different resonant frequencies in TE and TM polarization states. Figure 5a–c show the electric field distribution at frequencies of 5.93, 11, and 14 THz in TE mode. It is observed that the absorber exhibits a strong electric field at 3.5 THz at the left and right ends of the crossover, resulting in strong THz trapping and absorption. This means that the first absorption peak is caused by the coupling effect between two neighboring cells. When the frequency is 11 THz, the electric field is distributed in the horizontal gap between the upper and lower ends of the cross and the L-shaped dart. It is implied that the second absorption peak is caused by the interaction between the cross and the L-shaped dart. At 14 THz, the electric field strength of the vanadium dioxide resonant structure is particularly weak compared to the first two structures, indicating that the interaction between the vanadium dioxide surface structure and the incident terahertz wave is weak at this time. Most of the terahertz wave is lost in the dielectric layer at this time, thus producing absorption. Figure 5d–f show the electric field distribution of the absorber in TM mode at 1.93, 11, and 14 THz, which is similar to Figure 5a–c. Due to the perfect symmetry of the absorber designed in this paper, the absorber exhibits absolute polarization insensitivity. Therefore, the electric field distribution of absorber z is the same in both polarization states but with a 90° rotation.

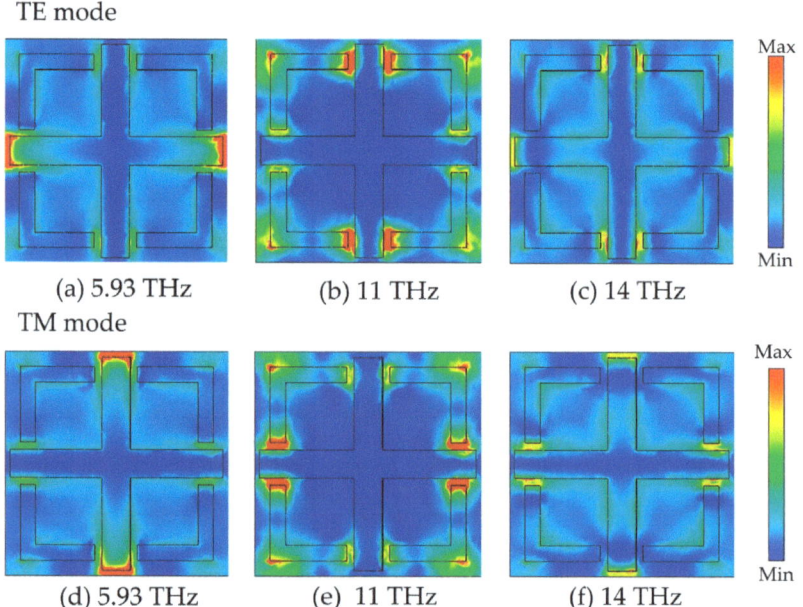

Figure 5. The electric field intensity distribution of the designed metamaterial wideband absorber at a frequency of 5.93 THz, 11 THz, and 14 THz; (**a**–**c**) for the TE mode and (**d**–**f**) for the TM mode.

Figure 6 shows the current distribution on the surface of the top VO_2 resonant layer and the bottom metallic layer at two resonant frequencies of 5.93 and 11 THz when VO_2 is in the metallic state. At 3.5 THz, the direction of the surface current on the top VO_2 resonance structure is inversely parallel to the direction on the bottom metal layer. The middle part of the dielectric layer is treated as a magnetic dipole, forming a strong magnetic resonance. At a frequency of 11 THz, a portion of the surface current direction on the top VO_2 resonant structure is parallel to that on the bottom metal layer, when an electrical resonance is also induced. Thus, under the excitation of magnetic and electrical resonances, the incident electromagnetic waves are extremely suppressed in the resonator, leading to broadband absorption properties.

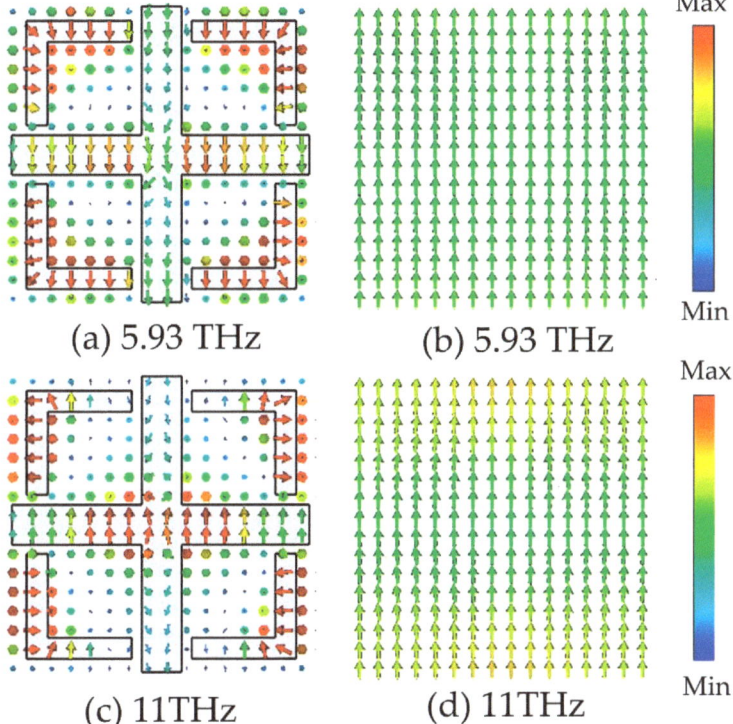

Figure 6. Distributions of the surface current on the top VO$_2$ resonant ring and bottom metal layer at the frequencies of 5.93 THz and 11 THz under normal incidence of TE incident wave.

Figure 7 shows the electric field and current distribution of the absorber as a function of the narrowband absorber at 13.88 THz when vanadium dioxide is in the insulating state. As shown in Figure 7a, at a frequency of 13.88 THz, the electric field is mainly concentrated in the middle of the horizontal gap between the left and right ends of the cross-shaped structure and between the upper and lower ends of the cross-shaped and L-shaped darts. The narrowband absorption in this case is due to the coupling effect between the components. When an external electromagnetic wave interacts with the vanadium dioxide dielectric layer, the vanadium dioxide layer acts as a resonator, causing charges to accumulate in the surface structure and forming electric dipole resonances. This induced electric dipole resonance couples to the underlying metal plate, resulting in the formation of a magnetic dipole resonance in the absorber, which leads to a solid magnetic resonance producing a resonant absorption peak at 13.88 THz. Figure 7b,c show the surface and bottom metal current distributions of vanadium dioxide in the narrowband absorption state and the insulating state. The current direction at the top of the vanadium dioxide is parallel to the current direction at the bottom of the metal, forming a loop. This current distribution further validates the theoretical analysis section. In addition, the vanadium dioxide layer acts as a dielectric layer, where the thickness of the dielectric layer increases, providing a useful space for electromagnetic wave propagation and results in a narrowband peak of near-perfect absorption.

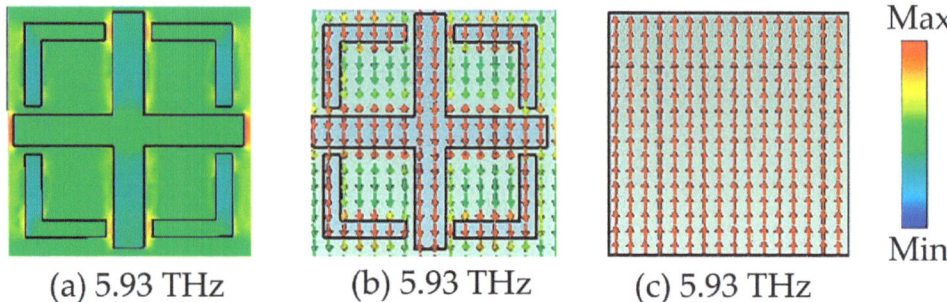

Figure 7. Distribution of electric fields and currents at 13.88 THz in narrowband absorption; (**a**) electric field distribution diagram; (**b**) surface current distribution diagram; (**c**) bottom surface current distribution diagram.

To investigate the effect of the geometric structure parameter on the results, it is necessary to verify the effect of the geometric structure parameter on the absorber. One parameter is analyzed for simplicity, while the additional parameters are fixed. Figure 8 shows the effect of the length of the cross-shaped structure and the length of the L-shaped dart on the absorption spectra, respectively. In Figure 8a, as the L-shaped dart length b keeps increasing, the absorption spectrum shifts slightly at low frequencies, but the absorption intensity gradually decreases. Finally, the x absorption spectrum with a maximum absorption more significant than 90% is obtained at $L_2 = 7.4$ μm. In Figure 8b, with the increasing length of the cross-type, the absorption peak at low frequencies gradually shifts blue, and the absorption intensity gradually decreases with the increasing length, which is caused by the coupling effect between adjacent unit structures. The effects of the cross-shaped structure k-width and L-shaped dart width on the absorption spectra are shown in Figure 9a,b, respectively. In Figure 9a, with the increasing length of W_1, the absorption peaks with >90% absorption at low and high frequencies are red-shifted and blue-shifted, respectively, but the absorption intensity gradually increases, and finally, the absorption broadband is at its maximum at $W_1 = 2.8$ μm. In Figure 9b, with the gradual increase in the L-shaped dart width, the position of the absorption peak at elevated frequency gradually shifts blue, and the absorption intensity gradually decreases. In summary, the best effect of the absorption bandwidth can be obtained by adjusting the geometric parameters of the structure. It will have important implications for practical fabrication.

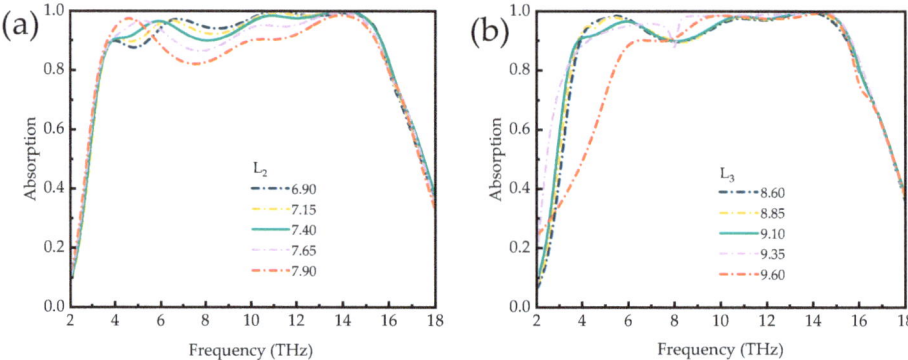

Figure 8. Influence of the (**a**) length L_2 of the cross-like structure and (**b**) length L_3 of the L-shaped, dart-like structure based on VO$_2$ on the absorption spectrum.

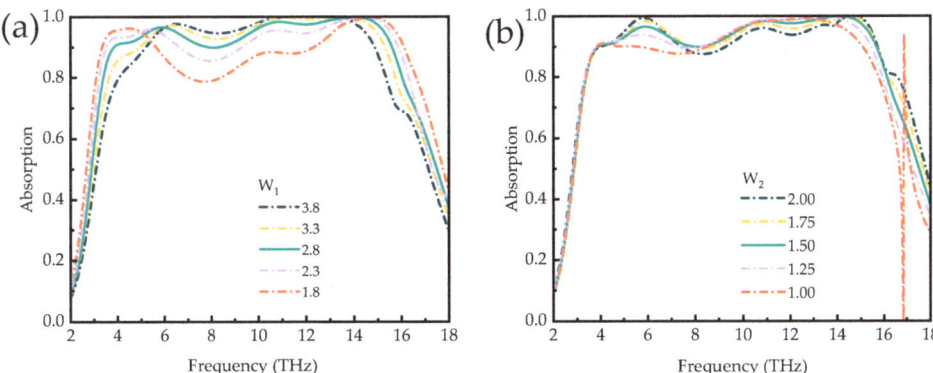

Figure 9. Influence of the (**a**) length W_1 of the cross-like structure and (**b**) length W_2 of the L-shaped, dart-like structure based on VO_2 on the absorption spectrum.

In practice, a major factor in evaluating absorber properties is the absorption properties at different polarization angles and incidence angles. Figure 10a,b show the absorption spectra of the proposed absorber for normal incident waves at broadband and narrowband with polarization angles ranging from 0° to 90°. As seen from the plots, the absorption spectrum remains constant; thus, this absorber has the advantage of polarization insensitivity. Figure 11a,b show the absorption spectra of the broadband absorber under TE and TM polarized waves, respectively. When the incidence angle is less than 25°, the absorber has superior absorption characteristics in the range of 3.8–15.6 THz. The blue shift occurs when the incidence angle is 25° to 52°, with superior absorption characteristics in the field of 3.8–18 THz. In the TM mode in Figure 12b, when the incidence angle is greater than 40°, the absorption rate will blue-shifted; however, when the incidence angle is less than 55°, the absorber shows superior absorption characteristics. The absorption rate of the absorber in the TM and TE modes will be blue-shifted with the increasing incidence angle. The main reason is that the tangential component of the electric field decreases as the incidence angle increases. Remarkably, the bandwidth of the incidence angle widens with increasing incidence angle. This is because the TE polarization produces different absorption peaks at extreme frequencies, resulting in the broadening of the absorption bandwidth. For TM polarization, the absorption peak is significantly blue-shifted at high frequencies, resulting in a wider absorption bandwidth. Figure 12 shows the absorption spectra of the narrowband absorber in TE and TM modes at different incidence angles when vanadium dioxide is in the insulating state. In Figure 12a, with the increase in the incidence angle, the absorption spectrum gradually blue-shifts, the absorption rate is lower at the low frequency, and the absorption rate is higher in the range of 60°. In Figure 12b, the absorption spectrum also blue-shifts with increasing incidence angle in TM mode and maintains a high absorption peak in the range of 50°. As a narrowband absorber, it has a high incidence angle to maintain a high absorption rate in both modes, which increases the application value of the device.

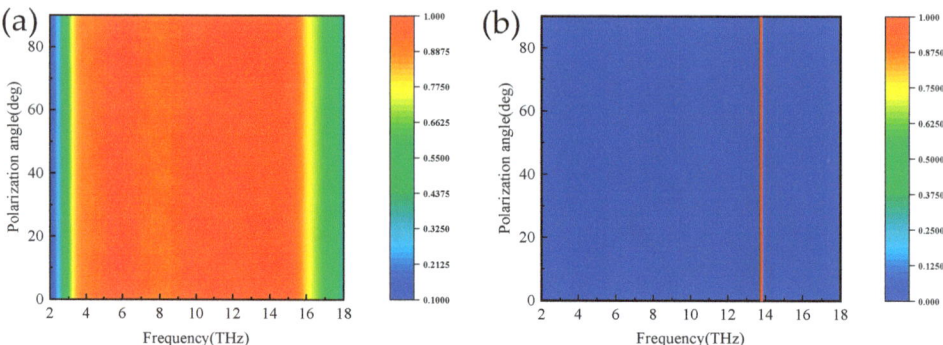

Figure 10. The absorption spectra of broadband absorbers in TM and TE modes; (**a**,**b**) different polarization angles.

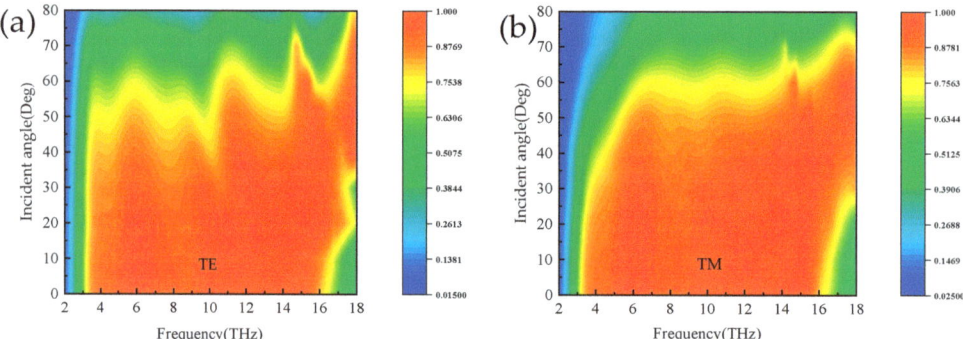

Figure 11. The absorption spectra of broadband absorbers in TE and TM modes; (**a**,**b**) different incident angles.

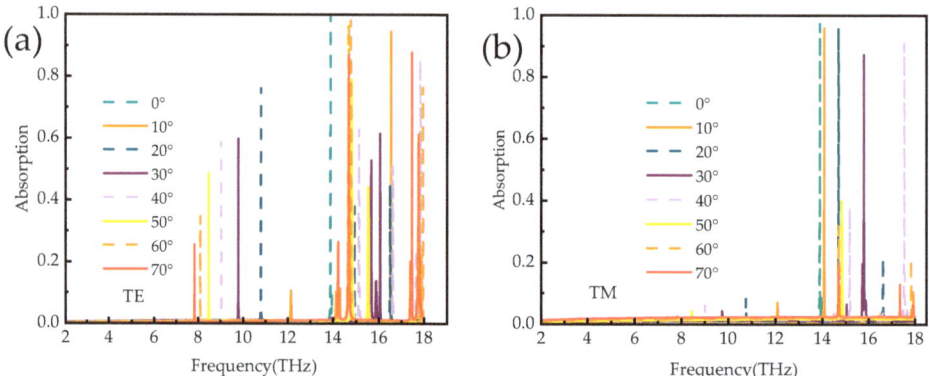

Figure 12. The absorption spectra of narrowband absorbers in TE and TM modes; (**a**,**b**) different incident angles.

Since the sensor designed in this paper is a broadband and narrowband tunable absorber, we additionally evaluated the performance of the narrowband sensor absorber to consider practical applications of the sensor. Figure 13 shows the absorption spectrum of the narrowband absorber as a function of the refractive index of the surrounding environment. Figure 13a shows that the position of the narrowband absorption peak is red-shifted with the increasing refractive index. Sensitivity (S) is commonly used to evaluate the sensing performance of narrowband absorbers. In general, the sensor sensitivity is defined as shown in Equation (6). In addition, the FOM value reflects the influence factor bandwidth, which is also one of the critical metrics to reflect the sensor performance and is calculated as shown in Equation (7).

$$S = \Delta f / \Delta n, \tag{6}$$

$$FOM = S/FWHM, \tag{7}$$

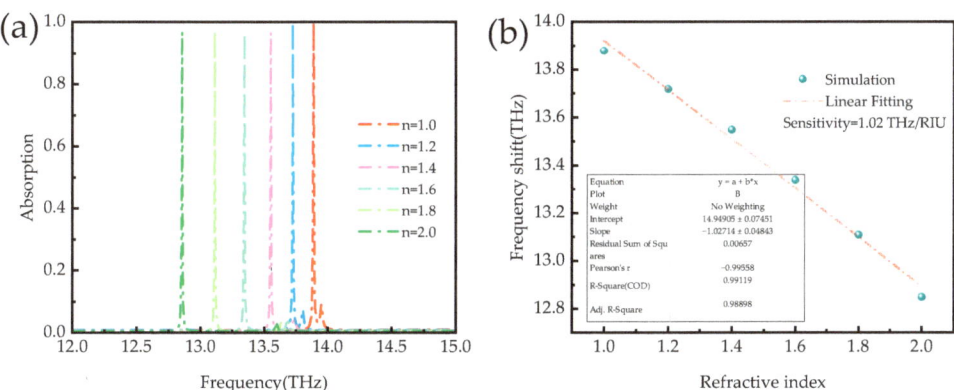

Figure 13. (a) Absorption spectra with different refractive index values of the surrounding dielectric environments; (b) the frequency shift against different refractive indices.

In Equation (6), Δf is the peak absorption shift due to the refractive index (Δn) change. From Figure 13, it can be seen that when the refractive index gradually increases, the absorption peak of the sensor gradually red-shifts. After fitting the sensitivity of the sensor calculated as 1.02 THz/RIU, and with the increase in the refractive index, the absorption of the sensor is above 95% and also has the performance of a high Q value. In Equation (7), the FWHM is the full width at a half-height maximum of the absorption peak, and the higher the FOM value, the better the performance of the representative sensor and the higher the sensing accuracy, which can be calculated from the FOM value of the designed narrowband absorber of 102. Therefore, the narrowband absorber proposed in this paper can be used for detection by detecting changes in the refractive index of the analyte and has shown superior performance in subsequent bio-detection and sensing.

This section presents the specific process and fabrication steps of the metamaterial absorber, which is shown in Figure 14. Due to the limited size of the terahertz band material units, most of them use micro-nano processing. These include the laser lithography process, inkjet printing process, MEMS process, etc. Due to the extreme accuracy of THz metamaterials, the resonant structure of the top unit cell of the metamaterial is fabricated in this work with laser etching. In Figure 14a, a layer of the metal film is sputter-grown on the back of the PMI layer after 0.2 μm using magnetron sputtering, and then a Topas insulating layer is superimposed on the PMI in the same way. Figure 14b shows a layer of photoresist uniformly spin-coated on the washed and air-dried prototype and dried at a suitable temperature. Figure 14c the photoresist is exposed to UV light under the premise that the sample is aligned with the mask plate, the exposed sample is carefully put into the developing solution and dried, and the pattern of the mask plate will appear on the developed sample. Figure 14d shows the sputtering of vanadium dioxide onto the developed sample in c using magnetron sputtering. Figure 14e shows the sample immersed in acetone solution to remove the excess photoresist, resulting in the designed vanadium dioxide resonance pattern. After the above process, a sample of the metamaterial absorber is obtained as shown in Figure 14f.

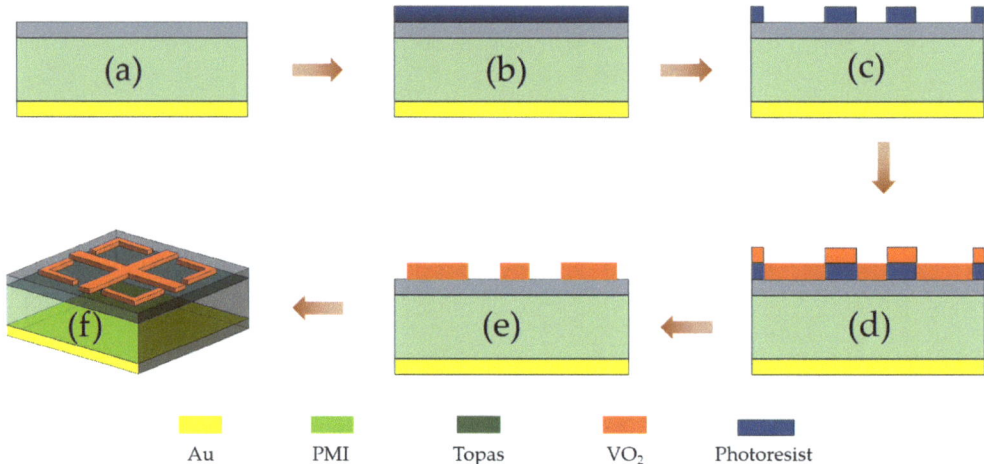

Figure 14. Absorber process flow diagram; (**a**) preparation of the bottom gold and PMI dielectric layers; (**b**) photoresist coating; (**c**) exposure, development; (**d**) sputtered precipitated vanadium dioxide layer; (**e**) release photoresist; (**f**) absorber samples.

In Table 2, we compare the performance of the designed metamaterial absorber with alternative absorbers. The metamaterial absorber designed in this paper not only enables the switchable functionality of both broadband and narrowband absorbers, but it also has excellent absorption performance when the absorber is in the broadband regime, with a wider range of absorption than previously reported. When the absorber is in the narrowband, not only does it have the advantage of high Q values, but it also detects the ambient refractive index. Moreover, the metamaterial absorber designed in this paper has the advantages of modest size, simple structure, and easy handling.

Table 2. The performance of the proposed absorber is compared with that of recent years.

Reported Year and Reference	FB > 90 (THz)	BW > 90 (THz)	Materials	Functions (Absorption Band)	Layers
2021 [49]	1.04–5.51	4.47	Graphene and VO_2	narrowband	5
2021 [50]	8.5–11	3.5	VO_2	narrowband and broadband	3
2022 [51]	4.5–10	5.5	VO_2	narrowband	4
2022 [52]	1.63–13.39	10.06	VO_2	narrowband	6
2023 [53]	3.3–5.62	3.32	VO_2	narrowband	6
2023 [54]	2.63–5.27	2.64	VO_2	narrowband	3
This work	3.8–15.6	11.8	VO_2	narrowband and broadband	4

4. Conclusions

In summary, we have designed an ultra-broadband and ultra-narrowband switchable, bi-functional THz absorber based on vanadium dioxide. The phase-shifted nature of vanadium dioxide, which modifies the conductivity with temperature, enables the switching of metamaterial device functions. When vanadium dioxide is in the metallic state, the absorber can be used as an ultra-wideband absorber, providing superior broadband absorption performance with an absorption rate of more than 90% in the 3.8–15.6 THz range and an absorption bandwidth of 11.8 THz. When vanadium dioxide is insulating, the designed absorber switches from an ultra-broadband absorber to an ultra-narrowband absorber. The designed sensor is a refractive index sensor to simulate different refractive indices in the ambient medium. The sensitivity of the proposed narrowband absorber was found to be 1.02 THz/RIU. Therefore, the proposed narrowband absorber can probe the refractive index by observation and sensing, and different devices exhibit excellent performance. The metamaterial absorber designed in this paper also has the properties of polarization insensitivity, wide incidence angle, small size, elevated Q value, and easy handling. The design has potential applications in terahertz imaging, electromagnetic stealth, and small optoelectronic switches.

Author Contributions: Conceptualization, X.W. and Y.L.; methodology, X.W. and Y.L.; software, X.W.; validation, Y.L.; formal analysis, Y.L.; investigation, X.W.; resources, Y.L.; data curation, Y.J.; writing—original draft preparation, X.W.; writing—review and editing, X.W. and Y.L.; visualization, Y.L.; supervision, Q.W. and N.S.; project administration, X.W.; funding acquisition, Q.W. All authors have read and agreed to the published version of the manuscript.

Funding: This research was funded by the Double First-Class Disciplines National First-Class Curriculum Construction, the National Future Technical College Construction Project, the Shanxi Province Postgraduate Education Reform Project, and the "173" Projects of China; grant numbers 11013168 and 11013410, 11013169, 11012103 and 11012133, 2020JCJQZD043 and 2021JCJQJJ0172.

Data Availability Statement: The data that support the findings of this study are available from the corresponding author upon reasonable request.

Acknowledgments: The authors thank the School of Instrument and Intelligent Future Technology of the North University of China for its support.

Conflicts of Interest: The authors declare no conflict of interest.

References

1. Hasan, M.; Arezoomandan, S.; Condori, H.; Sensale-Rodriguez, B. Graphene terahertz devices for communications applications. *Nano Commun. Netw.* **2016**, *10*, 68–78. [CrossRef]
2. Lyu, N.F.; Zuo, J.; Zhao, Y.M. Terahertz Synthetic Aperture Imaging with a Light Field Imaging System. *Electronics* **2020**, *9*, 830. [CrossRef]
3. Chen, P.Y.; Soric, J.; Padooru, Y.R.; Bernety, H.M.; Yakovlev, A.B.; Alu, A. Nanostructured graphene metasurface for tunable terahertz cloaking. *New J. Phys.* **2013**, *15*, 123029. [CrossRef]
4. Chen, M.; Singh, L.; Xu, N.; Singh, R.; Zhang, W.; Xie, L. Terahertz sensing of highly absorptive water-methanol mixtures with multiple resonances in metamaterials. *Opt. Express* **2017**, *25*, 14089–14097. [CrossRef]
5. Xiong, R.; Li, J. Double layer frequency selective surface for terahertz bandpass filter. *J. Infrared Millim. Terahertz Waves* **2018**, *39*, 1039–1046.
6. Mengual, T.; Vidal, B.; Marti, J. Continuously tunable photonic microwave filter based on a spatial light modulator. *Opt. Commun.* **2008**, *281*, 2746–2749. [CrossRef]
7. Fan, J.; Cheng, Y. Broadband high-efficiency cross-polarization conversion and multi-functional wavefront manipulation based on chiral structure metasurface for terahertz wave. *J. Phys. D Appl. Phys.* **2020**, *53*, 025109. [CrossRef]
8. Grady, N.K.; Heyes, J.E.; Chowdhury, D.R.; Zeng, Y.; Reiten, M.T.; Azad, A.K.; Taylor, A.J.; Dalvit, D.A.; Chen, H.T. Terahertz metamaterials for linear polarization conversion and anomalous refraction. *Science* **2013**, *340*, 1304–1307. [CrossRef]
9. Fan, Z.; Geng, Z.; Fang, W.; Lv, X.; Su, Y.; Wang, S.; Liu, J.; Chen, H. Characteristics of transition metal dichalcogenides in optical pumped modulator of terahertz wave. *AIP Adv.* **2020**, *10*, 045304. [CrossRef]
10. Berardi, S.-R.; Rusen, J.; Michelle, M.K.; Tian, F.; Kristof, T.; Wan, S.H.; Debdeep, J.; Lei, L.; Huili, G.X. Broadband graphene terahertz modulatorsenabled by intraband transitions. *Nat. Commun.* **2012**, *4*, 3.
11. Jin, J.L.; Cheng, Z.; Chen, J.; Zhou, T.; Wu, C.; Xu, C. Reconfigurable terahertz Vivaldi antenna based on a hybrid graphene-metal structure. *Int. J. RF Microw. Comput.-Aided Eng.* **2020**, *30*, e22175.1–e22175.8. [CrossRef]
12. Wang, B.X. Quad-Band Terahertz Metamaterial Absorber Based on the Combining of the Dipole and Quadrupole Resonances of Two SRRs. *IEEE J. Sel. Top. Quantum Electron.* **2017**, *23*, 4700107. [CrossRef]
13. Liu, X.; Fan, K.; Shadrivov, I.V.; Padilla, W.J. Experimental realization of a terahertz all-dielectric metasurface absorber. *Opt. Express* **2017**, *25*, 191–201. [CrossRef] [PubMed]
14. Landy, N.I.; Sajuyigbe, S.; Mock, J.J.; Smith, D.R.; Padilla, W.J. Perfect metamaterial absorber. *Phys. Rev. Lett.* **2008**, *100*, 207402. [CrossRef]
15. Liu, Y.; Huang, R.; Ouyang, Z.B. Terahertz absorber with dynamically switchable dual-broadband based on a hybrid metamaterial with vanadium dioxide and graphene. *Opt. Express* **2021**, *29*, 20839–20850. [CrossRef]

16. Huang, J.; Li, J.N.; Yang, Y.; Li, J.; Li, J.H.; Zhang, Y.T.; Yao, J.Q. Broadband terahertz absorber with a flexible, reconfigurable performance based on hybrid-patterned vanadium dioxide metasurfaces. *Opt. Express* **2020**, *28*, 17832–17840. [CrossRef]
17. Kumar, P.; Lakhtakia, A.; Jain, P.K. Graphene pixel-based polarization-insensitive metasurface for almost perfect and wideband terahertz absorption. *J. Opt. Soc. Am. B-Opt. Phys. B* **2019**, *36*, 1914. [CrossRef]
18. Liu, W.W.; Song, Z.Y. Terahertz absorption modulator with largely tunable bandwidth and intensity. *Carbon* **2020**, *174*, 617–624. [CrossRef]
19. Zhu, H.L.; Zhang, Y.; Ye, L.F.; Li, Y.K.; Xu, Y.H.; Xu, R.M. Switchable and tunable terahertz metamaterial absorber with broadband and multi-band absorption. *Opt. Express* **2020**, *28*, 38626–38637. [CrossRef] [PubMed]
20. Zhang, B.H.; Qi, Y.P.; Zhang, T.; Zhang, Y.; Yi, Z. Tunable multiband terahertz absorber based on composite graphene structures with square ring and Jerusalem cross. *Results Phys.* **2021**, *25*, 104233. [CrossRef]
21. Han, J.Z.; Chen, R.S. Tunable broadband terahertz absorber based on a single-layer graphene metasurface. *Opt. Express* **2020**, *28*, 30289–30298. [CrossRef]
22. Zhong, Y.J.; Huang, Y.; Zhong, S.C.; Lin, T.L.; Luo, M.T.; Shen, Y.C.; Ding, J. Tunable terahertz broadband absorber based on MoS_2 ringcross array structure. *Opt. Mater.* **2021**, *114*, 110996. [CrossRef]
23. Xu, Y.; Wu, L.; Kee, A.L. MoS_2-based highly sensitive nearinfrared surface plasmon resonance refractive index sensor. *IEEE J. Sel. Top. Quantum Electron.* **2019**, *25*, 4600307. [CrossRef]
24. Zamharir, S.G.; Karimzadeh, R.; Luo, X. Tunable polarization independent MoS_2-based coherent perfect absorber within visible region. *J. Phys. D Appl. Phys.* **2021**, *54*, 165104. [CrossRef]
25. Li, W.Y.; Cheng, Y.Z. Dual-band tunable terahertz perfect metamaterial absorber based on strontium titanate (STO) resonator structure. *Opt. Commun.* **2020**, *462*, 125265. [CrossRef]
26. Li, D.; Huang, H.L.; Xia, H.; Zeng, J.P.; Li, H.J.; Xie, D. Temperature-dependent tunable terahertz metamaterial absorber for the application of light modulator. *Results Phys.* **2018**, *11*, 659–664. [CrossRef]
27. Zou, H.J.; Cheng, Y.Z. Design of a six-band terahertz metamaterial absorber for temperature sensing application. *Opt. Mater.* **2019**, *88*, 674–679. [CrossRef]
28. Luo, H.; Cheng, Y. Thermally tunable terahertz metasurface absorber based on all dielectric indium antimonide resonator structure. *Opt. Mater.* **2020**, *102*, 109801. [CrossRef]
29. Chen, F.; Cheng, Y.Z.; Luo, H. Temperature Tunable Narrow-Band Terahertz Metasurface Absorber Based on InSb Micro-Cylinder Arrays for Enhanced Sensing Application. *IEEE Access* **2020**, *8*, 82981–82988. [CrossRef]
30. Luo, H.; Wang, X.; Qian, H. Tunable terahertz dual-band perfect absorber based on the combined InSb resonator structures for temperature sensing. *J. Opt. Soc. Am. B Opt. Phys.* **2021**, *38*, 2638–2644. [CrossRef]
31. Ren, Y.; Zhou, T.L.; Jiang, C.; Tang, B. Thermally switching between perfect absorber and asymmetric transmission in vanadium dioxide-assisted metamaterials. *Opt. Express* **2021**, *29*, 7666–7679. [CrossRef]
32. Qazilbash, M.M.; Brehm, M.; Chae, B.G.; Ho, P.C.; Andreev, G.O.; Kim, B.J.; Yun, S.J.; Balatsky, A.V.; Maple, M.B.; Keilmann, F.; et al. Mott transition in VO2 revealed by infrared spectroscopy and nano-imaging. *Science* **2007**, *318*, 1750–1753. [CrossRef]
33. Briggs, R.M.; Pryc, I.M.; Atwate, H.A. Compact silicon photonic waveguide modulator based on the vanadium dioxide metal-insulator phase transition. *Opt. Express* **2010**, *18*, 11192–11201. [CrossRef] [PubMed]
34. Chen, L.L.; Song, Z.Y. Simultaneous realizations of absorber and transparent conducting metal in a single metamaterial. *Opt. Express* **2020**, *28*, 6565–6571. [CrossRef] [PubMed]
35. Zhou, Z.K.; Song, Z.Y. Switchable bifunctional metamaterial for terahertz anomalous reflection and broadband absorption. *Phys. Scr.* **2021**, *96*, 115506. [CrossRef]
36. Meng, H.Y.; Shang, X.J.; Xue, X.X.; Tang, K.Z.; Xia, S.X.; Zhai, X.; Liu, Z.R.; Chen, J.H.; Li, H.J.; Wang, L.L. Bidirectional and dynamically tunable THz absorber with Dirac semimetal. *Opt. Express* **2019**, *27*, 31062–31074. [CrossRef] [PubMed]
37. Shan, Y.; Chen, L.; Shi, C.; Cheng, Z.; Zang, X.; Xu, B.; Zhu, Y. Ultrathin flexible dual band terahertz absorber. *Opt. Commun.* **2015**, *350*, 63–70. [CrossRef]
38. Wang, R.X.; Li, L.; Liu, J.L.; Yan, F.; Tian, F.J.; Tian, H.; Zhang, J.Z.; Sun, W.M. Triple-band tunable perfect terahertz metamaterial absorber with liquid crystal. *Opt. Express* **2017**, *25*, 32280–32289. [CrossRef]
39. Liu, S.; Chen, H.B.; Cui, T.J. A broadband terahertz absorber using multi-layer stacked bars. *Appl. Phys. Lett.* **2015**, *106*, 151601. [CrossRef]
40. Liu, L.; Liu, W.W.; Song, Z.Y. Ultra-broadband terahertz absorber based on a multilayer graphene metamaterial. *J. Appl. Phys.* **2020**, *128*, 093104. [CrossRef]
41. Huang, J.; Li, J.; Yang, Y.; Li, J.; Zhang, Y.; Yao, J. Active controllable dual broadband terahertz absorber based on hybrid metamaterials with vanadium dioxide. *Opt. Express* **2020**, *28*, 7018–7027. [CrossRef] [PubMed]
42. Song, Z.Y.; Wang, K.; Li, J.W.; Liu, Q.H. Broadband tunable terahertz absorber based on vanadium dioxide metamaterials. *Opt. Express* **2018**, *26*, 7148–7154. [CrossRef] [PubMed]
43. Song, Z.Y.; Wei, M.L.; Wang, Z.S.; Cai, G.X.; Liu, Y.N.; Zhou, Y.G. Terahertz absorber with reconfigurable bandwidth based on isotropic vanadium dioxide metasurfaces. *IEEE Photonics J.* **2019**, *11*, 4600607. [CrossRef]
44. Wang, S.X.; Cai, C.F.; You, M.H.; Liu, F.Y.; Wu, M.H.; Li, S.Z.; Bao, H.G.; Kang, L.; Werner, D.H. Vanadium dioxide based broadband THz metamaterial absorbers with high tunability: Simulation study. *Opt. Express* **2019**, *27*, 19436–19447. [CrossRef]

45. Dao, R.N.; Kong, X.R.; Zhang, H.F.; Su, X.R. A tunable broadband terahertz metamaterial absorber based on the vanadium dioxide. *Optik* **2019**, *180*, 619–625. [CrossRef]
46. Bai, J.J.; Zhang, S.S.; Fan, F.; Wang, S.S.; Sun, X.P.; Miao, Y.P.; Chang, S.J. Tunable broadband THz absorber using vanadium dioxide metamaterials. *Opt. Commun.* **2019**, *452*, 292–295. [CrossRef]
47. Song, Z.Y.; Jiang, M.W.; Deng, Y.D.; Chen, A.P. Wide-angle absorber with tunable intensity and bandwidth realized by a terahertz phase change material. *Opt. Commun.* **2020**, *464*, 125494. [CrossRef]
48. Wu, G.Z.; Jiao, X.F.; Wang, Y.D.; Zhao, Z.P.; Wang, Y.B.; Liu, J.G. Ultra-wideband tunable metamaterial perfect absorber based on vanadium dioxide. *Opt. Express* **2021**, *29*, 2703–2711. [CrossRef]
49. Zhang, C.Y.; Zhang, H.; Ling, F.; Zhang, B. Dual-regulated broadband terahertz absorber based on vanadium dioxide and graphene. *Appl. Opt.* **2021**, *60*, 4835–4840. [CrossRef]
50. Zheng, Z.P.; Zheng, Y.; Luo, Y.; Yi, Z.; Zhang, J.G.; Liu, L.; Song, Q.; Wu, P.H.; Yu, Y.; Zhang, J.F. Terahertz perfect absorber based on flexible active switching of ultra-broadband and ultra-narrowband. *Opt. Express* **2021**, *29*, 42787–42799. [CrossRef]
51. Peng, H.; Yang, K.; Huang, Z.X.; Chen, Z. Broadband terahertz tunable multi-film absorber based on phase-change material. *Appl. Opt.* **2022**, *61*, 3101–3106. [CrossRef]
52. Zhang, P.Y.; Chen, G.Q.; Hou, Z.Y.; Zhang, Y.Z.; Shen, J.; Li, C.Y.; Zhao, M.L.; Gao, Z.Z.; Li, Z.Q.; Tang, T.T. Ultra-Broadband Tunable Terahertz Metamaterial Absorber Based on Double-Layer Vanadium Dioxide Square Ring Arrays. *Micromachines* **2022**, *13*, 669. [CrossRef] [PubMed]
53. Niu, J.H.; Yao, Q.Y.; Mo, W.; Li, C.H.; Zhu, A.J. Switchable bi-functional metamaterial based on vanadium dioxide for broadband absorption and broadband polarization in terahertz band. *Opt. Commun.* **2022**, *527*, 128953. [CrossRef]
54. Gao, P.; Chen, C.; Dai, Y.W.; Luo, H.; Feng, Y.; Qiao, Y.J.; Ren, Z.Y.; Liu, H. Broadband terahertz polarization converter/absorber based on the phase transition properties of vanadium dioxide in a reconfigurable metamaterial. *Opt. Quantum Electron.* **2023**, *55*, 380. [CrossRef]
55. Li, Z.B.; Sun, X.A.; Ma, C.R.; Li, J.; Li, X.P.; Guan, B.O.; Chen, K. Ultra-narrow-band metamaterial perfect absorber based on surface lattice resonance in a WS$_2$ nanodisk array. *Opt. Express* **2021**, *29*, 27084–27091. [CrossRef] [PubMed]
56. Wu, X.L.; Zhao, W.C.; Yi, Z.; Yu, J.X.; Zhou, Z.G.; Yang, H.; Wang, S.F.; Zhang, J.G.; Pan, M.; Wu, P.H. High-performance dual-control tunable absorber with switching function and high sensitivity based on Dirac semi-metallic film and vanadium oxide. *Opt. Laser Technol.* **2022**, *153*, 108145. [CrossRef]
57. Xing, L.Y.; Cui, H.L.; Zhou, Z.X.; Bai, J.; Du, C.L. Terahertz scattering and spectroscopic characteristics of polymethacryl imide microstructures. *IEEE Access* **2019**, *7*, 41737–41745. [CrossRef]
58. Sun, P.; You, C.L.; Mahigir, A.; Liu, T.T.; Xia, F.; Kong, W.J.; Veronis, G.; Dowling, J.P.; Dong, L.F.; Yun, M.J. Graphene-based dual-band independently tunable infrared absorber. *Nanoscale* **2018**, *10*, 15564–15570. [CrossRef]
59. Li, J.K.; Chen, X.F.; Yi, Z.; Yang, H.; Tang, Y.J.; Yi, Y.; Yao, W.T.; Wang, J.Q.; Yi, Y.G. Broadband solar energy absorber based on monolayer molybdenum disulfide using tungsten elliptical arrays. *Mater. Today Energy* **2020**, *16*, 100390. [CrossRef]
60. Fan, F.; Hou, Y.; Jiang, Z.W.; Wang, X.H.; Chang, S.J. Terahertz modulator based on insulator-metal transition in photonic crystal waveguide. *Appl. Opt.* **2012**, *51*, 4589–4596. [CrossRef]
61. Tang, N.M.; Li, Y.J.; Chen, F.T.; Han, Z.Y. In situ fabrication of a direct Z-scheme photocatalyst by immobilizing CdS quantum dots in the channels of graphene-hybridized and supported mesoporous titanium nanocrystals for high photocatalytic performance under visible light. *Rsc Adv.* **2018**, *8*, 42233–42245. [CrossRef] [PubMed]
62. Zhao, Y.; Huang, Q.P.; Cai, H.L.; Lin, X.X.; Lu, Y.L. A broadband and switchable VO2-based perfect absorber at the THz frequency. *Opt. Commun.* **2018**, *426*, 443–449. [CrossRef]
63. Tittl, A.; Harats, M.G.; Walter, R.; Yin, X.H.; Schäferling, M.; Liu, N.; Rapaport, R.; Giessen, H. Quantitative angle-resolved small-spot reflectance measurements on plasmonic perfect absorbers: Impedance matching and disorder effects. *ACS Nano* **2014**, *8*, 10885–10892. [CrossRef] [PubMed]

Disclaimer/Publisher's Note: The statements, opinions and data contained in all publications are solely those of the individual author(s) and contributor(s) and not of MDPI and/or the editor(s). MDPI and/or the editor(s) disclaim responsibility for any injury to people or property resulting from any ideas, methods, instructions or products referred to in the content.

MDPI AG
Grosspeteranlage 5
4052 Basel
Switzerland
Tel.: +41 61 683 77 34

Micromachines Editorial Office
E-mail: micromachines@mdpi.com
www.mdpi.com/journal/micromachines

Disclaimer/Publisher's Note: The title and front matter of this reprint are at the discretion of the Guest Editor. The publisher is not responsible for their content or any associated concerns. The statements, opinions and data contained in all individual articles are solely those of the individual Editor and contributors and not of MDPI. MDPI disclaims responsibility for any injury to people or property resulting from any ideas, methods, instructions or products referred to in the content.